BREWER STATE JR. COLLEGE

W9-BQY-792

LIBRARY

TUSCALOOSA CENTER

DISCARDED

BC
177
.T58

Toulmin, Stephen
Edelston.

Human understand-
ing

DATE DUE			
FEB 1 '79			
JAN 5 '82			
APR 6 '82			
OCT 2 '84			
AUG 26 '85			

DISCARDED

Human Understanding

STEPHEN TOULMIN

Human Understanding

The Collective Use and
Evolution of Concepts

PRINCETON UNIVERSITY PRESS
PRINCETON, NEW JERSEY

Copyright © 1972 by Princeton University Press
ALL RIGHTS RESERVED
L.C. CARD 73-166391

ISBN 0-691-01996-7 (paperback edn.)
ISBN 0-691-07185-3 (hardcover edn.)

First PRINCETON PAPERBACK printing, 1977

PRINTED IN THE
UNITED STATES OF AMERICA
BY PRINCETON
UNIVERSITY PRESS

H. D. A.
W. J. C.
conlegis at amicis
d.d. auct.
S. E. T.

Preface

EVER since the ancient Greeks fell in love with geometry, philosophical thought about the nature of knowledge has been dominated by models derived from mathematics and theoretical physics. This fact has had two regrettable consequences. On the one hand—to stand Whitehead's epigram on its head— it has doomed the whole of subsequent philosophy to being a series of footnotes to Plato. On the other hand, it has tempted philosophers to concentrate on questions of logical form, to the neglect of questions about rational function and intellectual adaptation. The present 'critique of collective reason' is the first stage in a comprehensive reappraisal of our working ideas about rationality, designed to redress that balance. (The two succeeding instalments will be corresponding critiques of individual reason, and of judgement.) The idea of rationality, I shall be arguing, is concerned far more directly with matters of function and adaptation—with the substantive needs and demands of the problem-situations that men's collective concepts and methods of thought are designed to handle—than it is with formal considerations; and it is with the express intention of emphasizing this fact that I have introduced the phrase 'intellectual ecology'.

The central thesis of the present volume was first presented in my earlier book, *The Uses of Argument* (1958). The work I have done on the development of scientific thought and related topics during the subsequent decade makes it possible to expound it here at greater length and in a historical frame. This thesis can be summed up in a single, deeply held conviction: that, in science and philosophy alike, an exclusive preoccupation with logical systematicity has been destructive of both historical understanding and rational criticism. Men demonstrate their rationality, not by ordering their concepts and beliefs in tidy formal structures, but by their preparedness to respond to novel situations with open minds—acknowledging the shortcomings of their former procedures and moving beyond them. Here again, the key notions are 'adaptation' and 'demand', rather than 'form' and 'validity'. Plato's programme for philosophy took it for granted that the functional adequacy of geometrical forms was self-guaranteeing—so that there must, in the last resort, be only one Idea of the Good, and that a mathematical one. The philosophical agenda proposed here sets aside all such assumptions in favour of patterns of analysis which are at once more historical, more empirical and more pragmatic.

In the course of working on this book, I have incurred intellectual debts to many friends and colleagues. First and foremost, I am indebted to innumerable discussions with my wife, June Goodfield. Her lack of any strict professional training in academic philosophy has made her a more helpful and uninhibited critic; while her obstinate insistence on relevant and substantive issues has repeatedly compelled me to clarify my arguments at points which remained obscure. She has patiently read and re-read successive drafts of the chapters that follow, and her frank criticism has greatly contributed to improving them.

I have profited a great deal also from long and valued exchanges with Henry Aiken, whose broad humanity and feeling for the demands of real life have preserved in him all the intellectual virtues of candour and open-mindedness, in the face of philosophical experience and sophistication. With him I have repeatedly discussed most of the concepts analysed here, particularly those of 'authority' and 'disciplinability'.

Since moving to the United States in 1965, I have had the opportunity of learning much about the methodology of linguistics and anthropology from talking with Catherine Bateson; about the complexity and subtlety of theoretical evolution in science from the writings and conversation of Gerald Holton; and about population theory and the conditions of effective speciation from discussions with Ernst Mayr. More recently, I have had valuable exchanges with Donald Campbell about the wider significance of intellectual innovation and selection procedures, and with Dudley Shapere about the developing problematics of scientific 'enterprises'. (How well I have, in fact, succeeded in learning from the knowledge and wisdom of all these friends remains to be seen.) Others whose arguments, comments, criticisms, and ideas have played a part in the development of my argument include—among colleagues—Mark Adams, Bo Anderson, Robert Cohen, Norman Geschwind, Leonard and Donna Kasdan, Imre Lakatos, Eric Lenneberg, Henryk Skolimowski, Marx Wartofsky, and Marty Wishnatsky; and—among my own students—Nancy Baker, David Brighton, Yehuda Elkana, Allan Janik, Terry Perlin, and Robert Rice. Steve St. Clair drew my attention to the passage from Kierkegaard used as a motto for this first volume.

The research for this book was carried out, in part, with the help of grants from the National Science Foundation. Certain sections of the argument have been expounded more briefly in lectures and papers for the American Association for Advancement of Science (Section L), the American Philosophical Association, the Boston Colloquium for Philosophy of Science, the Philosophy of Science Association, the Society of Sigma Xi, and others: but in all material respects the argument is presented here for the first time.

Quite aside from these intellectual obligations, I have two further debts to acknowledge of a more practical kind. The first is to my secretaries—especially Constance Worthington and Joy Cull—who have deciphered and retyped my edited drafts with exemplary calm and good humour. The other, inestimable debt is to Bill Callaghan—a prince among department chairmen—and my other colleagues at Michigan State University, whose imaginative generosity gave me the opportunity I needed, during the last two years, to concentrate on the completion of this book.

STEPHEN TOULMIN

July 1971

Concepts, like individuals, have their histories, and are just as incapable of withstanding the ravages of time as are individuals.

Kierkegaard.
The Concept of Irony

A man demonstrates his rationality, not by a commitment to fixed ideas, stereotyped procedures, or immutable concepts, but by the manner in which, and the occasions on which, he changes those ideas, procedures, and concepts.

Contents

Human Understanding

General Introduction

1: *The Theory and Practice of Knowledge*

THE problem of human understanding is a twofold one. *Man knows, and he is also conscious that he knows.* We acquire, possess, and make use of our knowledge; but at the same time, we are aware of our own activities as knowers. In consequence, human understanding has developed historically in two complementary ways: it has grown, but at the same time it has deepened, so becoming at once more extensive and more reflective. Looking 'outside ourselves' and mastering the problems posed by the world we live in, we have extended our understanding; looking 'inwards' and considering how it is that we master those problems, we have deepened it. And throughout the history of thought these twin activities have gone on continuously in parallel.

The relations between these 'outward-looking' and 'inward-looking' activities have changed from one period of intellectual history to another, and ever since the beginning of rational speculation this duality has been a source both of opportunities and perplexities. There have been times—including some of the most fruitful periods in human thought— when these two activities were seen as being intimately related; and when natural philosophers made it their business to push the boundaries of our understanding outwards, while at the same time striving to recognize more clearly and analytically the nature of the intellectual processes and procedures in which they were themselves engaged. There have been other times—including much of our own twentieth century—when the two activities were seen as being quite separate and independent, and so the concern of different intellectual professions. At these times, natural scientists have kept their eyes turned resolutely outwards, so as to avoid becoming entangled in philosophical word-splitting; while professional philosophers have been equally preoccupied with pursuing their own analytical debates on an autonomous basis, and have set aside the new discoveries of science as being merely empirical. And one question that we must face here is whether the current separation of the practice of knowledge from its theory—of 'natural science' from 'epistemology', and of scientists from philosophers—cannot easily be carried too far. There are in fact good reasons, both historical and substantial, for re-establishing links between the scientific extension of our knowledge and its reflective analysis, and reconsidering our picture of ourselves as knowers in the light of recent extensions to the actual content of our knowledge.

This relationship between our knowledge of nature and our own self-knowledge has always been a tricky one to describe and discuss. The spatial metaphors implicit in terms like 'introspection' and 'the external world' need to be handled with extreme care: if we take them too literally, we can easily misconceive the relationship between the content of our knowledge and our awareness of it. Men's recognition of their own status as knowers has, accordingly, always tended to lead them into a certain hampering self-consciousness. The price we pay for articulating our experience of moral good and evil was captured long ago in the myth of Eden; and throughout philosophy we are at continual risk of being trapped in an intellectual hall of mirrors, whose multiple reflections destroy confidence in our own posture and capacities.

In principle, a proper grasp of our epistemic situation should reinforce confidence in the best-founded of our beliefs; in practice, it often leaves us in a bewildered and universal scepticism. Yet this trap can be avoided. During the constructive phases of human thought, the realistic appraisal of human understanding has often been an instrument for its systematic improvement. So, in the fundamentals of natural science—particularly where theoretical physicists work on the very boundary of physical thought—extensions of knowledge have often sprung from reflective reappraisal of beliefs and principles previously claimed as 'known'. (Despite twentieth-century claims for the autonomy of philosophical epistemology, men like Einstein and Heisenberg have unavoidably trespassed into the theory of knowledge.) And more generally, whenever

we have had occasion to sit down in a cool hour, and to reconsider what our present patterns of thought might be and
might do for us, the reflective reanalysis of our knowledge
has often been a step towards its improvement.

These two extremes of scepticism and pragmatism delimit
the area within which philosophically-minded men must
construct an adequate theory of human understanding. Even
in natural science the standards for reappraising our existing
stock of ideas are not self-evident; rather, our practical choice
of intellectual standards shows what results we feel entitled to
demand on theoretical grounds in our intellectual dealings with
the world, and what deficiencies we are prepared to recognize
in our current ideas—in short, what conception we actually
work with of our own epistemic situation. A realistic grasp
of that situation, is the best protection against moral or intellectual vertigo. Equally, it is the deeper basis on which all well-
founded criticism must rest.

We shall do well, therefore, to consider a man's practical
ideals of intellectual method in the light of his theoretical ideas
about intellectual activities and higher mental functions. By
making explicit the arguments underlying his conceptual
ambitions and dissatisfactions, we bring to light his own
epistemic self-portrait: the particular picture of human beings
as active intelligences which governs his stance towards the
objects of human understanding. The general problem of
human understanding is, in fact, to draw an epistemic self-
portrait which is both well-founded and trustworthy; which is
effective because its theoretical basis is realistic, and which is
realistic because its practical applications are effective. So
considered, our practical standards of rational judgement are
one face of the coin whose other face is philosophical. The
practical ('outward-looking') task is to apply these standards
of judgement to decide what ideas, concepts, or points of view
in fact have the strongest current claim to intellectual authority
over our thoughts and actions. The theoretical ('inward-
looking') task is to give an analytical account of the considerations
by which the authority of those very standards is to be judged.

On the face of it, then, the intellectual procedures which
the philosopher analyses, as a matter of theory, are applied in
practice in science and elsewhere. True, this convergence is

rarely complete. The practical refinement of our rational procedures and the theoretical analysis of their status lead towards a common goal only in the long run. Yet, if our theoretical grasp of human understanding ever were complete, exact and explicit, the critical procedures of science and other rational enterprises could be regarded as practical applications of the very explanatory principles that were brought to light in this theory. In the meantime, any excessive separation between the theory of knowledge and the practice of knowers should prompt us to ask, 'Are scientists becoming unreflective? Or are philosophers losing their sense of relevance?'

The search for an adequate theory of understanding has a history as long as the critical reason itself. Originally, physics and metaphysics came into existence together. The rationalist cosmologies of sixth-century Ionia and their precursors in the Middle-Eastern creation-myths carried double meanings: they were concerned not just with Nature, but with Intelligible Nature—not just with a World that Man might or might not understand, but with this World as an object of Man's understanding. In fourth-century Athens, this connection between the theory of knowledge and its practice became quite explicit. For Plato and Aristotle, all metaphysical theses had methodological morals, and every maxim of scientific procedure had a justification in philosophy. In arguing for the existence of Independent Forms, Plato was also working his way towards a physical theory based on abstract mathematical principles; in arguing against the existence of Independent Forms, Aristotle was developing a natural science whose explanatory concepts were empirical and taxonomic, rather than abstract and geometrical.[1] In forming his conception of 'true knowledge' as the product of science—i.e. in deciding what a completed theory of Nature would have to be like—each of the two natural philosophers committed himself also to a corresponding

---

[1] As to the connection between Plato's philosophical doctrines and his views about scientific methodology, see especially his arguments about the relation between mathematical theories and the phenomena they are used to explain, e.g., in the *Republic*, Book VII, 521c–531c, and the *Timaeus*, 27c–29d. As to Aristotle, see J. H. Randall jr., *Aristotle* (New York, 1960), especially chapters 4, 8, 10, and 11, which discuss particularly the relevance of biological theory to Aristotle's philosophical theory of 'essences'.

conception of 'true knowing' as the activity of science—i.e. to an account of what was involved in coming to understand Nature. In basing their respective scientific theories on different types of explanatory concept, or Ideal Form, they presupposed equally different modes of understanding. Process and product were correlative.

In seventeenth-century Europe the theory and practice of knowledge were again in harmony. Men's scientific expectations once more dovetailed with their ideas about their own epistemic situations—their ideas about 'true knowledge' with those about their status as 'knowers'. As the ultimate elements of his physics, René Descartes chose certain minute geometrical figures, devoid of density, colour, and other essentially non-geometrical properties; and he did so because his philosophical ideal of true knowledge was also derived from geometry. Scientific concepts could be fundamentally trustworthy, only if they shared the clarity and certainty of Euclid's mathematical concepts. Similarly, Isaac Newton and John Locke saw certain sensory impulses transmitted up the nerves to the depths of the brain as the ultimate sources of our knowledge about Nature; and they did so because the basic physical structure of the material world was for them 'corpuscular' also.[1] Any adequate account of human knowing had to fit with a corpuscular picture of the known world. In the rationalism of Descartes and the empiricism of Locke alike, explicit ideals for judging the products of intellectual enquiry were associated with explicit ideas about the manner in which these products were achieved. Men worked with both scientific and philosophical confidence just because they had a unified conception of human knowledge and human knowing.

So the philosopher-scientists of classical Greece and seventeenth-century Europe faced the central epistemic issues of their respective cultures: neither as philosophical problems

---

[1] For a fuller discussion of the connection between the seventeenth-century conception of 'mechanistic' explanation and the *sensorium* theory in epistemology, see my essay, 'Neuroscience and Human Understanding', in the collection, *The Neurosciences*, edited by G.C. Quarton *et al.* (New York, 1967), pp. 822ff. John Locke makes the connection between his epistemological arguments and the 'corpuscular' view of matter explicit at many points in his *Essay concerning Human Understanding*. For Newton's view, see the final paragraph to the General Scholium, added as a conclusion to the second (1713) edition of his *Principia*.

alone, nor as scientific problems alone, but by stating questions, doctrines, and ideas operative equally in all relevant domains of thought. Each man's account embraced both what Man knows and what there is for him to know. It was, at once, an account of the bodily processes and mental procedures employed in arriving at knowledge; of the World of Nature as comprising the objects of the verb *to know*, and of Man as its grammatical subject: and of the languages, notions, concepts, or ideas through which human knowledge finds an expression and crystallization.

If those were the epistemic problems for the two most profoundly original periods of human thought, they can serve once again to define 'the problem of human understanding' for us today. Judged by twentieth-century standards, the epistemic arguments of Plato and Descartes, Aristotle and Locke, may appear to overlap half a dozen academic disciplines, while belonging exclusively to none. They drew upon contemporary psychological ideas, but they were not 'psychological hypotheses'. They took account of the physical properties of material things, but they were not 'physical theories'. They had direct connections with the neurology, the linguistics, the formal logic of the times, but they never became subordinate parts of those subjects. Nor were their concerns—in our own jargon—narrowly 'epistemological'. They did not analyse our knowledge of the external world by exercises in 'logical construction' alone; instead, they pieced together conceptions of knowledge from the best available insights in all related areas of enquiry. The resulting pictures might be incomplete, but one had to do the best with what was available. For how could any man judge the intellectual authority of his own ideas, concepts and points of view adequately, unless he were prepared to exploit the separate, fragmentary lights coming from all available directions?

In previous generations, then, the rational appraisal of human understanding has normally involved many disciplines, without being 'in' or 'of' any one of them. So, if we are seriously to revive this central philosophical aim, we must ignore contemporary attempts to divide off the various epistemic disciplines by academic frontiers with professional checkpoints.

By its very nature, the problem of human understanding—the problem of recognizing the basis of intellectual authority—cannot be encompassed within any single technique or discipline. For the very boundaries between different academic disciplines are themselves a consequence of the current divisions of intellectual authority, and the justice of those divisions is itself one of the chief questions to be faced afresh.

Like cosmogony and matter-theory, the field of *epistemics* is necessarily an area of interdisciplinary enquiry. Half a dozen disciplines have epistemic aspects, sectors, or implications—e.g. the physiology of perception, the sociology of knowledge, and the psychology of concept formation. Like cosmogony and matter-theory, epistemics is also a very long-lived subject. For twenty-five centuries, the problem of human understanding has been an enduring preoccupation: a permanent feature of men's speculative thought, ramifying in many directions, but preserving a recognizable unity and continuity. Each new generation has reformulated the detailed, specific questions of epistemics in its own terms; but the central, guiding problems have remained the same.

> What sort of things do we know? What kinds of certainty can our knowledge have? How do we acquire that knowledge, or the concepts in terms of which it is framed? And what part does the evidence of our senses play in this process?

> How far are our concepts—even the most basic ones—derived from sense-experience? Must our claims to knowledge be backed, in every particular, by sensory evidence? Or do our concepts and categories, rather, predetermine our abilities to perceive and recognize? Do people with different concepts or languages even see the world differently?

> Either way, how can we compare the merits of rival concepts? And how, in their turn, can our claims to knowledge—moral or mathematical, scientific or practical—be justified and appraised?

Though posed afresh in a dozen different terminologies, these central questions have remained alive throughout, and continue to face us in our own time.

Just how far these problems are from definitive solution will be clear, if we look for a moment at the crucial term in which we naturally express them. The term *concept* is one that every-body uses and nobody explains—still less defines. On the one hand, the word has a familiar currency in twentieth-century history and sociology, psychology and philosophy alike. For many twentieth-century philosophers, indeed, concepts provide their central subject-matter, their very bread and butter. They raise questions about the concept *good* or the concept *number*, or even about the concept *red*; they recognize certain necessities and impossibilities as holding in virtue of our concepts, and mark off a special resulting class of 'conceptual' truths; they describe the structure of natural science as comprising a number of 'conceptual systems', which can be more or less adequately represented by formal or axiomatic systems; and they speculate about the origins of the common 'framework' of concepts presupposed in our everyday knowledge of the world. Many of them would even describe the central task of philosophy itself as being that of conceptual analysis. Yet, despite all their scrupulous care in the actual practice of conceptual analysis, the precise meaning of the terms 'concept' and 'conceptual' is rarely made explicit and frequently left quite obscure. As it was for St. Augustine with time, so it is for them with concepts: they know perfectly well what concepts are, just so long as nobody asks them to *say*. A concept? . . . it has something to do with how we use language; or with how we structure our experience, or with how we categorize the objects we have to deal with . . . something, but—we must now insist on asking—what exactly?

Suppose, instead, that we turn to professional psychologists for an account of concepts. We shall find that they leave us almost equally in the dark. Psychologists, too, study concept-formation in the intellectual development of young children, or the part played by conceptual interpretation in determining what we perceive in any situation; they too recognize that the acquisition and employment of concepts is closely associated—but not identifiable—with the learning and use of language; and they too speculate about the possibility that certain concepts, or at any rate certain conceptual capacities, may be an inborn part of our genetic inheritance—operating, perhaps through hereditary patterns of brain structure. Neurophysiologists, in their turn, raise similar questions, about whether there may not be separate 'centres' within the brain, serving respectively as the store for language, and as the seat of conceptual information-processing, or thought. 'As time passes there is formed within the brain the *ganglionic equivalent of a word* and the *ganglionic equivalent of a concept.*'[1] Yet psychologists and physiologists as much as philosophers tend to assume that, for practical purposes, we already know whàt concepts are, and how they differ from words on the one hand and from intelligent non-verbal behaviour on the other. They too are content to take the term for granted, and do not trouble to give it any formal definition.

Perhaps their instinct is correct. For strictly scientific purposes, the current colloquial uses of terms like 'concepts' and 'conceptual' may be clear enough to be going on with; so that nothing is gained by insisting on more explicit definitions and distinctions at the present stage. But, for our purposes, something more is required. The term 'concept' is in danger becoming an irredeemably vague catch-all, for the same reasons as its forerunners, 'impression' and 'idea', 'notion', 'essence', and 'substance'. Even supposing we set aside popular corruptions of the term—such as 'a new concept in packaging'—it already carries as much intellectual load as it can safely bear and possibly more. Between concepts as discussed in developmental psychology (Piaget, Vygotsky, Bruner) or in cybernetics (Wiener and Rosenblueth); in the history of physics (Mach, Koyré, Holton) or in the foundations of mathematics (Frege's *Grundgesetze*); between concepts regarded as shaping the thought of individual thinkers, and as the collective possession of communities of concept-users; between concepts as connected together into the conceptual systems of Viennese inductive logic, as forming the everyday conceptual framework of Strawson's descriptive metaphysics, or as giving rise to the conceptual questions of Wittgenstein's logical grammar . . . between all these things, what coherent pattern of relations can be recognized and described? We can answer that question only within the framework provided by a systematic reanalysis

[1] Wilder Penfield and Lamarr Roberts, *Speech and Brain-Mechanisms* (Princeton, 1959), p. 230.

of the problems of human understanding, and of the role of 'concepts' in the growth and expression of knowledge: only, that is, by bringing our philosophical analysis of concepts back into relation with historical and scientific discoveries about men's conceptual evolution and development, and so giving our intellectual judgements a realistic basis.

From the philosophical point of view, this last problem is the crucial one: *In what scales are our own concepts and judgements themselves to be weighed?* The central issues of epistemic philosophy—justification and appraisal, judgement and criticism—have never been concerned with factual matters alone. Questions about the processes, procedures, and mechanisms by which our concepts are developed, acquired, used, and/or improved may be topics for particular sciences or disciplines: neurophysiology, experimental psychology or logic, cultural anthropology or the sociology of science. The philosophical question in epistemics is, by contrast, the question from what sources our concepts ultimately derive their intellectual authority. Like the ultimate source of moral and political authority, this question remains a matter for judgement and evaluation, as well as factual knowledge. All the same, while the nature and origin of intellectual authority may not itself be a straightforwardly factual issue, each generation of men has always had to reconsider this question in the light of its own best factual knowledge; and, in this respect, we in the late twentieth century are no differently placed from our predecessors. Suppose we wish to develop our understanding, both of the world with which we have to deal and of our own dealings with that world; and suppose that we wish to do this, reflectively, and with self-awareness. In that case, it will not be enough for us to rely on our existing concepts unthinkingly—either as they are, or with minor modifications. We shall need, in addition, to come to terms with our own intellectual creations: recognizing them for what they are, and facing once again, in terms operative for our own time and context, the central epistemic questions about rational judgement and appraisal.

What Warren McCulloch asked about number, we must ask about the World:[1]

[1] Cf. Warren S. McCulloch, 'What is a Number that a Man may know It, and a Man that he may know a Number?' *General Semantics Bulletin* (1961), pp.

What is Man that he may understand the World? And what is the World that Man may understand it?

In particular, so as to focus on the central element in human understanding, we must ask:

What are the skills or traditions, the activities, procedures, or instruments of Man's intellectual life and imagination—in a word the *concepts*—through which that human understanding is achieved and expressed?

The final, philosophical goal of the inquiries that follow is, accordingly, to give an adequate account of the intellectual authority of our concepts, in terms of which we can understand the criteria by which they are to be appraised. 'Adequate by what standards?' it may be asked, and here two considerations will carry weight. First, our philosophical account must be *relevant* to the actual practice of rational criticism. We cannot be satisfied to lay down canons of intellectual appraisal as a matter of abstract epistemological theory, without asking what bearing they have on the practical tasks actually facing the critical intellect. Secondly, it must be given in terms which are *operative* in the light of all our present knowledge. We must take care not to beg empirical questions for which well-founded answers are already available in the epistemic sectors of other disciplines—whether they have to do with the historical evolution of concepts in different communal activities, or with their acquisition and employment in the lives and experience of individuals. So far as possible, indeed, our account should be worked out in the light of those empirical insights and answers, and stated using terms that dovetail with those of all the associated fields. And, if we enter those other fields, it will simply be because our own philosophical problems concern the mutual bearing and relevance of those subjects—and these mutual relations are the sole property of no one discipline.

The central question for all our subsequent arguments can be stated in three parts, and the three succeeding groups of enquiries will tackle each of these in turn:

Suppose we consider, first, our current ideas about the historical evolution of human knowledge and understanding

7–18; reprinted in the collection, *Embodiments of Mind* (Cambridge, Mass., 1965), pp. 1–18.

—i.e. the *growth* of concepts—and, secondly, those about the development of such understanding within the lifespan of human individuals—i.e. the *grasp* of concepts —what can we then learn about the *worth* of concepts—i.e. the foundations on which their intellectual authority rests, and the standards against which it is to be appraised?

In the first phase of our argument, we shall consider concepts as entering into the conceptual aggregates, systems, or populations that are employed on a collective basis by communities of 'concept-users'. We can concern ourselves with the patterns of events through which such aggregates of concepts are created, develop historically, and fall into disuse; and ask in what respects the resulting changes—historical or sociological, anthropological or whatever—have a bearing on the intellectual authority of the concepts concerned. In the second phase of the argument, we shall consider the skills and abilities through which an individual displays his personal grasp of concepts. We can concern ourselves with the patterns of events by which such conceptual abilities are acquired, exercised, and lost; and ask in what respects the resulting changes—physiological or psychological, linguistic or whatever—are relevant to the appraisal of the concept themselves. By the end of these first two phases, we shall have reconstructed an outline picture of the actual empirical matrix within which the human understanding operates; and one which attempts to take full account of recent changes in our knowledge of that matrix. In the final phase of the argument, we shall return to the underlying issues of judgement and evaluation, and ask explicitly what general account of intellectual authority or rational criticism is consistent with our current picture of concepts and understanding, both communal and individual. Only when this has been done, shall we recognize clearly just how little philosophical epistemology can afford to claim entire intellectual autonomy, and just how far philosophers have impoverished themselves by declaring their independence from all the natural and human sciences.

Once we do reach that point, however, we shall free ourselves from the temptation to think of epistemology as a genuine, self-contained discipline, having an authentic subject-matter and distinctive problems of its own. On the contrary, the proud self-sufficiency of philosophical epistemology will turn out to be, in actual fact, a mark of its sheer irrelevance. If twentieth-century epistemologists have debated questions on which natural science and human history have no bearing, that is for one reason only. The reflective, philosophical theory of knowledge—as so interpreted—has lost touch with the scientific and historical procedures by which our practical knowledge is extended. Questions of pure epistemology have ceased to be operative outside the boundaries of formal philosophy, because their intellectual roots in the rest of human thought have withered entirely away.

2: *The Three Axioms of the Seventeenth-Century Tradition*

If so much in twentieth-century epistemology lacks an organic connection with the natural and human sciences, that is not a mark of emancipation from empirical presuppositions. The truth is more curious. If we remind ourselves briefly how the philosophical theory of knowledge became divorced in the first place from the maxims of intellectual practice, we shall appreciate the real irony of the present situation, and free our hands to deal with the consequent problems. For the questions of twentieth-century epistemology do still rest on scientific and historical presuppositions. It is just that these presuppositions are some three hundred years out of date. Philosophical ideas about human understanding today are tacitly shaped by axioms from seventeenth-century debates in science and history.

René Descartes and John Locke were candid and critical men, yet they were men of their age, who discussed the principles of human knowledge in the light of current ideas both about the Order of Material Nature (i.e. physics), and about the Mental and Bodily Powers of Man (i.e. psychology and physiology). Their theories of human understanding took careful account of the best contemporary scientific opinions and, being 'natural philosophers', they would scarcely have been satisfied with anything less. Specifically, they and their epistemological followers took three commonplaces for granted. These can be put in capsule form:

(1) The Order of Nature is fixed and stable, and the Mind of Man acquires intellectual mastery over it by reasoning

in accordance with Principles of Understanding that are equally fixed and universal.[1]

(2) Matter is essentially inert, and the active source or inner seat of rational, self-motivated activity is a completely distinct Mind, or Consciousness, within which all the highest mental functions are localized.[2]

(3) Geometrical knowledge provides a comprehensive standard of incorrigible certainty, against which all other claims to knowledge must be judged.[3]

To most seventeenth-century thinkers, these three axioms appeared common form—even 'common sense'—and supporters of the 'new, mechanical philosophy' in particular treated them as above question. At the time, those few men who challenged them could not produce counter-arguments conclusive enough to put them in serious doubt, still less to enforce conviction. And, before their doubts became serious and active, epistemology was well launched on the main road that it has followed right up to the present century. By now, its original framework of historical, scientific, and mathematical presuppositions may long since have disintegrated. Yet the questions at the centre of epistemological debate, even today, too often remain direct descendants of those which Descartes and Locke put into circulation in a quite different—and forgotten—intellectual context.

(1) Consider, first, the influence on epistemology of the belief in a Fixed Order of Nature. Until almost 1800, most scholars and scientists were restricted within an ahistorical world-view, which telescoped without discomfort philosophical doctrines inherited from the Greeks and an orthodox time-scale based on Holy Scripture. No matter how forward-looking they might be, rationalists and empiricists, corpuscularians and vorticists found it all but impossible to break out of this ahistorical straitjacket. Trapped within the few thousand years of Biblical tradition, they were in no position to grasp the true antiquity

---

[1] Cf. Stephen Toulmin and June Goodfield, *The Discovery of Time* (London and New York, 1965), especially Chapter 4. Referred to below as *Discovery*.

[2] Cf. Stephen Toulmin and June Goodfield, *The Architecture of Matter* (London and New York, 1963), especially Chapter 14. Referred to below as *Architecture*.

[3] This was one of Descartes' chief debts to Plato, and has been perpetuated in our own time by Bertrand Russell.

of the natural world; and this fact in turn concealed from them the full mutability of all natural things. Where we see the whole World of Nature as in slow but continuous flux, to their eyes it possessed a permanent and unchanging Order, established at the Creation only a few thousand years back.[1] About the details of the Creation, men were free to disagree: Descartes and Leibniz, for instance, rejected the corpuscularian belief in 'atoms'. But no leading seventeenth-century thinker seriously questioned the existence either of fixed laws or of motion in physics, or of fixed principles of understanding in epistemology. For Protestants particularly, these were only two features of a larger God-given Order, which embraced also the planetary system, the physiological structures of man and animals, and even the very numbers, shapes and distributions of the elementary 'atoms' or 'particles'. Only the Epicureans seriously questioned the assumption that the Order or Structure of Nature were unchanging; but Epicureanism, being associated with atheism, had no serious public following in seventeenth-century Europe.

Until men became accustomed to a time-scale millions of years in length, the functional patterns in the present state of Nature remained unanswerable evidence of a Fixed Order. In turn, this Order was interpreted as displaying the Rationality and Forethought of its Creator, since how should it be understood, except as an outcome of deliberate Design? So the first duty of the devout scientist (or 'Christian Virtuoso') was to master its Fixed Laws. In this way, four distinct conceptions were compressed into a single article of scientific faith: the moral legitimacy of scientific enquiry rested on the discovery of rational Laws of Nature, and the Divine Creation of the World was a standing argument for its historical stability.

Given this immutable Order of Nature—including the human body and brain along with the rest of the material world—it was a short step to assuming that Human Nature was similarly fixed and permanent. So the fundamental epistemological question was simple:

By what principles or processes does the Human Mind acquire intellectual mastery over the Order of Nature?

[1] *Discovery*, chapters 4, 5, and 6.

Since both terms in this relationship—the Intelligence of Man, and Intelligible Nature—presumably operated on stable, unchanging principles, the relationship between them was presumably stable and unchanging also. The philosopher's task was then to analyse and explain the fixed and universal principles of Human Understanding by which ideas were formed and rational thought directed. Once again, there might be disagreements over matters of detail, but over the basic point all philosophers were united. Fixed Mind masters Fixed Nature according to Fixed Principles. The operative question was, '*What* fixed Principles?'

(2) Consider, next, the conception of Matter lying at the heart of seventeenth-century physics. Matter was intrinsically inert, and its essential properties were—on the majority view—either geometrical or mechanical, or both. All material objects comprised small three-dimensional parts having different geometrical shapes: these moved about the world, exchanging motion when and only when they pressed against, or collided with, one another. The World of Matter had been given a certain initial quantity of motion at the Original Creation. This apart, the types of activity open to material things were restricted to three: contact, collision, exchange of motion— that was all.[1] It appeared quite problematic, for instance, whether Matter could of itself sustain Life. Certainly, it was incapable of mental or rational activities, since these involved kinds of choice or spontaneity which must be the outcome of 'non-material' agencies. Any suggestion that material things with a sufficient degree of complexity could display genuine mentality or rationality was dismissed as ridiculous, even blasphemous. Once Matter had been defined as extended and inert, the activities of Mind and the attributes of Reason could be explained only in other, incommensurable terms.

But this was only a beginning. Belief in an intrinsically-inert Matter also landed seventeenth-century thinkers with a physiological model of the central nervous system far more exact and 'centralized' than the actual evidence warranted. In this way difficult but genuine questions about knowledge and perception were transformed into insoluble puzzles about the interaction between those two separate 'substances', Mind

[1] *Architecture*, chapters 7, 8, and 14.

and Matter. For, how could a nervous system made up of material components interacting causally serve functions which were—in human beings, at least—mental or rational?[1]. To protect the fundamental distinction between mental functions and material structures, the overlap between Mind and Matter must be reduced to an absolute minimum. So all genuinely rational functions were confined within a limited region deep inside the brain, and the non-material realm of Mind or Consciousness was banished into an 'inner' *sensorium commune*. (That locus, Descartes speculated, Was it perhaps in the pineal gland?) As a result, seventeenth-century physics pushed philosophers towards a centralist model which separated epistemological issues entirely from the empirical facts of neurophysiology. All questions about perception and cognition presumably referred to events within Consciousness—i.e. within the 'inner, mental world' of the *sensorium*—rather than to occurrences in the 'external, material world' of the nervous system.

This centralist model has not exhausted its philosophical influence. Bertrand Russell, for instance, has described how he was himself led into philosophy. The process was purely Cartesian. At the age of fifteen, he tells us, he 'became convinced that the motions of Matter . . . proceeded entirely in accordance with the laws of dynamics', so that 'the human body is a machine'; whereas, 'I came to the conclusion that consciousness is an undeniable datum, and therefore pure materialism is impossible.'[2] When Russell was a boy in the 1880s, however, the authoritative view of Matter was still that represented by the mechanics of Newton and the chemistry of Dalton; so, in dismissing 'materialism', he was still taking the essential inertness of Matter for granted. On this point, Russell's views never radically changed. This was his position in 1914 when he posed the central epistemological problems for so many of his successors, in his lectures on *Our Knowledge of the External World*. And his later writings on the subject still took the same centralist neurophysiology for granted. One quotation from *The Analysis of Matter* (1927) will make the point well enough:

What the physiologist sees when he looks at a brain is part of his own brain, not part of the brain he is examining . . . The percept

[1] See above, p. 5, n. 1.
[2] B. Russell, *Autobiography*, Vol. I (London, 1967), p. 47.

must . . . be nearer to the sense-organ than to the physical object, nearer to the nerve than to the sense-organ, and nearer to the cerebral end of the nerve than to the other end.[1]

Here we see Russell still pursuing the 'inner seat' of conscious perception and cognition up the causal chain of the sensory nerves to its supposed 'non-material' terminus.

(3) The third of the seventeenth-century presuppositions was the Euclidean model of certainty. Descartes hoped to defend the epistemic claims of natural science against the sceptics in the same way as Plato two thousand years earlier. Michel de Montaigne had questioned whether mere humans could ever have rational grounds for deciding between opposed systems of physical theory, and—like Socrates—had confined our understanding to matters of humane experience and concern: virtue and friendship, beauty and statecraft.[2] Descartes was less ready to accept scientific defeat. Inspired in youth by Galileo, he met Montaigne's challenge with the same device that Plato used to counter the cosmological scepticism of Socrates. The legitimate demands of reason in scientific theory could be met—at a price. A properly formulated system of scientific concepts could claim intellectual authority, on condition that it measured up to the standards of rigour and certainty set by geometry.

Before long, epistemologists were accepting this ideal as past question. There was much controversy over the question, in what respects our empirical knowledge of nature could achieve geometrical certainty. On this topic, Descartes's own rationalist successors differed sharply from the empiricist followers of John Locke. Yet the ideal itself kept its charm; and most philosophical epistemologists have continued to regard mathematical necessity as the epitome of knowledge and certainty. Claims to true knowledge must be backed either by incorrigible, self-authenticating data, or by arguments as complete and rigorous as those of pure mathematics, and

---

[1] B. Russell, *The Analysis of Matter* (London, 1927), p. 383.

[2] For Montaigne, see *Apologie de Raimond Sebond, Essais* II, 12: for Descartes, see his *Principia*, Book III, 3. On the connection between Montaigne and Descartes, see A. O. Lovejoy, *The Great Chain of Being* (Cambridge, Mass., 1936), Chapter 4, and the Introduction by A. Koyré to Descartes, *Philosophical Writings*, ed. Geach and Anscombe (Edinburgh, 1964).

preferably by both. In return the seeming deficiencies of our actual empirical knowledge provoked perplexity, hesitation, and scepticism—sometimes to the extreme of solipsism. Considered in these terms, knowledge lacking mathematical certainty was—philosophically speaking—no better than ignorance. Rather than confess universal ignorance, epistemologists attempted to discover general conditions on which other kinds of knowledge might achieve an intellectual authority comparable to those of the 'formal, demonstrative sciences'. With the passage of time, indeed, this became for many the very defining task of epistemology. Their agenda now was, to find additional principles or premises which would bring arguments from substantial fields of enquiry—scientific, historical, or ethical—up to the mathematical ideal. The alternative course, of challenging the formal ideal itself, seemed to them a betrayal of philosophy: abandoning all philosophical claims to 'rational certainty', and opening the gates to sceptics.[1]

Betrayal; abandonment; opening the gates . . . this is no longer the language of a pure, disinfected, and value-neutral science. Nor had Descartes and Locke used the established framework of science to define the empirical matrix of their theories solely out of piety towards 'up-to-date scientific facts'. On the contrary, each basic presupposition protected a seemingly essential philosophical demand. The belief in universal principles of human knowledge defined a standpoint for impartial rational judgements between men from different backgrounds. In a world whose material processes slavishly obeyed causal laws, the separation of Matter from Mind, and Brain from Consciousness, likewise defended the rational claims of our higher mental procedures. And, meanwhile, the geometrical ideal of knowledge provided the universal criterion which was needed to underpin our rational confidence in all branches of practice.

Having been accepted in the first place as satisfying these philosophical demands, the presuppositions have survived for lack of any alternative. Without the guarantees so provided, it has appeared that the authority of rational thought itself would be gravely impoverished. Unless, at a deep enough level, the

[1] B. Russell, *Human Knowledge: its Scope and Limits* (London, 1948), pp. 439.

principles of human understanding are applicable universally, how could the soundness of ideas from different epochs or contexts ever be compared? Failing this guarantee, we are seemingly threatened with an intellectual relativism that we should much rather avoid. So the historical invariance of rationality—i.e. the existence of universal principles of human understanding—has always appeared to be a precondition of rational judgement, without which cross-cultural and cross-historical comparisons are inconceivable. Once one accepted the notion of inert Matter, again, it scarcely appeared possible to state—still less defend—a thoroughgoing materialist explanation of rational activities. The processes of 'thought' and the operations of 'machines' seemed flatly incommensurable. Wollaston put the point ironically, in reply to Priestley: 'Who can imagine Matter to be moved by arguments, or ever ranked syllogisms and demonstrations among levers and pulleys?'[1] Without a central *sensorium* absolutely distinct from Matter, what room could be found, in the world as we know it, for conceptual thought and rational judgement? The men of the seventeenth century asked this question in a rhetorical spirit, confident that no real alternative existed. As for the geometrical ideal of knowledge: given 'clear and distinct' concepts, conforming to timeless and self-validating mathematical principles, scepticism could be avoided, and some basis for rational agreement guaranteed. Failing this final and unarguable criterion, what rationally defensible substitute could be found for the comfortable certainty it formerly provided?

The broader presuppositions of seventeenth-century philosophy may have remained influential by default, but in the natural and human sciences the position has changed drastically. There, the older axioms have long since been called in question, and their implications discounted. (1) During the past 250 years, physical science has broken its earlier ties with theology, and the familiar scientific phrase 'laws of nature' is no longer thought of as implying the 'sovereign will' of a Creator. By now, even the long term invariance of those 'laws' is merely a working assumption. Meanwhile, the other, structural

[1] Quoted by Joseph Priestley in his *Disquisitions relating to Matter and Spirit* (London, 1777), Section VIII, Objection III.

invariants of the seventeenth century world-picture have all crumbled under the touch of history and been transformed into historical variables. From the ultimate unchanging particles of matter, through the planetary system and animal species, to the timeless imperatives of morality and social life: every aspect of Nature is considered today as historically developing, or 'evolving'. If a theory of human understanding is to follow the rest of twentieth-century science and history, then, it must be based not on unchanging principles and guarantees, but on the developing interactions between Man, his concepts, and the world in which he lives. Human variability is restricted only within the slowly changing limits of our genetic constitution and cultural experience. The problem of human understanding in the twentieth century is no longer an Aristotelian one, in which Man's epistemic task is to recognise the fixed Essences of Nature; nor is it an Hegelian one, in which Human Mind alone develops historically against a static background of Nature. Rather, it is a problem that requires us to come to terms with the developing relationship between Human Ideas and a Natural World, neither of which is invariant. Instead of Fixed Mind gaining command over Fixed Nature by applying Fixed Principles, we should expect to find variable epistemic relationships between a variable Man and a variable Nature.

(2) At the same time, physics has been stripping Matter of just those 'essential' characters which earlier scientists saw as cutting Matter and its causal processes off from Mind and its rational activities. In place of clear geometrical boundaries, impenetrability, strictly-localized influence and purely mechanical contact-action, today's fundamental particles have acquired characteristics reminiscent less of Democritean 'atoms' than of the active 'harmonies' in the Stoic view of nature. In physiology, likewise, centralist models of the nervous system have also been losing ground in favour of newer systemic conceptions. Rather than being sharply localized in a non-material *sensorium*, the higher mental activities of Man now call into play elaborate physiological systems—involve, indeed, the co-operative interaction of those systems with appropriate elements of the 'external' environment. These changes have once again shifted the philosophical burden of proof. From the

start, thoroughgoing critics of Descartes and Locke saw that any effective challenge to their theories must call in question the definition of Matter assumed by all the New Mechanical Philosophers.[1] By now, however, the scientific basis of the *sensorium* theory has been completely eroded, and the constraints it formerly placed on epistemic speculation have been removed. Current neurological ideas pose problems about human understanding quite unlike those that faced Descartes and Locke. This is not to suggest that Descartes' ideas are, in every respect, inconsistent with twentieth-century science: still, having divorced his dualism from its roots in seventeenth-century physics, we must at the very least restate his central points in a less dramatic form. Released from this straitjacket any weakened form of Cartesian dualism will—necessarily—be very different from the dogmatic dualistic systems current during the period following Descartes.

(3) Within mathematics itself, the Euclidean ideal had already lost ground before the end of the eighteenth century. The charm of Euclid's geometry, as the supreme exemplar of rational understanding, sprang from the belief that the Greek geometers had brought off a spectacular double feat: viz. that they had at one and the same time established a rigorous logical network of deductions and also given a correct description of the actual World of Nature. The last two centuries have seen the death-blow to any such belief. Saccheri's vain attempt to demonstrate the mathematical uniqueness of Euclid's system by a *reductio ad absurdum* was followed by the construction of non-Euclidean systems, beginning with Lambert and Gauss.[2] Yet, once non-Euclidean geometries became as acceptable, by any formal standards, as Euclidean, a wedge was driven between the two halves of Greek geometry, i.e. between questions about formal consistency and questions about empirical relevance. Given alternative formal systems of spatial, or quasi-spatial relations, men had now to decide the actual degree of relevance and applicability of any one formal 'geometrical' system to actual experience and practice. The problem of empirical geometry was one starting-point for

[1] See Priestley, op. cit., Sections III and IV.

[2] See Gottfried Martin, *Kant's Metaphysics and Theory of Science* (Manchester, 1955), especially Chapter I, secs. 2–5, esp. p. 18.

Immanuel Kant's critical inquiries. As a young man, he had taken it as certain that the empirical relevance of Euclid's and Newton's systems was demonstratively guaranteed. Disabused of this idea partly by his friendship with Lambert, partly by his reading of Hume, he was obliged to find some other source to justify the special claims of Euclidean geometry and Newtonian dynamics on our intellectual allegiance. If we consider geometry today as a formal system, much—if not all—of our confidence in it springs from the fact that it is our own creation. Its rigour is the inner articulation of a man-made calculus; while its pragmatic applicability reflects, not verbal definitions, but our practical procedures for identifying 'straight lines', 'circles', and the rest among the objects and figures of the real world.

Giambattista Vico's motto, *certum quod factum*—'An agent can understand fully only what he has himself made'— evidently applies also to pure geometry and to formal logic.[1] Many of the arguments by which Vico himself vainly tried to hold back the tide of Cartesianism have been integrated into twentieth-century thought. Outside formal philosophy, Cartesian knowledge and certainty would nowadays be claimed only on behalf of our own intellectual artefacts, and, outside the man-made fabric of pure mathematics, the question now arises whether the demand for formally demonstrative arguments was not always a will-o'-the-wisp. Perhaps the idea of timeless, eternal standards, applicable to arguments-in-general in abstraction from their practical contexts, was always (as Vico claimed) a Cartesian delusion. Over-reliance on the model of Euclidean geometry has led philosophy into dead ends before now; since the mathematicians themselves have reappraised the status of their knowledge, philosophers too should reconsider their own standards of certainty.

We have seen how the seventeenth-century presuppositions imposed on philosophy a certain epistemic picture, of Man the Rational Knower facing Nature the Unchanging Object of Knowledge, and so limited the options between which epistemologists were at liberty to choose. These presuppositions were

---

[1] See Isaiah Berlin, 'The Philosophical Ideas of Giambattista Vico', in *Art and Ideas in Eighteenth-Century Italy* (Rome, 1960), and also *Giambattista Vico: An International Symposium*, ed. G. Tagliacozzo and H. V. White (Baltimore, 1969), Part IV.

of course 'axiomatic', only in a loose sense: viz. beliefs too fundamental to be currently called in question. Correspondingly, if we wish to relate a late-twentieth-century theory of knowledge to contemporary ideas in other fields, the task is to substitute a new epistemic self-portrait for the old, and so reopen intellectual options closed off throughout the heyday of Cartesian epistemology.

For the moment, not many contemporary philosophers yet see their epistemic task in quite these terms. The older epistemic self-portrait is unthinkingly familiar to anybody who immerses himself in the ironically named 'modern philosophy' of the universities, and has continued to direct much professional philosophy well into the twentieth century. Early in *The Longest Journey*, E. M. Forster introduces us to a group of Cambridge undergraduates, talking late into the night. They are discussing a cow in a field—'Does it really exist? Can we know that it exists? If so, *how* do we know? How, above all, can we *prove* that it exists? . . . ' Here Forster is delicately capturing the undergraduate debates of his own Cambridge contemporaries, particularly Bertrand Russell and G. E. Moore: about such things as sense-data and material objects, inner consciousness and the external world. To compress these questions into a nutshell:

> How can Man acquire a reliable understanding of Nature? By what permanent principles can the human observer (i.e. Mind), peering out from an inner vantage-point deep within the body (i.e. Matter), transcend those Inner Sensations which alone serve him as evidence, and achieve authentic knowledge of an External World, on the far side of his sensory receptors? And how can Man's empirical arguments about that External World ever possess the rigour and certainty which formal logic and mathematics entitle him to demand?

So at the beginning of the twentieth century, Russell and Moore presented English philosophers with a set of epistemological questions based squarely on the traditional epistemic self-portrait. They saw the Human Mind as approaching the External World in a historically fixed posture; as confronting it through the medium, or obstacle, of the Five Senses; and as demanding of all its epistemic arguments nothing less than Euclidean rigour and certainty. Since their time, many of the weaknesses in their analyses have been recognized, and nobody would accept their statement of the problem of epistemology in just this form. Still, the task of paring away from philosophy all the last consequences of the seventeenth-century assumptions will evidently compel us to cut very near the bone. In this situation, we have two options. We may start again from scratch undistracted by the seventeenth-century legacy, and attempt to frame problems for our own theory of human understanding within the framework of late-twentieth-century beliefs about man and history, ideas, and nature—which will mean abandoning the philosophical autonomy of epistemology, and re-opening closed doors between formal philosophy and more substantive disciplines. Or we may recoil from this whole task and retreat into a philosophical tradition that originated—confessedly—in an intellectual context quite unlike our own. Yet, in this latter case, the question will then have to be faced: 'What shall we achieve in this way, except to keep the lights in students' rooms burning into the small hours?'

3: *Programme for a New Theory of Human Understanding*

Let me summarize the tasks that face us in the following inquiries. The general aim is to piece together a new 'epistemic self-portrait': that is, a fresh account of the capacities, processes, and activities, in virtue of which Man acquires an understanding of Nature, and Nature in turn becomes intelligible to Man. This account must sit as comfortably with the questions, conceptions, and commonplaces of twentieth-century thought as the older epistemic picture of Descartes and Locke did with seventeenth-century ideas. It must accordingly carry conviction beyond the frontiers of formal philosophy into all these other disciplines which are concerned with perception and cognition, knowledge and ideas. This is so for two reasons. We shall be working in an area where empirical considerations can—at best—only suggest directions in which both science and philosophy might attempt to move; and we can never wholly disentangle the scientific aspects of human understanding from its philosophical aspects.

Even now, the 'scientific facts' are not philosophically unambiguous, and we are therefore concerned with a delicate task of readjustment. While we reconsider traditional philosophical issues in the light of contemporary scientific ideas, we may equally find the results of our reappraisal reacting back in certain respects on the sciences themselves. Thus, any philosophical account of conceptual knowledge presupposes something—however crude—about the mechanisms by which 'sensory inputs' are 'processed' in the nervous system; yet, in return, neurophysiology must bear in mind the central nervous system's functions as an instrument in the improvement of knowledge.[1] Similarly, we can understand the intellectual authority of our concepts clearly, only if we bear in mind the socio-historical processes by which they develop within the life of a culture or community; yet, in turn, a clearer analysis of that intellectual authority gives us the means of developing more exact ideas about those very processes.[2]

With this final qualification, our agenda is complete, and we can see how its elements dovetail together to form a single sequence of questions and arguments. The central topic of Part I will be the evolution of human understanding, as represented by the historical growth of concepts. Its subject-matter will be the changing populations of concepts and procedures characteristic of collective intellectual activities: the manner in which these function within communities of concept-users, and the various factors that play a part in the processes of conceptual change. In positive terms, this first group of inquiries will be focused around the question:

> Through what socio-historical processes, and intellectual procedures, do populations of concepts and conceptual systems—the methods and instruments of the collective understanding—change and develop in their transmission from each generation to the next?

In posing this question, we must deliberately avoid confining ourselves to the traditional range of epistemological problems. Our first step will be to set aside the seventeenth-century

---

[1] Cf., for example, J. Lettvin *et al.*, 'What the Frog's Eye tells the Frog's Brain', *Proc. Inst. Radio Engineers*, 47 (1959), 1940–51.

[2] Cf. the issues in Part I, Chapter 4 below.

question, 'What universal principles of understanding govern the operations of the Human Mind, in its task of mastering the Fixed Order of Nature?' and so to reopen philosophical options closed off by the traditional approach. But we shall do so only at a price. For this will land us at once with the further question:

> In the absence of immutable Principles of Human Understanding, how can intellectual positions or arguments from different historical and cultural contexts be rationally compared?

The aim of our first group of inquiries will be to show that this second, philosophical question can be answered satisfactorily, only in the light of the prior, empirical question; by locating precisely, within the socio-historical matrix of human understanding, the points at which rational appraisal and criticism find their operative niches.

The central topic of Part II will be the development of human understanding, as expressed in a psychological grasp of concepts. Its subject-matter will be the changing skills and abilities through which individual men acquire and exercise that grasp: the manner in which these abilities function in the lives of particular concept-users, and the different factors which play a part in their acquisition and use. In positive terms, this second group of inquiries will be focused on the question:

> Through what physiological processes, and psychological sequences, are conceptual skills and abilities—the methods and instruments of the individual understanding—acquired, employed, and sometimes lost, within the lifetime of individual concept-users?

In posing this question, we must again ignore the restrictions of traditional epistemology. We shall deliberately begin by setting aside the older question, 'How does the perceiving Mind, from its standpoint deep within the Brain, acquire information about the External World?'; so reopening forgotten options. But, in doing so, we shall again be faced with a further question:

> If we take the results of twentieth-century neuroscience seriously, how can we characterize rational thought in terms

consistent with our causal understanding of brain physiology and psychology?

In this second part, our aim will again be to show that the philosophical question can be answered satisfactorily, only in the light of the prior, empirical question; by locating precisely, within the psycho-physiological matrix of human understanding, the points at which rational thought and judgement find their functional correlates.

In Part III, we shall tackle directly the consequential questions for philosophy. Our central topics will there be the rational appraisal of human understanding and the intellectual worth of concepts. Any epistemic self-portrait that carries conviction in other contemporary sciences must also give us a sound grasp of our epistemic situation, and so an understanding of our proper canons of rational criticism. In positive terms, therefore, the final group of enquiries will be organized around the question:

> Given the operative niches and functional correlates of concepts within the communal and individual matrices of human understanding, how must we—at any rate, for our own generation—sift, compare, and so justify critical confidence in our own best-founded concepts and beliefs?

In tackling this final philosophical question we shall set aside the Euclidean ideals of certainty and rationality, so ignoring the philosophical constraints imposed by the traditional demand that true knowledge be both indubitable and incorrigible. In return, however, we shall have to attack head-on the consequential question:

> Failing the absolute certainties of the Euclidean ideal, by what alternative standards of conceptual judgement can we decide where our intellectual trust is rationally to be placed?

We shall therefore embark on our final group of inquiries with two linked aims. The first is to relate rationality and its associated categories to the actual niches within which rational appraisals are in practice made: the second is to show how the intellectual authority of our concepts finds an ultimate source within the empirical matrices of understanding itself.

*   *   *

A last remark can help to set this whole enterprise into its historical perspective. From one point of view, our agenda seems designed, Penelope-like, to unravel the entire fabric of epistemology—problems, questions and methods alike—that philosophers have woven so patiently over the last 350 years. From this point of view, we are indeed disavowing all the substantive doctrines for which Descartes stood, along with the tradition of epistemological enquiry which he initiated. Yet, from another and deeper point of view, our investigations are strictly faithful to Descartes's own methodological principles. Whatever the changes in a philosopher's questions, conclusions, and techniques of argument, the general spirit of his philosophy may still remain the same; and, for Descartes, that spirit was defined by his fundamental maxim of 'systematic doubt'. As he repeatedly insisted, a philosopher is rationally obliged to doubt whatever he can consistently call in question, and our enquiries here will merely carry this programme one stage further. If in the late twentieth century we can pursue our own systematic doubts beyond the extreme point of seventeenth-century thought—abandoning now the three basic presuppositions which even Descartes and Locke took for granted—then, on Descartes's own standards of philosophical rationality, it is only rational for us to do so. And if, despite all his prodigies of doubt, Descartes nevertheless made assumptions which we ourselves can now 'consistently call in question', it becomes our Cartesian duty to investigate the consequences of these more radical doubts.

Our agenda may therefore compel us to abandon Descartes' beliefs; but it will not require us to betray his ideals. On the contrary, it would be a worse betrayal of Descartes' programme if we neglected to challenge, at all its dubitable points, the epistemological tradition of which his work was the origin. Rather, we must harden our hearts and swear in his name to be, in our own twentieth-century terms, *plus Cartesien que Descartes même.*

Part I

The Collective Use and Evolution of Concepts

SECTION A   Rationality and
Conceptual Diversity

Introduction

Each of us thinks his own thoughts; our concepts we share with our fellow-men. For what we believe we are answerable as individuals; but the language in which our beliefs are articulated is public property. To understand what concepts are, and how they play a part in our lives, we must come to terms with the central relationship between our thoughts and beliefs, which are personal or individual, and our linguistic and conceptual inheritance, which is communal.

In this respect, the problem of human understanding—the problem of accounting for the intellectual authority which our collective methods of thought exert over individual thinkers —displays some little-remarked parallels to the central problem of social and political theory: that of accounting for the corresponding authority which our moral rules and customs, our collective laws and institutions, possess over the individual members of a society. These parallels are implicit in the arguments of the idealist philosophers. Strictly speaking, they argued, the exercise of individual rights presupposes the existence of society, and is possible only within the framework of social institutions; and equally, we could add, the articulation of individual thoughts presupposes the existence of language, and is possible only within the framework of shared concepts. So the paradox of political freedom, as enunciated by Jean-Jacques Rousseau, faces us in the epistemic field also. *Man is born free, and everywhere he is in chains;*[1] yet, on closer inspection these chains turn out to be the necessary instruments of effective political freedom. Intellectually, also, Man is born with the power of original thought, and everywhere this originality is constrained within a particular conceptual inheritance; yet, on closer inspection, these concepts too turn out to be the necessary instruments of effective thought.

[1] J. J. Rousseau, *Du Contrat Social* (1752).

It would be easy enough, then, to discuss the philosophical problems of rationality in terms of quasi-political imagery. Individualists or anarchists will see the shared concepts which are the instruments of our thinking, and the communal language in which we express our thoughts, as shackles or constraints imposed by a tyrannous collectivity, so warping the free consciousness of the self-moved, creative individual. Collectivists will retort, with equal plausibility, that an individual's inherited concepts represent neither a prison nor a set of stocks. Rather they are an intellectual structure or platform raised above the bloody, improvident chaos of brutish existence, on which alone he can live a genuinely human life. He owes this platform to the efforts of his forefathers, and his own true creativity consists in working responsibly and effectively to improve it for his successors. Yet, in either case, the picturesque imagery of 'shackles', 'platforms', and the rest will, in the long run, be acceptable only as a preface to a more serious, literal-minded analysis. Its only merit is that of focusing attention on the complex relations to be elucidated in any convincing epistemic self-portrait. Concepts play their parts in the lives of individuals, and scarcely have any actuality apart from these roles. At the same time, individual concept-users acquire the concepts they do within a social context, and the sets of concepts which they employ play identifiable parts also in the lives of human communities—whether societies, congregations, or professions. In the intellectual as in the political sphere, an individual's initiatives—whether social or conceptual—are an expression of his personal thought about collective problems. The conceptual innovations of the individual physicist (say) are judged in relation to communal ideas which he shares with the rest of his profession; and he thinks creatively when he makes his contribution to the improvement of this communal 'physics'.

In intellectual affairs and in politics alike, then, even personal originality has a social or collective dimension. And, furthermore, these two roles are clearly correlative. Collective understanding is realized through the intellectual performances of individuals; the individual's understanding applies concepts taken from a communal stock, or modifies them in ways that represent potential improvements of that stock. So the epistemic self-portrait to be built up in these enquiries must first be drawn from two distinct points of view, in two separate dimensions—one individual, the other collective—and the two resulting pictures must afterwards be put together into proper perspective. Which of these dimensions shall we tackle first? In principle, one might analyse the two empirical aspects of understanding quite independently, but there will be certain advantages here in looking at the collective aspect first. To begin with, this reverses the customary procedure. From Locke to Russell, and from Descartes to Chomsky, orthodox epistemologists have interpreted the problem of knowledge as requiring them to explain, at the outset, how an individual thinker or observer can arrive single-handed at valid ideas, truths, or grammatical forms; and this choice of priorities has given rise to grave difficulties, by distracting attention from the social character of language and the communal criteria of validity. Only recently, e.g. in the later work of Ludwig Wittgenstein, has the scale tipped decisively the other way, showing us the crucial connections between concept-acquisition and 'enculturation'. (To anticipate: many contemporary developmental psychologists, e.g. Jean Piaget and Jerome Bruner, have also missed the full significance of this connection.)[1]

There is another reason, too, for considering the collective aspects of concepts first. I have here spoken of the 'personal' and 'communal' aspects of concepts, rather than of their 'private' and 'public' aspects, deliberately. Traditionally, the notion of *privacy* has been associated with a view that we must treat with the greatest caution: namely, the view that mental states and activities are *essentially* 'internal', and so 'hidden'. One scarcely needs to emphasize the central part played by the dichotomy, Inner Mind/External World, in shaping the problems of orthodox epistemology. So it will be worthwhile here to see just how far we can carry our own enquiries, without resorting to any of the familiar spatial metaphors, which equate the physical with the 'external', and the mental with the 'internal'. By choosing to consider the communal aspect of concepts first, we shall find it possible to defer until much later the whole perplexing problem of 'thought' as 'inner life'.

[1] On Piaget, see below, pp. 417–46. For Bruner, see the report on his 'conservation' experiments, in J. S. Bruner *et al.*, *A Study of Thinking* (New York, 1956).

In explaining this decision, we must add only one qualification. Our discussion (in Part I) of scientific disciplines and other such rational enterprises will depart from the practice of men like Ernst Mach and Bertrand Russell in one additional respect. They regarded the particular task of justifying scientific theories as merely one special case of a larger philosophical problem: viz., that of explaining 'our knowledge of the external world' in general terms.[1] For Mach and Russell, as for Locke and Hume before them, the central question was, 'How can we penetrate the inner veil or screen of private experience, so as to describe, infer, and/or logically construct an external world of public objects beyond it?' In these enquiries, I shall attempt to break the link between these two problems, by considering the collective use of concepts—in scientific theory and elsewhere—apart from all questions about the reliability of sense-perception. My idea is not to beg any crucial questions against Mach and others like him; at most, I shall be trying to show how far his equation of these two problems was premature. So, for the purposes of Part I, we may treat questions about the communal aspect of concept-use independently of all questions about the roles of concepts in individual sensation and perception. True: in Part II, I shall in fact be going on to criticize any assumption that 'sensory experience' is somehow 'interposed between' the Mind and the External World. But for the moment I shall be concerned to show only how much we can usefully achieve without even bringing that contentious issue into consideration.

We acquire our grasp of language and conceptual thought, then, in the course of education and development; and the particular sets of concepts we pick up reflect forms of life and thought, understanding and expression current in our society. In certain respects, the patterns so developed are—demonstrably—products of cultural history and prehistory. They differ from country to country, they may change quite strikingly within a few years, and any normal human readily learns or relearns them in their characteristic local forms. In these

---

[1] For Mach, see E. Mach, *Beiträge zur Analyse der Empfindungen* (Jena, 1886), or the English version, *The Analysis of Sensations* (Chicago, 1914): popularized in England by Karl Pearson, *The Grammar of Science* (London, 1892). Cf. also B. Russell, *Our Knowledge of the External World* (London, 1914).

respects, our conceptual inheritance is re-created in each new generation by all the processes of 'enculturation', whether imitation or interaction, instruction or formal education. In other respects, of course, these very forms of life and thought are merely cultural expressions of capacities and sensitivities common to all men, or even to all higher animals: features 'built into' the human brain and body, during the organic evolution of our species from its progenitors. In those other respects, our conceptual skills may still need to be learned— just as *walking* needs to be learned—but the effect of that learning is simply to evoke and develop capacities whose primary basis is physiological. The consequent patterns will vary between cultures or societies only in minor details, and will be correspondingly resistant to change.

It remains an open question, today, exactly which features of our conceptual inheritance have a predominantly physiological or genetical basis, and which features call for explanation in predominantly cultural or educational terms. To some extent, no doubt, every aspect of collective intelligence has both physiological and cultural aspects, with different cultures representing alternative expressions of 'native capacities' spread widely, or even universally, through the human species. The capacity to create language-in-general (*langage*) thus seems to be a 'species-specific' human capacity, requiring a particular constellation of neuro-anatomical and physiological correlates; while the development of particular languages (*langues*) represents so many alternative expressions of this general human capacity. In our own generation, the task of disentangling the contributions of genetical inheritance (nature, native capacities) and of environment (nurture, enculturation) to intelligent human behaviour promises to be as tricky as the corresponding task facing our grandfathers in general physiology and medicine.

To begin with, however, our analysis of concepts, language, and collective understanding need not be hampered by this particular difficulty. We need only recognize that the conceptual abilities we exercise as adults are, primarily, those that we have inherited; and for many purposes it will not matter —within limits—just how far this inheritance is transmitted physiologically, just how far by enculturation. We may (e.g.) criticize the particular forms of life and understanding into which we have grown up, seeking to improve on them and working beyond them to better forms; so our individual reflective thought may innovate on, modify, and eventually replace those inherited concepts. In this case, both the original concepts and their replacements will be not merely products of a cultural process but also expressions of our native capacities. Yet that duality will make no difference to the operative questions in the case: namely, what considerations play a part in conceptual innovation, and how novel conceptual variants are to be judged. For these purposes, the earlier forms of concept remain the starting-points for all subsequent innovations, and the new 'reformed' concepts will be the potential property of all our fellows quite as much as their predecessors. Neither old nor new concepts will be manifestations of universal genetical properties, or of our private experiences, alone. So we come back to the first, inescapable point. Our personal beliefs find expression only through the use of communal concepts. The new moulds in which our individual thoughts are cast acquire a definite form only when they become—at any rate, potentially—the collective intellectual instruments of an appropriate community.

The collective aspects of understanding are, accordingly, the general topic of Part I. The individual aspects of concept-use will concern us chiefly in Part II. We may reserve until Part III our attack on the special philosophical problems that arise, when we set the personal and communal aspects of concept-use alongside one another, and reappraise the intellectual authority of our concepts and beliefs within an integrated account of human understanding.

CHAPTER 1

The Problem of
Conceptual Change

1.1 : *The Recognition of Conceptual Diversity*

CERTAIN dates in human history acquire a retrospective significance that no one could have recognized at the time. Looking
back from a later age, we see in them the point at which some
new factor, influence, or idea entered—imperceptibly but
irreversibly—into the course of historical development. A seed
was planted in men's minds or affairs, whose growth we can
trace from beginnings which were too slight to recognize. One
such date is 13 April 1769.

On that day, Captain James Cook arrived at Tahiti in
H.M.S. *Discovery*, so completing the first half of a voyage around
the world.[1] The circumnavigation had been planned as a
scientific expedition: Cook brought with him Joseph Banks and
a party of scientists assembled under the auspices of the Royal
Society. Tahiti had been selected as a convenient observation
point on the far side of the globe from Greenwich for recording
the transit of the planet Venus across the sun's disc due to
take place on 3 June 1769; so Cook's first concern on arriving
at the island was to choose a suitable headland—named Point
Venus, in honour of the planet—on which to instal the necessary astronomical instruments. These transit observations were
needed to make good one last key detail hitherto missing from
the orthodox Newtonian picture of God's astronomical Creation. By comparing the lengths of time taken by Venus to
transit the sun, as observed from a number of different points
scattered across the face of the earth, astronomers hoped to
calculate at last, in absolute measure, the actual sizes of the
planetary orbits—triangulating the heavens in the same way

[1] J. C. Beaglehole (ed.), *The Journals of Captain James Cook*, Vol. I: *The Voyage of
the 'Endeavour', 1768–1771* (Cambridge, Eng., 1955), especially pp. clxxii–cxcii; also
H. Woolf, *The Transits of Venus: a study of Eighteenth Century Science* (Princeton, 1959)

that a navigator triangulates the ocean, or a surveyor the land.

This premeditated part of the expedition's mission was delicate but comparatively straightforward, and it was completed without any serious hitch. On her return home in July 1771, however, the *Discovery*'s voyage became the talk of Europe for quite other and unforeseen reasons. The problem of astronomical distances might stir the imagination of scientists and theologians, but most educated Europeans shared Alexander Pope's belief that 'the proper study of mankind is Man'. So the most forcible impression left by the expedition's reports came not from its intended astronomical results, but rather from some curious anthropological by-products. The customs of the Tahitians proved far more intriguing than the distances of the planets. Life on Tahiti, though tranquil enough, was surprisingly free and easy: the inhabitants got along happily, even though disregarding many of the taboos and commandments regarded in Europe as essential elements in God's Order. Eighteenth-century Europeans read accounts of the voyage with a mixture of amazement and envy. The amiability of the Tahitian women became a by-word, and the name of Point Venus acquired quite unintended associations. The new, romantic ideal of the Noble Savage, as the incarnation of a Human Nature undistorted by the demands of society, convention, and prejudice, had—it seemed—sprung to life on the islands of the South Seas.

This confrontation was just one index of a conflict between the traditional assumptions of philosophy and theology and the facts of historical and cultural variety, which the subsequent 200 years have only sharpened. Cook's voyage was planned to establish finally the eternal structure of God's Creation; instead, its outcome was to concentrate attention on the variety and apparent inconsistency of men's moralities, cultures, and ideas. From this point on, it became progressively more difficult to square men's factual knowledge about the theories and practices of their own fellow humans with accepted views about the impartial standpoint from which rational judgement and appraisal were to be conducted.

Looking back, we can see that these accepted views had telescoped several independent demands, and the immediate effect of new discoveries in history and anthropology was to drive a wedge between those demands. As a theoretical problem, the search for an 'impartial rational standpoint' was one of the starting-points for the whole tradition of Western philosophy. If the testimony of the senses referred—as Heraclitus had insisted—only to particular moments and places, we needed some more permanent theoretical principles to adjudicate between the contradictions in their evidence. If the same mutability and contingency undercut the basis of language also— as Cratylus next inferred—we needed, in addition, some more enduring criteria to guarantee the accepted meanings of words. And, if 'justice' were not to be merely a name for 'the will of the politically stronger'—as Thrasymachus argued in Plato's *Republic*—philosophers must show how social and political disagreements could be resolved by appeal to general principles, rather than by resort to naked power. Indeed, one implication of the Gospel According to Socrates had been the importance of 'keeping the talk going'; and keeping it going in terms that did not discriminate, in principle, between the parties to any issue. So, from the beginning, the problem of rationality was equivalent to the problem of keeping man 'open to reason'. This meant establishing both an impartial forum or court of reason, in which all men would have the same intellectual standing, and also impartial methods and procedures, whose even-handedness they could all alike acknowledge.[1]

The alternative was what Socrates called *misology*: that is, a contempt for rational debate. Failing an impartial forum and procedures, rationality would end by going the same way as justice. Truth would yield to the belief of the loudest-mouthed, soundness to the ideas of the most respectable, validity to the intellectual methods of the most persuasive. In the theoretical as in the practical realm, disagreements would be decided by the balance of power rather than of principles; and the pursuit of well-founded intellectual positions would be replaced by a verbal clobbering-match—a matter (in Lenin's pungent phrase) of 'Who whom'. Stated in these terms, the problem of identifying and characterizing an impartial standpoint for rational discussion, comparison, and judgement has been a

---

[1] About the idea of the 'Court of Reason', let me acknowledge a debt to conversations with Peter Herbst at Oxford in the early 1950s.

permanent preoccupation of philosophy. And the obstacles to the rational resolution of men's disagreements—which are both the symptoms of misology, and a measure of the practical difficulties of countering it—are the same today as they were in Socrates' time: a refusal to listen to your opponent's case, an obsession with power rather than principles, and a willingness to impose, by violence or threats, opinions which have failed to carry conviction on their merits.

From early on, however, all the philosophical theories proposed as solutions to this problem began to develop in a single direction. The need for an impartial forum and procedures was understood as calling for a single, unchanging, and uniquely authoritative system of ideas and beliefs. The prime exemplar of such a universal and authoritative system was found in the new, abstract networks of logic and geometry. In this way, 'objectivity', in the sense of impartiality, became equated with the 'objectivity' of timeless truths; the rational merits of an intellectual position were identified with its logical coherence; and the philosopher's measure of a man's rationality become his ability to recognize, without further argument, the validity of the axioms, formal entailments, and logical necessities on which the claims of the authoritative system depended. Yet this particular direction of development—which equated *rationality* with *logicality*—was never compulsory. On the contrary (as we shall shortly see) accepting this equation made an eventual clash with history and anthropology inevitable.

For the time being, however, this equation held the field; and, so long as it did so, it limited the options considered in the philosophical debate. The impartial forum of reason was defined as requiring an unchanging system of axioms, or principles; the only question for discussion was, how one should explain the *source* of those universal principles. Plato and Aquinas, Descartes and Kant each looked for their ultimate source, or ground, in a different direction. For Plato, rationality was ultimately associated with certain 'ideas' external to the human mind, whose validity was independent of our individual opinions, but which we could—so to say—come to 'see' with the 'eye' of the mind. The philosopher helped men use their intellects in such a way that the 'timeless truths' about these ideas made themselves intuitively apparent to them. For the medievals, the objective ground of rational knowledge lay, rather, in the Divine Mind. Human insight into the permanent principles of rationality relied on God's grace, not on man's skill. Descartes combined these two positions: the ultimate ground for confidence in the unchanging principles of rationality lay, for him, in a harmony or correspondence established by God between those 'ideas' which the human mind found totally 'clear and distinct' (e.g. the basic concepts of Euclid's geometry) and those structures in the external world to which they supposedly corresponded. For Kant, this compromise was not good enough. How could we ever, in the nature of the case, establish such an isomorphism between our 'clear and distinct' ideas and any 'external reality'? In the last resort, all we could really claim to know about with certainty was the ideas themselves. Our confidence in their rational claims must be internal, lying within the rational organization of thought itself, and the principles of rationality must be those through which we ourselves give structure to our experience . . . And so the debate has continued.

Still, despite their detailed disagreements, all these philosophers worked within the same, self-imposed general limitations. Whatever the ultimate source of rationality, all concerned assumed that its principles were—and must be—historically invariant. For Kant, this invariance was especially important. By his time, philosophers were beginning to recognize the growing challenge from historical and anthropological discovery, and Kant took care to guard himself against misinterpretation. The principles of human understanding (he insisted) could not just be empirical generalizations about the actual thinking habits of all human beings, past and present; they were not a matter of 'mere anthropology' alone. Universality, in the factual sense, could give the rational principles of human knowledge no intellectual *authority*: on that basis, they would never become, in Kant's word, 'apodeictic'. Rather, an effective critique of reason and judgement must find the source of its principles *a priori*, in a form that could impose itself on all rational thinkers, independently of their cultural and historical differences. So, despite his Copernican Revolution in philosophy, Kant ended by re-emphasizing the historico-cultural invariance of all valid rational procedures; and he went so far as to claim for the whole of Newtonian mechanics (including even the inverse-square law of gravitation) the same uniquely rational status that had originally belonged to logic and geometry alone.[1] So for Kant, as for Plato, the rationality of a man's thoughts was still to be judged by universal, *a priori* principles; for Kant, as for Plato, one particular natural philosophy was uniquely sound both in form and in content; and for Kant, as for Plato, the supreme intellectual merit of this natural philosophy lay in its systematicity and coherence. It was this commitment to a single universal system of intrinsically rational principles which, during the nineteenth century, finally came into head-on collision with the discoveries of history and anthropology.

From very early on, however, this dominant tradition in philosophy had turned itself into a recipe for intellectual authoritarianism rather than genuine impartiality. Here as elsewhere, the belief in 'universally valid' and 'intuitively self-evident' principles encouraged a certain self-righteousness and parochialism. When the Hellenes contrasted themselves with the barbarians, or eighteenth-century Europeans prided themselves on their superiority to the ancient Britons and other 'primitive' peoples, they could hardly avoid supposing that the ultimate, uniquely valid truths would turn out to support their own basic concepts and methods of thought. So the initial impact of historical and geographical discovery was somewhat muffled.

After all, at a certain level, the sheer variety of men's tastes, habits, and social structures had been familiar to European thinkers since well before the 1770s; and its full significance took some time to be recognized. In his treatise on *L'Esprit des Lois*, Montesquieu had contrasted human people and cultures existing in different geographical locations, with differing climates and soils, resources and traditions, and had used these environmental differences to explain why the societies in question ended up with correspondingly varied systems of law, institutions, and economic techniques—as well as varied national temperaments and artistic tastes.[2] Yet for Montes-

---

[1] I. Kant, *Prolegomena to any Future Metaphysics* (1783), especially Section 38.

[2] For Montesquieu, see *L'Esprit des Lois* (Geneva, 1748); Montesquieu's analysis established a tradition of static, structural ideas which has kept its hold in France

quieu these differences still amounted to something less than outright inconsistencies. They still represented so many expressions of the same fixed social laws and principles, resting on the same permanent 'nature of things'; or, alternatively, so many structural variations on a common theme, reflecting different equilibria between the relevant conditions of social life. And even a writer as liberal and unorthodox as Voltaire could treat the variety of human customs as a superficial matter, behind which the fundamental demands of Reason and Nature remained as fixed and ineluctable as ever: 'There is only one Morality, as there is only one Geometry.'[1]

Montesquieu and Voltaire, however, knew little about cultures and societies outside the Middle-Eastern and European traditions. China and India had barely emerged from the realms of the fabulous; the tribal societies of Central Africa and North America could still be ignored as barbarous and primitive; while, about the lost civilizations of the Incas and Aztecs, the testimony of the Spanish *Conquistadores* was both sketchy and suspect. There was adequate evidence about the classical Romans, Greeks, and Persians, about the various European nations and, to a lesser extent, about the Jews and the Muslims; all of whose cultures were comparatively homogeneous, being based alike either in the Mediterranean Basin or in the cultural traditions of the Bible. So, until almost 1800, the surface variety of laws, customs, and practices found in known human societies could still be interpreted as different manifestations of a single static Human Nature, operating everywhere according to the same fixed, quasi-geometrical principles.

It took the circumnavigations of such men as Cook and Bougainville and the reports of European travellers to South and East Asia to open up the last and remotest parts of the globe, so displaying the full spectrum of human variability. The fillip which zoological and botanical discoveries were giving to speculations about taxonomy and the origins of organic species, the resulting anthropological observations gave also

---

up to the present day, as evidenced by (e.g.) Claude Lévi-Strauss's *Structural Anthropology* (Paris, 1958), English translation by C. Jacobson and B. G. Scheop (New York, 1963). See also *Discovery*, Chapter 5.

[1] Voltaire, *Dictionnaire Philosophique* (1767), s.v. 'Morale': ed. J. Benda, Vol. II (Paris, 1943), pp. 160-2.

to speculations about morality and the origins of society. In Darwin's crucial arguments for organic variability, evidence about the present geographical distribution of species was to dovetail with geological and palaeontological evidence about the past. So too, around 1800, the discoveries of ethnology and anthropology had united with a more profound understanding of the historical past into a powerful argument for cultural variability; and subsequent nineteenth-century developments only gave strength to this argument.

The moment men acknowledged the full range and variety of customs and moral ideas accepted in different cultures and epochs, they could no longer see these, at all plausibly, as nothing more than alternative effects of eternal and universal causes, alternative expressions of static social laws, or alternative solutions of timeless social equations. Rather, the temptation was to fly in the opposite relativist direction, and to conclude that moral ideas and practices were a matter for local option, like codes of law and administrative procedures; so that one must say of moral principles, too, *cuius regio eius lex*. At first, however, the appeal of relativism was confined to matters of practical conduct: e.g. to questions of morality. Within the social fabric of Tahiti, free love and infanticide seemed as accepted and respectable as monogamy and parental affection in eighteenth-century Europe; and these evident differences raised the doubt, whether all such arrangements were not quite arbitrary.[1] But while the customs and morals of the Pacific Islanders might challenge the ethical presuppositions of Europeans, there seemed at first no corresponding reason to regard (say) the numbering-system, cosmology, and animistic beliefs of the Australian aborigines as genuine alternatives to European mathematics, astronomy, and physics. So, for some 100 to 150 years, men avoided extending the arguments for ethical relativism to mathematical and scientific disciplines or concepts. The intellectual claims of Euclidean geometry and Newtonian physics—which Kant had attempted

---

[1] Beaglehole, op. cit, pp. clxxxviii–clxxxix. As the work of Moerenhout was later to show, the actual situation in Tahiti was more complex than first reports had suggested: the scandalous but thrilling behaviour that titillated the imaginations of eighteenth-century Europeans turned out to be characteristic largely of a single, specialized sub-group of Tahitians, having semi-priestly functions.

to establish by transcendental, *a priori* arguments—still set them apart in man's minds from the rest of European thought and from the proto-scientific ideas of other cultures.

A few idealist philosophers (it is true) took historical change and variety seriously enough to argue that all the basic categories of human thought and theory—mathematical and physical, as well as ethical and theological—were the current products of an historical sequence, rather than being 'universal and apodeictic' features of the 'pure reason'. But most nineteenth-century men did not find this argument compelling. Only since the year 1900—particularly since 1905—has it become finally evident that intellectual judgements and concepts are exposed to an historico-cultural variety, or relativity, comparable to that of legal practices, moral beliefs, and social institutions. As we now realize, the numbering-procedures, colour-nomenclatures, cosmogonies, and technologies of different societies rest on principles that differ as radically as those underlying different moral attitudes and social organizations. Meanwhile, within the citadel of physical science itself, the classical system of Euclid-cum-Newton no longer claims the unique intellectual authority it preserved for much of the nineteenth century. So the chief barrier to extending the relativist argument from practical conduct to intellectual concepts has been removed. By now, the challenge of historico-cultural variety confronts us as sharply in intellectual as in practical life.

As a result, we find ourselves today in a new and perplexing situation. If the eighteenth century placed its ultimate reliance on Reason or Nature, and the nineteenth found its intellectual confidence in the providential workings of History, the twentieth century has been plagued by the unsolved problem of Relativity. Over the last seventy years, men have finally become aware that the relativity of human judgements affects not merely morals, religion and personal relations, but all other types of concepts—including even our most fundamental scientific ideas—as well. And this relativity holds not only between cultures separated by distance—between Rome and Florence, China and Japan, North America and Southern Asia—but between different communities and individuals within a single

city. It sets apart not only men and epochs divided by time—
the nineteenth from the eighteenth century, modern from
medieval society, the twentieth century A.D. from the eighth
century B.C.—but the men of successive decades and generations. What concepts a man employs, what standards of
rational judgement he acknowledges, how he organizes his
life and interprets his experience: all these things depend—
it seems—not on the characteristics of a universal 'human
nature', or the intuitive self-evidence of his basic ideas alone,
but also on when he happened to be born and where he
happened to live.

This discovery puts us in a quandary. If all men's concepts,
interpretations, and rational standards—in morals or in practical life, in natural science or even mathematics—are historical
and cultural variables, so that our habitual modes of thought
are as much reflections of our particular time and place as our
habitual modes of social behaviour, then the same fundamental
problem arises in each case. What solid claims can *any* concepts
and modes of thought have on our intellectual allegiance? If
every culture has its own fundamental ideas, and all its judgements must be interpreted with an eye to those particular conceptions, would it not be as much of an anachronism or failure
of understanding to judge the intellectual beliefs and concepts
or one culture or epoch entirely from the standpoint of another
as it is in the case of their moral or social attitudes? How are
we, in that case, to decide what concepts have a genuine intellectual authority over us, or any serious right to our attention?
In this new situation, indeed, what meaning can the idea of
'rational authority' have for us, beyond the local and transitory
claims of some one community and epoch? Not surprisingly,
our century has seen the growth of a healthy political scepticism towards absolute claims for national sovereignties and
ideologies; yet, at the same time, the very rapidity of scientific
development has encouraged many non-scientists to depreciate,
and to undervalue, even the qualified intellectual authority to
which those ideas are properly entitled. (Surely that's just how
we think about the world *now*?) Thus the old Socratic problem
of finding an 'impartial standpoint' for rational judgement
arises today in a new and less tractable shape.

We are brought up with certain ideas about society and

morality, about geometry and algebra, about matter and the universe; we learn to regard certain methods of investigation and types of argument as rational or scientific, and others as superstitious or muddle-headed—only to find that elsewhere, and at other times, quite different ideas, methods, and arguments have carried equal conviction and authority. What claim can then be made, beyond that of custom, on behalf of (say) the geometrical ideas of Euclid or Riemann or Minkowski, as compared with the spatial conceptions of the Australian aborigines or the classical Chinese? On what rational foundation can we base a preference for the political theories of Locke or Marx or Marcuse over those of the Aztecs or Bismark or de Maistre? And what scope remains in technology, aesthetics, and social thought, for rational judgements about our tastes, choices, and patterns of life? In every sphere, the recognition of conceptual diversity gives the problems of rationality and authority an embarrassingly sharper cutting-edge.

In one respect, however, the twentieth-century situation is also quite new. The facts of history and anthropology have, by now, finally driven a wedge between the Socratic problem and its Platonist solution. The rational demand for an impartial standpoint remains pressing and legitimate. The choice is still one between the exercise of superior power and respect for even-handed discussion, between the authoritarian imposition of opinions and the intrinsic authority of well-founded arguments. But we can no longer afford to assume that our rational procedures, however impartial, find a guarantee in *unchanging principles* mandatory on all rational thinkers—still less, in some uniquely valid system of natural and moral philosophy.

The belief that human knowledge should be governed by fixed principles may retain a certain attraction as a philosophical dream; but when it comes to understanding and appraising the actual basis of our claims to knowledge, that belief is no longer of any help to us. In Kierkegaard's words, 'Concepts, like individuals, have their histories, and are just as incapable of withstanding the ravages of time as are individuals.'[1] Our recognition of intellectual variety makes it impossible for us to

---

[1] S. Kierkegaard, *The Concept of Irony* (1841; Eng. tr. London and New York, 1966), Introduction, para. 2. This book was in fact Kierkegaard's doctoral dissertation.

ignore any longer the present-day version of the Socratic problem. To put that problem in a single sentence:

> Failing the guarantees formerly provided by the assumption of fixed principles of human understanding, how else can the impartial forum of rationality—with its even-handed procedures for comparing alternative sets of concepts and methods of thought—find a philosophical foundation that is acceptable in the light of our other twentieth-century ideas?

1.2: *Frege, Collingwood, and the cult of Systematicity*

From the classical Greeks right up to the eighteenth century, then, every 'new instauration' of philosophy was provoked by a similar quandary, and this quandary was in each case the occasion for a similar critique. In these respects the twentieth-century problem—of reconciling the claims of rational impartiality with the diversity of men's actual modes of thought—is simply a sharpened-up version of a dilemma already familiar to Socrates and Montaigne. Still, each time that philosophy has completed this intellectual *ronde*, the questions at issue have become that little bit more specific and precise; and it is especially worthwhile here to examine the reactions of twentieth century philosophers to this particular dilemma.

The recognition of conceptual and intellectual diversity has in fact had a strongly polarizing effect on twentieth-century positions. Most philosophers have reacted as though two possibilities alone were open, in direct opposition; and we can learn as much by asking why these particular alternatives so often appear the only possible ones, as we can by scrutinizing either position separately. As so often happens with polar extremes, we shall find that—for all their surface incompatibility—the two positions are united by one common presupposition. Both still accept the familiar assumption that rationality must be equated with logicality, and that different concepts and beliefs can be compared 'rationally', only so long as they can both be referred to a single 'logical system'.

The two dominant reactions are, firstly, to deny history; secondly, to bow to it. On the one hand, there are those philosophers who acknowledge the facts of conceptual and historical variety simply as facts, but allow them no relevance to the central questions of philosophy. On the contrary (they retort) the greater the diversity in men's actual concepts and beliefs, the more important it is to define an 'objective' standpoint in terms of 'absolute' standards of rational judgement; these standards should, for preference, be stated in abstract general terms, even at the price of losing touch with the actual complexities of historical change. On the other hand, there are those who are so deeply impressed by the actual diversity of human ideas that they go to the opposite extreme: abandoning any demand for a universal, objective standpoint as no longer tenable, and falling back on local, temporary, or 'relative' standards. The former, absolutist reaction repeats, in twentieth-century terms, the same move towards an abstract formalism modelled on mathematics by which Plato short-circuited the Socratic problem, and Descartes evaded the scepticism of Montaigne. The absolutist treats the actual diversity of men's concepts and beliefs as a superficial matter, behind which the philosopher must find fixed and enduring principles of rationality, reflecting the pure, idealized forms of concepts. By contrast, the relativist takes the historico-cultural variety of concepts too seriously. Instead of ignoring the diversity of conceptual systems, he yields entirely to it, abandons the attempt to judge impartially between different cultures or epochs, and treats the notion of 'rationality' as having no more than a local, temporary application.

At this point, let us consider the position taken up by one distinguished representative of each approach, and use their two arguments to demonstrate the general characters—and common assumptions—of the rival positions. Our representative absolutist is Gottlob Frege, whose writings did as much as anything to revive the 'mathematicizing' approaches of the Platonist tradition around 1900, and did so—quite explicitly —as a means of protecting philosophy from subordination to the facts of history and psychology. Our relativist is R. G. Collingwood, a philosopher of talent and enthusiasm who also wrote first-class professional history, but who let himself be tempted into relativism, as the price which must be paid in order to rescue philosophy from the historical irrelevance into which Frege's very success had led it.

Neither Frege nor Collingwood would happily have accepted

the other as a philosophical ally, and on the surface their central theses were in direct contradiction. Yet even contradictory statements have something in common: they both accept the validity of the question to which they give or imply incompatible answers. So, in considering the arguments of Frege and Collingwood, we must take particular care to identify the shared question over which their positions contradict each other. If we can bring this question—and the assumptions on which it rests—into the open, we may then find a middle way which undercuts those assumptions and avoids the embarrassing consequences into which both absolutism and relativism drive us.

To start with the absolutist argument: suppose that Descartes had been told about the cosmology of the Australian aborigines, or the geometry of the ancient Chinese. What bearing would these have had on his own philosophical position? The answer is, None. In Descartes's eyes, history had no more intellectual significance than foreign travel—it broadened a man's range of experience, but it did nothing to strengthen his reason. The sheer multiplicity of human conceptions and misconceptions did not make them philosophically any the more interesting. Though the varieties of human error were limitless, the truth was one and unitary; and the philosopher's task was to see past all these variations to the inner core of rationality by which all men are united. 'The power of forming a good judgement and of distinguishing the true from the false, which is properly speaking what is called Good Sense, or Reason, is by nature equal in all men.'[1] The exercise of this universal, divinely-implanted power alone gives intellectual authority to its products, and we can properly speak of rationality only where that Universal Reason has been put to work. As a result, the standards of rational judgement must be formulated in timeless, a-historical terms, and be relevant equally in all historical and cultural contexts.

The Platonist strand in Descartes's philosophy was revived, at the end of the nineteenth century, by Gottlob Frege, who

---

[1] R. Descartes, *Discourse on Method*, Part I, second sentence. See (e.g.) *Descartes' Philosophical Writings*, ed. Haldane and Ross (Cambridge, Eng., 1911; rept. New York, 1955), p. 81.

promulgated the original programme of 'conceptual analysis' in his *Foundations of Arithmetic*. During the nineteenth century, (Frege argued) the discussion of *concepts* had become intolerably muddled, through the failure of philosophers to make it clear whether their interest in concepts had to do with necessary logical relations, or with merely empirical considerations. As a result philosophers had come to use the word 'concepts' in several inconsistent ways: 'Its sense is sometimes psychological, sometimes logical, and sometimes a confused mixture of both.'[1]

Frege himself was rebelling particularly against the tendency to telescope formal and prescriptive 'laws of thought', which were the proper concern of logic, with empirical and descriptive 'laws of thinking', which were the business of cognitive psychologists; and his fire was concentrated against any account of logic and mathematics which might make those pre-eminently rational subjects dependent on the empirical laws governing actual 'mental processes'. His opposition to adulterating philosophy with psychology extended also to history. Having distinguished the philosophical (or logical) analysis of concepts from the psychological (or empirical) study of thought-processes, he went on to distinguish it with equal sharpness from the history of ideas.

The historical approach, with its aim of detecting how things began and arriving from these origins at a knowledge of their nature, is certainly perfectly legitimate; but it has also its limitations. *If everything were in continual flux, and nothing maintained itself fixed for all time, there would no longer be any possibility of getting to know about the world, and everything would be plunged into confusion.*[2]

It would be hard for Frege to make his own essential Platonism —his conviction that all true knowledge must be founded ultimately on changeless, ahistorical properties, relations, or principles—more explicit than he does in these last words.

As philosophers (Frege goes on) we should ignore all merely empirical discoveries, whether about the development of

[1] G. Frege, 'On Concept and Object' (1892), Eng. tr. in *Translations from the Philosophical Writings of Gottlob Frege*, ed. P. Geach and M. Black (Oxford, 1966), p. 42.

[2] G. Frege, *The Foundations of Arithmetic* (Breslau, 1884), tr. J. L. Austin (Oxford, 1950), p. vii. (My italics in this and the two following quotations.)

understanding in the individual mind or about the historical evolution of our communal understanding.

We suppose, it would seem, that concepts grow in the individual mind like leaves on a tree, and we think to discover their nature by studying their growth; we seek to define them psychologically, in terms of the human mind. But this account makes everything subjective, and if we follow it through to the end, does away with truth. *What is known as the history of concepts is really a history either of our knowledge of concepts or of the meanings of words.*[1]

Philosophers must concern themselves with 'concepts' only as timeless, intellectual ideals, towards which the human mind struggles, at best, painfully and little by little. 'Often it is only after immense intellectual effort, which may have continued over centuries, that humanity at last succeeds in achieving *knowledge of a concept in its pure form*, by stripping off *the irrelevant accretions which veil it from the eye of the mind.*'[2] For a Platonist like Frege, as for Descartes earlier, the actual 'conceptions' current in any existing community are philosophically significant only as an approximation to the eternal system of ideal 'concepts'. Philosophical questions about rationality, intellectual allegiance and authority arise only in terms of that ideal system of concepts; and any actual, historical set of conceptions has a legitimate intellectual claim on us, only to the extent that it approximates to that ideal.

Given Frege's initial preoccupation with the philosophy of pure mathematics, his approach was not unreasonable and soon led to some significant mathematical results. He had set himself the task of defining the fundamental concept of arithmetic—i.e. *number*—in terms that succeeded in 'stripping off' all the 'irrelevant accretions which veil it from the eye of the mind', and so displaying it 'in its pure form'. For his purposes, it was beside the point to ask how men's actual use of number-conceptions had developed historically, or what differences anthropologists had found between the methods of counting and figuring used in different cultures; such factual studies merely chronicled the changing meanings of number-words in our historical gropings towards fully adequate or 'pure'

[1] G. Frege, *The Foundations of Arithmetic*, loc. cit.
[2] Ibid.

number-conceptions. A rationally based arithmetic, by contrast, must concern itself with the ideal and final system of number-concepts, and this will provide a unique intellectual standard, or template, for judging all men's earlier and cruder proto-arithmetical creations. The analysis of number-concepts must therefore be undertaken using the instruments of logic alone. It calls for the construction and interpretation of a rigorous network of formal definitions and relations, to supplement the axiomatic system already being worked out for arithmetic by the Italian mathematician, Peano.

Frege's *Foundations of Arithmetic* served as a philosophical example which was soon followed by others. The programme he first enunciated in the 1890s became a model for Bertrand Russell's work on philosophical logic, and for half a century's research on philosophy of science, especially in Vienna and the United States. Just how Frege's successors put his design for philosophy into execution in detail, we need not examine here; it is enough to demonstrate their continuing commitment to his original absolutist principles. In the case of Russell, this is not hard to do. From his early paper 'On Denoting', by way of *Principia Mathematica* and on up to his classic lectures on *The Philosophy of Logical Atomism*, Russell conceived his philosophical task in terms that remained strictly within Frege's original limits.[1] He defined that task by appeal to a central distinction between the 'grammatical forms' of the *sentences* used in different 'natural' or historically existing languages—which might, but more often did not, display in a perspicuous manner the common meaning which those different sentences were intended to convey—and the 'logical form' of the underlying *proposition* which truly carried that meaning. The philosophical task was to peel away all the intellectual confusions which the idiosyncrasies of grammatical form and custom impose on our thought, and bring to light the underlying propositions whose logical forms and relations alone are unambiguously significant.

So, in his early writings, Russell explained Frege's 'irrelevant

---

[1] B. Russell, 'On Denoting', *Mind*, 14 (1905), 479–93. 'The Philosophy of Logical Atomism', *Monist*, 28 (1918), 495–527, and 29 (1919), 32–63, 190–222, 345–80; also B. Russell and A. N. Whitehead, *Principia Mathematica* (Cambridge, Eng., 1910, 1912, 1913).

accretions' as produced by the misleading aspects of ordinary language and everyday usage. This explanation added one detail to Frege's position, without changing its essentials. For the early Russell, as for Frege, concepts and propositions remained ideal, timeless entities which were captured at best incompletely by the colloquial words and sentences employed at one or another moment in history. The true character of those timeless entities could be displayed only in logical terms, as a system of necessary relations; this meant that philosophers must develop a logical symbolism and calculus, by which to extend Frege and Peano's treatment of arithmetical concepts, first to mathematics as a whole, and then to the remaining concepts of natural science and practical life. In this way, one might finally separate off the philosophical analysis of concepts proper, which aimed at a formal system of necessary relations, both from the historical study of changes in our collective conceptions and word-meanings, and from the psychological study of intellectual development in the individual. In this way alone we could be sure of escaping the twin heresies of 'psychologism' and the 'genetic fallacy'.

The task of realizing Frege's intellectual programme completely has, of course, proved much larger than he and Russell alone could carry through. Still, their work provided the first indispensable foundation, and the formal calculi of Frege's *Begriffsschrift* and Russell and Whitehead's *Principia Mathematica* soon acquired the same sovereign authority for twentieth-century mathematical philosophers that Euclid's geometry had possessed for Descartes. At its most ambitious, their philosophical goal was then to integrate the entire body of positive scientific knowledge: this goal was referred to in the 1920s and 1930s as a Unified Science. Where Klein had mapped pure geometry on to arithmetic, where Peano and Russell had founded pure mathematics on pure logic, where Hamilton and Hertz had converted physical dynamics into rational mechanics, supporters of the Unity of Science movement planned to turn the whole of natural science into a single logical system. Having added further primitive terms, postulates, and correspondence rules to Russell's symbolic logic, they hoped to incorporate all genuine branches of science into the resulting axiomatic edifice; and, since the essential mathematical portions of this edifice were best displayed in the existing logical formalism—particularly, the so-called 'lower functional calculus'—the same notation should be extended and adapted to serve the purposes of the whole. As a result, the symbolism of mathematical logic became, as Euclid's geometrical ideas had been earlier, the obligatory medium for expounding a coherent and unified scientific theory or Weltbild.[1]

So much for programme and manifestos; when put into actual effect, however, Frege's design for a universal, quasi-mathematical system of 'pure' concepts has run into difficulties. By analysing our standards of rational judgement in abstract terms, we avoid (it is true) the immediate problem of historical *relativism*; but we do so only at the price of replacing it by a problem of historical *relevance*. To see how these difficulties arise, we must begin by drawing a distinction that Frege's own successors do not always bear in mind, between the questions:

(1) whether the concepts in any field of study lend themselves to Frege's style of formal analysis at all, and
(2) what light this formal analysis throws on the rationality of intellectual changes in the field concerned.

Sharing Frege's commitment to the mathematization of philosophy, Russell and the Vienna Circle philosophers naturally accepted also the traditional Platonist faith in the special virtues of logical relations and logical systematicity. Demonstrating the *formal possibility* of analysing any set of concepts in logical symbolism has appeared to them all the evidence required that the resulting system is *applicable in practice*. They have not sufficiently faced the question, how any abstraction can be self-validating, or can guarantee its own relevance. In pure mathematics, where questions of empirical relevance are set aside, this approach may again have some justification. Once we go beyond Frege's own chosen realm of arithmetic, however, we must ask afresh how far this method is in fact helpful. Leaving mathematics aside for the moment,

---

[1] See my essay, 'From Logical Systems to Conceptual Populations', in *Boston Studies in Philosophy of Science*, Vol. VIII, ed. R. S. Cohen and R. Buck (Dordrecht and New York, 1972).

is it safe to extend this Platonizing approach to the concepts used in other fields? In those other fields of intellectual activity, does the historical and cultural variety of conceptions and judgements, standards and criteria once again overlay universal abstract ideas of a formalizable kind? Can the philosopher hope in those cases, too, to strip away 'irrelevant accretions' from the historico-anthropological variety of local conceptions and so reveal, in their 'pure form', ideal timeless concepts valid for all human beings alike, regardless of their cultural backgrounds, professional interests and preoccupations? Or is the ambition to extend Frege's method beyond arithmetic delusory?

Something of the kind can perhaps be attempted in parts of formal logic, but there are special reasons why this is so. Though different speech-communities clothe their logical operations in contrasted linguistic garb, all languages contain some locutions, tones of voice, or other devices whose functions are those of negation, conjunction, etc. In this case, therefore, it is reasonable to search behind differences in usage, for standard forms of operations which embody in each case the same universal 'laws of deduction'. Since even the simplest human languages allow for some elementary logical and arithmetical operations, one can see how Frege and his successors could regard Man's historical conceptions of *negation* or *number* as gropings towards the ultimate formulation of 'pure concepts' and feel that they could, in these cases, safely ignore the complexities of historical and anthropological fact.

Outside logic and pure mathematics, this Olympian stance is not so easily maintained, nor is history so readily escaped. Consider, for instance, physics and political theory. There we may wish to compare (say) the dynamical concepts of Buridan, Aristotle, and Einstein, or the political ideas of Machiavelli, Plato, and Marx. If we do so, it becomes highly doubtful whether Frege's fundamental assumptions still have any relevance; for how can we now contrast the mere historical facts about men's local and temporary 'knowledge of concepts' with the true philosophical authority of 'concepts in their pure forms'? In substantive and developing fields of enquiry, like dynamics and political theory, the philosopher's central task is no longer to recognize how, 'after immense intellectual effort', humanity at last 'stripped away irrelevant accretions' from the 'pure concepts' in question, and so arrived at the perfect idealizations which alone have philosophical interest or authority. Rather, it is a step-by-step task, of recognizing the considerations which justify replacing one set of theoretical conceptions by another within the historical sequence (*weight* and *impetus* by *mass* and *momentum*, *polis* by *nation-state*, *rank* by *class*) and of finding impartial procedures for comparing the merits of the concepts actually employed in different contexts.

For these purposes—namely, for comparing the merits of alternative concepts within an historical sequence—the method of formal idealization is in fact of little value even in arithmetic itself. For suppose that we follow Frege's recommendations in such a case: confining our attention to the 'pure form' of the concept *number*, and dismissing all men's actual numbering-procedures as mere historical gropings irrelevant to philosophy. This will prevent us from even asking about the rational adequacy (or inadequacy) of the successive steps in these preliminary 'gropings'. Yet that is a perfectly reasonable question to ask. It is just as significant to ask whether, and in what respects, the counting-procedures of the ancient Greeks were an improvement on those of the Egyptians, or the computational methods of modern Europeans on those of the medievals, as it is to ask the corresponding questions about physical political concepts.

By focusing exclusively on the internal structure of idealized logical systems, moreover, Frege's method distracts us not only from the process of conceptual change, but also from questions about the external application of conceptual systems, as put to practical use. Even supposing that we could find an elegant and consistent way of presenting (e.g.) the standards of intellectual judgement used in natural science as a timeless, ideal system—as Frege's logical empiricist successors hoped to do in their 'inductive logic'—the formal construction of such an abstract schema would still be only the first step in a larger task. In addition, we must show how the resulting ahistorical standards bear on the actual practice of scientists working in all the varied fields and milieus over which this inductive logic claims authority. Generalizing Frege's abstract, Platonizing approach (in a word) does not free us from the problem of historico-cultural relevance; once we insist on being told how such formal analyses apply to real-life arguments, framed in historically existing concepts, that problem recurs in full force. Do the dynamical, or zoological, arguments of all cultures and epochs really share a common, underlying quasi-mathematical form, despite all their apparent variety and diversity? How, then, does this universal abstract form throw light on (say) the specific transition from Buridan and Oresme to Galileo and Newton, or on the differences between the dynamical concepts of medieval Europe and classical China? We can scarcely be expected to take the universal applicability of the Fregean approach for granted without examination; rather, this ought surely to have been demonstrated explicitly, with profuse historical illustrations.

Confronted with this problem, Frege's successors have consistently turned away from it. Where scientists like Galileo and Newton, Hertz and Einstein, have recognized that the empirical scope and relevance of their abstract theoretical systems had to be shown, the logical empiricists have taken the applicability of their formal artefacts for granted. Rudolf Carnap's system of inductive logic, for instance, was expounded not in terms of real-life scientific examples, but in a formalized logical symbolism whose relevance to actual scientific languages was always assumed, never demonstrated. When scientifically knowledgeable readers complained that the resulting formal system was too general and abstract to have any manifest bearing on the arguments used in contemporary theoretical physics (say), Carnap's answer was uncompromising. If that was so, he replied, it was because quantum mechanics had never been formulated according to the rigorous standards of modern logic—a reply which ignored where the real burden of proof lies.[1] Similarly, Carl Hempel has developed a formal analysis of confirmation in scientific theory, and speaks of this analysis as being expressed in 'the language of science'. But he goes on to explain that he means, by this last phrase, 'the lower functional calculus with individual constants ... universal

[1] See R. Carnap, *Logical Foundations of Probability* (London, 1950), p. 243. 'The structure of the new physical theory ... is so comprehensive and complicated that no physicist at any stage in the development has given a complete and exact formulation of it according to the rigorous standards of modern logic ...'

quantifiers for individual variables, and the connective symbols of denial, conjunction, alternation and implication'.[1] Once again, the philosopher's version of 'the language of science' turns out to be, not a mode of discourse ever employed in the actual work of professional scientists, but that very symbolism of twentieth-century formal logic whose relevance needs to be demonstrated.

We can now press this point home. Universal authority may be claimed for an abstract, timeless system of 'rational standards', only if it has first been shown on what foundation that universal and unqualified authority rests; but no formal schema can, by itself, prove its own applicability. Until the problem of authority is dealt with, our capacity to construct alternative logical systems is limited only by our formal ingenuity.[2] Given these alternatives, we must then face the additional question, 'Why are we to accept this formal analysis, rather than that?'—a question which is, evidently enough, concerned less with the internal consistency of the rival systems than with their power to throw light on the merits of substantive arguments. At this point, the full irony of the twentieth-century situation becomes apparent. Frege's philosophical successors have repeatedly chosen between rival systems of standards on the basis, not of their external relevance, but of considerations as formal and abstract as the systems themselves: above all, for their conformity to the 'perspicuous symbolisms' developed by Frege and Russell for the purposes of mathematical logic. Basing one's whole analysis of 'rationality' on a formal or aesthetic preference for the symbolism of the lower functional calculus (say) is, indeed, to mistake the trappings of intellectual authority for its substance!

At this stage, we can at last see the fatal weakness in Frege's reaction against history and psychology. An abstract, timeless analysis of the criteria of rationality must be shown to apply, not merely to judgements within a given family of concepts, but also to comparisons between different sets of concepts or

[1] C. G. Hempel, 'Studies in the Logic of Confirmation', *Mind*, Vol. 54 (1945); reprinted in *Aspects of Scientific Explanation* (New York, 1965), p. 35, n. 44.
[2] See the exchange of letters between Ernest Nagel and myself in *Scientific American*, April 1966, provoked by my review of Hempel's *Aspects* (see previous note) in the February 1966 issue.

conceptual systems. Where our task is to judge alternative hypotheses within the scope of a single scientific theory, the formal procedures of logic and probability-theory—significance-tests, techniques of curve-fitting, and so on—may serve us well; but the moment we have to compare hypotheses framed in terms of different theories, we go outside the scope of those procedures.[1] Any attempt to judge conceptual novelties in science, or to make comparisons across the intellectual boundaries between rival theories, soon drives us beyond the range of a purely formal analysis.

A very few contemporary logicians have attempted to come to terms with this crucial distinction, between the formal (or 'logical') considerations relevant within a given theoretical system, and the informal (or 'dialectical') considerations relevant as between successive theories. Willard van Orman Quine, for instance, deals with this distinction in his own way, by restricting 'logic' to the formal relations within given systems of concepts and propositions; the moment he begins to consider (e.g.) how we justify replacing one set of terms or concepts by another, he drops the mathematical idioms of symbolic logic for the pragmatic idioms of expediency. What 'entities' we shall accept in our account of the world, for instance, depends on what formal language we decide to use in describing it—in this sense, we are condemned to an 'ontological relativity'—but our best choice of language is never arrived at on a 'logical' basis alone, i.e., as the necessary conclusion of a formal inference. It can be justified only informally, in utilitarian terms, on the grounds of economy, simplicity, and convenience.[2] Yet, as every historian of ideas knows, this pragmatical approach to conceptual change yields—at best—

[1] Cf. Stephen Toulmin, Critical Notice of R. Carnap, 'Logical Foundations of Probability', *Mind*, 62 (1953), 86–99.

[2] See W. V. Quine, 'On What there Is', in *From a Logical Point of View* (Cambridge, Mass., 1953), especially pp. 18–19; also *Ontological Relativity* (New York, 1969), especially pp. 89–90. For a more stereotyped formalist reaction, which takes refuge from the threat of relativism in a quasi-mathematical absolutism, see I. Scheffler, *Science and Subjectivity* (New York, 1967), Chapter I, 'Objectivity Under Attack'; and also Mary B. Hesse, 'Positivism and the Logic of Scientific Theories', in *The Legacy of Logical Positivism*, ed. P. Achinstein and S. F. Barker (Baltimore, 1969), p. 114, which makes it crystal-clear that the author regards 'a logic of scientific theories' as the only conceivable alternative to 'a kind of historical relativism in which theoretical inference is regarded as being fundamentally irrational'.

a crude simplification of complex trains of reasoning. Pragmatists and positivists may tell us that Copernicus's theory of the planetary motions superseded Ptolemy's because it was 'simpler' or 'more convenient', but these words do less than justice to the issues actually facing the astronomers of the Renaissance. If Kepler and Galileo preferred Copernicus's new heliostatic system, their reasons for doing so were far more specific, varied, and sophisticated than are hinted at by such vague terms as 'simplicity' and 'convenience': especially at the outset, indeed, the Copernican theory was by many tests substantially less simple or convenient than the traditional Ptolemaic analysis. When we consider the conceptual changes between (say) successive physical theories, therefore, the rationality we are concerned with is neither a merely formal matter, like the internal articulation of a mathematical system, nor a merely pragmatic matter, of simple utility or convenience. Rather, we can understand on what foundations it rests, only if we look and see how, in practice, successive theories and sets of concepts are first applied, and later modified, within the historical development of the relevant intellectual activity.

The absolutist reaction to the diversity of our concepts, thus, emancipates itself from the complexities of history and anthropology only at the price of irrelevance. Even where an idealized system serves as a template in criticizing earler conceptions, as in arithmetic, such an analysis can tell us about the formal shortcomings of earlier ideas only as judged from a later, perfectionist point of view: it still throws no light on the rational adequacy of the individual steps by which, bit by bit, earlier men moved towards the idealized system. The progressive historical transformation of our ideas—involving the displacement of one proto-arithmetic, physical theory, or political doctrine by another—remains to be analysed and judged in other, less formal terms; and what those terms might be, Frege's absolutist analysis gives us no means of telling.

We may now turn to the arguments of R. G. Collingwood, which illustrate the opposite, relativist reaction to the facts of conceptual diversity. This approach takes good care to avoid the defects of historical irrelevance, but in doing so (as we shall see) it ends in equal difficulties, by denying itself any impartial standpoint for rational judgement.

From our point of view, Collingwood's *Essay on Metaphysics*—first published in 1940—has some outstanding merits, and we shall study the arguments for relativism as presented in that book.[1] In recent decades, philosophical relativism of one kind or another has been extremely popular, yet nobody has presented the case for the position in an explicit and general form with the same care as Collingwood took in this *Essay*. In vaguer terms, relativism has had obvious attractions for some time. In ethics, the rejection of absolutism led initially to the recognition of multiple moral authorities, each claiming its own local validity. The notions of 'right' and 'wrong' so came to be culture-dependent, and there no longer seemed to be any appeal against the collective moral standards of a particular time and place. (When in seventh-century Rome, one could only approve and disapprove as the seventh-century Romans did.) Subsequently, the arguments of the existentialists fragmented moral choice still further, with the individual's ethical judgement claiming supreme authority, at the expense of any communal code whatever. On this view, the only truly defensible choice is one in which the individual agent confronts an historically unique situation to the best of his character and ability, and accepts all the implications of the consequent 'agonizing' choice on his own personal responsibility. Each of us must make up laws binding on himself as he goes along.

Yet it is essential to recognize, as Collingwood did, that the fundamental arguments for relativism place no weight whatever on the subjective or sentimental character of ethics and aesthetics—the features traditionally relied on as distinguishing those areas of choice from (say) mathematics and natural science. Rather, those arguments arise immediately out of the facts of conceptual diversity and, as such, they apply to all areas of thought and action equally. Being 'enculturated' by our upbringings as we are, we think and act in terms of

---

[1] The argument discussed here is taken in the form set out in R. G. Collingwood, *An Essay on Metaphysics* (Oxford, 1940). See also M. Mandelbaum, *The Problem of Historical Knowledge* (New York, 1938); A. Shalom, *R. G. Collingwood: Philosophe et Historien* (Paris, 1967); and Louis O. Mink, *Mind, History and Dialectic* (Bloomington, Ind., 1969).

intellectual and moral presuppositions characteristic of our own culture; these presuppositions determine not only what kinds of conduct we consider right and wrong, but also what kinds of phenomena we regard as puzzling or self-explanatory, what sound or established picture of the world we use to interpret our experience, what types of scientific argument and evidence we find cogent or plausible, and so on; and, since men's intellectual standards have varied between different historical and cultural milieus in just the same way as their ethical and aesthetic preferences, we are faced with the inescapable question, what intellectual authority can be claimed —in principle—for one set of standards rather than another. How, indeed, can any rational standard carry weight at all, outside its original context? The only safe position (the relativist declares) is to concede final authority within any milieu to the particular intellectual standards current in it, while denying those standards any relevance or authority outside their original contexts. So a generalized relativism abandons the search for an absolute standpoint, not merely over the subjective questions of ethics and aesthetics, but also over intellectual matters. How we think or what we can understand depends entirely on the general presuppositions with which we are brought up; the search for a rationality transcending one particular milieu is the pursuit of a will-o'-the-wisp.

More recently, parallel arguments have penetrated to the very heart of philosophy, by way of the theory of language. Kant's claims for the universal authority of the 'pure and practical reason' were questioned almost immediately by Herder, who suggested that our fundamental categories were relative to a particular historico-cultural milieu, and by now the resulting critique of Kant's position has been carried through to the bitter end.[1] For, if Wittgenstein's final philosophical position is at all sound, this means that the very language through which our enculturation is achieved is itself intelligible only to men who share enough of our own modes of life. Any particular 'natural language'—he argues—comprises a variety of 'language games', whose significance is derived from the 'forms of life' of the communities in which the language in

---

[1] See the essay on Herder by A. O. Lovejoy, included in the collection, *Essays in the History of Ideas* (Baltimore, 1948), especially pp. 169–70.

question is learned, spoken, and put to practical use. In the absence of shared 'forms of life', linguistic communication must break down: 'What has to be accepted, the given, is (so one could say) forms of life.'[1] The philosophy of language thus becomes part of a wider 'natural history of man'. We shall find universal concepts and categories, represented in all languages and intelligible to the men of all cultures, only to the extent that equally universal patterns of life, thought, and conduct provide a shared framework for their use. And the actual existence of such universal patterns of life cannot be philosophically guaranteed.

So much about the background to relativism, considered as a general position: next, we must scrutinize Collingwood's own presentation of it.[2] In order to appreciate the force of his arguments, we must first look at his idiosyncratic account of the formal relations holding within conceptual systems—e.g., the logical connections between the different terms, questions, and propositions of a scientific theory. In any such theory or system (Collingwood tells us) the most comprehensive principles are located at the summit of a logical hierarchy or structure, while the progressively more specific propositions that are logically dependent on them occupy successively lower levels; and up to this point there is nothing unusual about this account. Beyond this point, however, his account is unorthodox in one significant respect. Most recent philosophers had thought of our concepts as organized into 'axiomatic' systems: that is, into systems the *truth* of whose general principles implies—and is in turn reinforced by—the *truth* of the specific propositions deduced from them. (For the empiricists, truth has flowed primarily 'upwards' from particular observation-statements to the general theoretical statements for which they were the supporting evidence; while, for the rationalists, it has flowed 'downwards', from general laws and principles to the particular statements of which they provided an interpretation; but on either account the relations concerned were truth-relations.)[3]

[1] L. Wittgenstein, *Philosophical Investigations* (Oxford, 1953), p. 226. Cf. also p. 82: '*The common behaviour of mankind* is the system of reference by means of which we interpret an unknown language', and para. 241, p. 88.

[2] Collingwood, *Metaphysics*, Chapters 4 and 5.

[3] I. Lakatos, 'Criticism and the Methodology of Scientific Research Programmes', *Proceedings of the Aristotelian Society*, 69 (1968), 149–86.

Collingwood rejected all such views equally. A formalized axiomatic structure, he argued, is at home only in those branches of pure mathematics whose basic concepts have been definitively settled: most typically, in an intellectually fossilized system like Euclidean geometry. Elsewhere, specific statements and questions depend on more general ones in a different manner. Our concepts form not axiomatic systems, but systems of 'presuppositions'; and the logical relations between propositions on different levels of generality are not truth-relations but meaning-relations. Thus, specific questions either 'arise' or 'do not arise' depending on what more general principles are assumed, and broader assertions are related to narrower ones, not as axioms to theorems, but rather as presuppositions to consequential questions. So it is not the truth of general principles that determines—or is determined by—the truth of particular statements; instead, specific statements rely on the validity and applicability of more general doctrines for their very *meaning*. (In Collingwood's terminology, the 'logical efficacy' of narrower, more specific concepts is referred to, and made dependent on, the 'logical efficacy' of broader, more general concepts. Or, to put the point in our own terms: the narrower, more specific concepts and questions are operative only where the broader, more general concepts and principles are relevant and applicable.)

This logical structure, Collingwood argued, characterizes conceptual systems in all their parts and at every level; so that any specific element in a 'presuppositional system' is, in the last resort, operative at all only provided that its most general principles are relevant and applicable. In a natural science like physics, for instance, the most general presuppositions determine what patterns of thought are employed in recognizing and interpreting physical phenomena; and, in doing so, they determine also what questions that are accepted as operative in that field of inquiry. Thus, classical nineteenth-century physics relied on a whole string of tacit assumptions: such as that the movement of inanimate bodies can normally be explained quite independently of their colours and smells, and that determinate actions or forces can be identified to account for all changes in linear momentum. Unless these presuppositions hold good, the specific concepts and questions of classical physics are no longer operative; in this sense, their very meaning depends upon the validity of those presuppositions. And certainly—from an historian's point of view— such an account throws a worthwhile light on Newtonian mechanics. For what would be the effect of abandoning the general axioms of Newton's dynamics entirely? To do so would not merely falsify a large number of statements about 'forces', and their effects on the 'momenta' of bodies, that were previously supposed to be true. It would actually strip these terms of meaning, so that the statements in which they were employed would cease to arise, be operative, or even make sense.

At this central point in his argument, Collingwood introduces a convenient and attractive pair of technical terms. Within the body of a presuppositional system, concepts and questions are related to other concepts and questions on adjacent levels, both above and below, in such a way that the logical efficacy of those on each level depends upon the logical efficacy of those on the next more general level.[1] Thus, physical questions about optical dispersion will have a determinate meaning, only within the scope of the broader concept of 'refractive index'. Our use of the term 'optical dispersion', that is to say, presupposes the relevance and applicability of the normal laws of refraction. (This is so, because the dispersion of a transparent substance is defined as the manner in which its refractive index varies with the wavelength of the refracted light.) Yet this presupposition remains—in Collingwood's terminology—merely a 'relative' one. For the concepts of 'refraction' and 'refractive index', in their turn, presuppose the applicability of other, still broader terms—such as 'light-ray'. Questions about refraction have a straightforward meaning, only as applied to optical phenomena which can be described straightforwardly in terms of light-rays; so that the whole terminology of refraction presupposes, in turn, yet more general ideas. In this way the same concepts which are presupposed at one level will, at another, be dependent on yet more general presuppositions. (That is what makes them, for Collingwood, merely relative presuppositions.)

When we arrive at the summit of a conceptual hierarchy, we

[1] Collingwood, *Metaphysics*, pp. 27ff.

face a family, or constellation, of general presuppositions which depend on no others of a yet more general kind. On this final level, our concepts and principles represent what Collingwood calls 'absolute' presuppositions, and the validity of a whole mode of thought can depend on the relevance and applicability of such general principles. Whereas relative presuppositions have logical relations in both directions, absolute presuppositions stand on their own feet. Once an argument reaches this final level, no more basic assumptions remain to be brought into the open. Abandoning a set of absolute presuppositions means giving up the corresponding pattern of thought in its entirety. This distinction between absolute and relative presuppositions is essential, Collingwood argues, for a proper understanding of the part played by conceptual systems, both in systematic science or philosophy and in the history of thought. In every field, the beginning of wisdom lies in recognizing how the patterns of conceptual change reflect the presuppositional structures of conceptual systems; and every intellectual discipline develops through a succession of distinct historical phases, characterized by a different set of concepts, questions, and propositions, and forming a self-contained hierarchy of presuppositions.

If we take Collingwood's picture of conceptual history in general terms, we can make it attractive and plausible; though, if we press him beyond a certain point, difficulties of interpretation arise. One can see, for instance, how a term like 'light-ray' serves as a fundamental notion for the whole of geometrical optics, so that abandoning this notion will mean doing without geometrical optics as we know it. (In the same kind of way, it would be hard to imagine a dynamics which totally lacked the concept of inertia.) When we examine Collingwood's account in detail, however it becomes unclear at just what level our presuppositions finally cease in practice to be 'relative'—i.e. dependent on other, still broader principles—and instead become 'absolute', or self-sustaining. What exactly does Collingwood regard as a specific example of an absolute presupposition? Over this question, his own illustrations do not help us much. For instance, he formed a highly personal conception of the development of physical science between 1680 and 1930, which divided that development into three successive stages: these were governed respectively—he said—by a Newtonian presupposition to the effect that 'Some events have causes', a Kantian presupposition to the effect that 'All events have causes,' and an Einsteinian presupposition to the effect that 'No events have causes.'[1] Now, the concept of *cause* has certainly played significantly different parts in physics at different times in the last three centuries, but Collingwood's three-stage formula does only the roughest of justice to this story; and, in general, his own applications of the theory of presuppositions to conceptual change in the natural sciences do his case more harm than good.

All the same, bad illustrations should not be allowed to distract us from the crucial feature of Collingwood's account. When he argued that certain fundamental concepts are—so to say—constitutive of the sciences within which they are used, he is on strong ground. Without the concepts 'light-ray' and 'inertia', geometrical optics and dynamics as we know them would be at an end. Other neighbouring sciences might perhaps step in and take over their fields of enquiry, as the collateral heirs and successors of those earlier sciences; but that is another matter. And Collingwood was also on strong ground in his next philosophical claim: namely, that the crucial intellectual choices in a science involve changes in its most basic concepts, of a kind that are intelligible only if we study them against their historical background. In the last resort, we can produce an adequate conceptual history of an intellectual discipline, only if we scrutinize the processes or procedures by which entire conceptual systems—'absolute presuppositions' and all—displace one another in the course of its historical development.

So Collingwood's *Essay on Metaphysics* confronted philosophers, directly and in precise terms, with the problem of conceptual change. This problem was presented not just as a local difficulty about (say) ethics or physics alone; instead, it was an entirely general difficulty, applying equally to any conceptual system characterized by a coherent structure of concepts and presuppositions. The understanding of conceptual change accordingly involved a three-stage enquiry, of an essentially historical kind. In order to map the intellectual development of a discipline, we must first 'identify different

[1] Collingwood, *Metaphysics* pp. 51–5; cf. also pp. 325, 328, 333.

constellations of absolute presuppositions' in the area of our concern; secondly, we must 'study their likenesses and unlikenesses', so as to discover which features do—and which do not—change in the transition from one historical phase to the next; and finally, we must go on 'to find out *on what occasions and by what processes* one such constellation has turned into another.'[1]

This threefold historical enquiry was, in Collingwood's view, the only legitimate task left for 'metaphysics'—but that was an idiosyncratic usage of the term, which we do not need to pursue here. For the moment, we must reconstruct the steps by which Collingwood's account landed him irrevocably in historical relativism. Put concisely, the argument goes as follows:

(1) The intellectual content of a discipline, at any given stage in its development, comprises a system of concepts and principles operating on different levels of generality.

(2) Our acceptance of concepts and propositions on the lower levels of generality is 'relative to' those on the higher levels, and such lower-level concepts and propositions are presupposed only 'relatively' to those on the more general levels.

(3) When we reach the most general level of all, our reasons for accepting concepts and principles cannot be explained in terms 'relative to' any more general considerations; so that those upper-level concepts and propositions are presupposed not relatively, but 'absolutely'.

(4) At any stage in the development of a discipline, different propositions and concepts can be rationally compared, to the extent that they are both operative 'relative to' the same constellation of absolute presuppositions.

(5) But no common, agreed principles or procedures of judgement are available for comparing propositions or concepts 'relative to' different constellations of absolute presuppositions, or for comparing different constellations of absolute presuppositions in their entirety.

(6) So propositions and concepts can be rationally appraised only 'relative to' one particular constellation of absolute presuppositions, viz. that within which they are operative;

[1] Ibid., p. 73.

and, once we leave the scope of one particular framework, we leave also the scope of rational comparison and judgement.

In the course of this argument, the undoubted fact that our rational standards depend in part on the historical context of judgement—what we have called the 'diversity' or 'relativity' of concepts and judgements—is taken as a reason for confining rational comparisons within one particular historical context. Historical relativity (in a phrase) is taken as entailing historical relativism; the need to remember the differences between intellectual contexts, when making comparisons between them, is made a ground for limiting rational judgement to relations holding within a single context.

So, where Frege replaces traditional metaphysics by a formal analysis of 'pure concepts', Collingwood replaces it by the historical analysis of 'absolute presuppositions'. When we try to put his new programme into effect, however, we quickly run into difficulties as grave as those facing Frege's programme. Those difficulties arise—remarkably enough—at just the same point as before: viz., over the problem of conceptual change. In particular, when it comes to explaining the rational considerations justifying men's actual transitions from one set of basic concepts (absolute presuppositions/patterns of explanation/calculating procedures) to its historical successor, neither man can give us the critical tools we need. Frege had dismissed all such historical questions as 'merely empirical', and concerned himself only with 'pure concepts' in their final, perfected forms. Collingwood for his part recognizes the importance of the question, but leaves himself no way of answering it.

His difficulties in this respect originate at two different levels. On the surface, they reflect certain ambiguities in his private terminology. The very rigour of his dichotomy between absolute and relative presuppositions—his own absoluteness, we might say—deprives him of the rational standards he needs for appraising changes in our fundamental concepts. Yet, at a deeper level, this very absoluteness springs from his commitment to the same ideal of logical systematicity that prevented Frege's successors from dealing with the problems of conceptual change.

To begin with the special difficulties created by Collingwood's terminology: the text of the *Essay on Metaphysics* formulates, in precise terms, the problem that will be our own concern in later chapters, viz. the 'dynamics' of conceptual change and choice. To state this in our own words, 'On what sorts of occasion, and by what processes and procedures, are the fundamental concepts or constellations of presuppositions characteristic of the modes of thought current in one human generation discredited and abandoned in favour of other successor-concepts or presuppositions?'[1] Collingwood succeeds in posing this question, but he does oddly little towards answering it. Having formulated the crucial problem, he stops. The remainder of the *Essay* describes a number of static historical cross-sections from the development of scientific and philosophical thought, but it leaves the problem of conceptual dynamics entirely without a solution. About the occasions on which 'constellations of presuppositions' succeed one another, Collingwood says something, though not very much. About the processes and procedures by which one constellation 'turns into another', he says nothing at all.

This gap in Collingwood's argument was so glaring that a colleague to whom he shewed his draft manuscript advised him to explain how he conceived of such transformations: 'I have hinted . . . that absolute presuppositions change. A friend thinks readers may credit me with the opinion that such changes are merely "changes of fashion" and asks me to explain what, otherwise, I believe them to be.'[2] Collingwood's reply to this question—added to the text in a substantial footnote—is worth quoting in full:

A 'change of fashion' is a superficial change, symptomatic perhaps of deeper and more important changes, but not itself deep or important. A man adopts it merely because other men do so, or because advertisers, salesmen, etc., suggest it to him. My friend's formula, 'If we like to start new dodges, we may', describes very well the somewhat frivolous type of consciousness with which we adopt or originate these superficial changes. But an absolute presupposition is not a 'dodge' and people who 'start' a new one do not start it because they 'like' to start it.

People are not ordinarily aware of their absolute presuppositions,

---

[1] Cf. Collingwood, *Metaphysics p. 73.*   [2] Ibid., p. 48n.

and are not, therefore, thus aware of changes in them: such a change, therefore, cannot be a matter of choice. Nor is there anything superficial or frivolous about it. It is the most radical change a man can undergo, and entails the abandonment of all his most firmly established habits and standards for thought and action. Why, asks my friend, do such changes happen? Briefly, because the absolute presuppositions of any given society, at any given phase of its history, form a structure which is subject to 'strains' of greater or less intensity, which are 'taken up' in various ways, but never annihilated. If the strains are too great, the structure collapses and is replaced by another, which will be a modification of the old with the destructive strain removed; a modification not consciously devised but created by a process of unconscious thought.[1]

This reply tells us a lot about Collingwood's own position; yet it sidesteps the crucial issue and leaves the basic difficulty unresolved. For it is given entirely in metaphorical terms, and—as with all metaphors—we can interpret it literally in alternative ways. We may, of course, speak of a system of presuppositions as a 'structure' subjected to 'strains'; provided only that we are prepared to explain through what sorts of effects these 'strains' reveal themselves, and by what criteria we can recognize when their 'destructive' effects have been 'removed'. When such questions are pressed, however, Collingwood hedges. He vacillates uneasily between two possible answers, and is unable in good conscience to handle these questions consistently in either of the two ways. Do we make the change from one constellation of absolute presuppositions to another because we have reasons for doing so; or do we do so only because certain causes compel us to? Are questions about the 'modifications' in our intellectual 'structures' to be answered in terms of reasons, considerations, arguments, and justifications —that is in terms of 'rational' categories? Or must they be answered, rather, in terms of forces, causes, compulsions, and explanations—that is, in terms of 'causal' categories?

Given Collingwood's own previous argument neither kind of answer can entirely satisfy him. He cannot answer in consistently rational terms, because his own analysis forbids it. If we advance 'reasons' to justify replacing one constellation of absolute presuppositions by another, the validity of this

---

[1] Collingwood, *Metaphysics*, p. 48n.

further argument will then have to be judged in terms of some yet more general principle. This will imply that neither constellation was fully 'absolute' or self-sustaining, in the first place; so we must go on and introduce a 'super-absolute' presupposition, for deciding when it is 'rationally justified' to step from one set of presuppositions to another; and both rival sets of presuppositions—though initially supposed to be 'absolute'—will then be 'relative' to this new, super-absolute presupposition. At this stage, the elimination of the 'strains' from a conceptual 'structure' will become, once again, a standard intellectual operation within a single basically-unchanging theory. We can therefore account for conceptual change at the fundamental level in terms that Collingwood can accept as 'rational', only at the price of giving up his central thesis, that 'absolute' presuppositions are self-supporting and serve as the last court of rational appeal. Such an interpretation (that is to say) involves abandoning Collingwood's distinctive contribution to the philosophical argument.

Yet what alternative is there? Under pressure, Collingwood hints—though without explicitly asserting—that we must give up all attempts to describe 'the removal of strains' in rational terms. Yet he is clearly unhappy about this prospect. If no possibility remains of justifying such transitions rationally, all we shall have left to do is explain their occurrence causally: and then how will 'thought' come into the matter? Even in the crucial footnote, Collingwood is already trying to dodge this conclusion, as when he argues that, since people are 'not ordinarily aware' of changes in their absolute presuppositions, these changes 'cannot be a matter of (rational) choice'; but he still credits such 'modifications'—in quasi-rational language— to a 'process of unconscious thought'. Elsewhere in the *Essay*, he goes further in a causal direction. For instance, he likens the 'strains' in conceptual systems to the 'social strains' that arise within a culture, society, or civilization; and he suggests that intellectual 'strains' within systems of ideas may sometimes be associated with—may even be epiphenomena of—broader socio-historical crises. Thus, he describes the history of the steam engine from James Watt up to Daimler and Parsons 'as a parable of the time', and he finds in it a direct parallel to 'the history of the English Parliamentary System, as worked

out by John Locke before the end of the seventeenth century':
'That theory . . . became the official doctrine of European
politics in the nineteenth century, when parliamentary
constitutions on Locke's model were manufactured with as
much regularity and as much self-satisfaction as steam-engines
on the model of Watt'.[1]

Certainly, if we consider the overall development of Collingwood's own position, we can see how a liberal form of 'historical
'materialism' was a natural sequel to his earlier philosophical
idealism. Yet Collingwood never took the final step of replacing
reasons by causes completely. At the absolute level, conceptual
change might come about through 'unconscious' thought;
but it remained a matter for 'thought', not for anonymous
'forces' or 'compulsions'. What then—we are left asking—is
the exact nature of this 'unconscious thought', and how is it
supposed to operate on us? (Given Collingwood's open scorn
for psychology, it seems unlikely that he had Freud's ideas in
mind; yet what alternative interpretation are we to put on his
phrase?)[2] How far is the 'thought' that goes on at this absolute
level analogous to the rational argument taking place at
less elevated levels in a presuppositional system, and how far
does it operate rather as some kind of causal agency? . . . We
ask these questions in vain.

This was a peculiarly tantalizing moment for Collingwood to
lapse into silence. Yet his very hesitation, between a rational
and a causal account of conceptual change, is itself the most
revealing thing about his whole argument. This is the price he
pays for keeping the presuppositional systems of different
historical epochs self-contained and self-sufficient: once he
commits himself to preserving this absolute self-sufficiency, he
lands himself in a relativism from which he cannot retreat. If
he had been little less devoted a historian, he might in fact
have done something to weaken the absoluteness of his views,
and so have rescued himself from these intolerable difficulties.
He could, for instance, have taken an intermediate position,
half-way between the ahistorical abstraction of Frege's approach
and the downright relativism in which he in fact ends: arguing

---

1 Collingwood, *Metaphyisics* p. 94.

2 Ibid., Chapter 9, especially p. 118: see also Collingwood's book, *The Principles
of Art* (Oxford, 1938), pp. 62–4 and 77n.

that every field of inquiry, or discipline, takes for granted certain basic concepts, principles, or presuppositions, but that these are characteristic not so much of a particular epoch or cross-section from its development as of the entire discipline. At the level of substantive theory, the intellectual content of such disciplines might then change in a discontinuous manner— with one family of concepts being displaced by another based on contrary, and even incongruous, theoretical principles. But at a deeper methodological level all our intellectual disciplines would display a far greater continuity; however radical the changes in their intellectual content, the argument after each such transformation would continue—methodologically speaking —in very much the same manner as before.

Given this weakened version of Collingwood's position, it might still be impossible in some cases to find *theoretical* principles to justify replacing one complete theory by another, but this need not rule out the possibility of justifying that step by appeal to more general *disciplinary* reasons. For the parties to such a debate—both those who cling to the older theory, and those who put forward a newer one—would still share some common ground: not any common body of theoretical notions, perhaps, but rather certain shared disciplinary conceptions, reflecting their collective intellectual ambitions and rational methods, selection-procedures and criteria of adequacy. Both parties to the debate would (in a word) be composed of physicists working as 'physicists', of neuro-physiologists working as 'neurophysiologists', or students of jurisprudence sharing a common concern with the general aims and procedures of the law.

Much of Collingwood's formal argument would, therefore, still hold good as applied, not to the basic theoretical principles current within a science at any particular time, but instead to the longer-term disciplinary principles constitutive of such a science. If he had left this latter interpretation open—if he had treated absolute presuppositions as defining the intellectual boundaries between distinct intellectual disciplines, rather than between historical epochs within the same discipline—he would indeed have ended up in a far less damaging position. Unfortunately, his own examples rule out this interpretation. Recall, for instance, his contrast between the physics of the

1680s, the 1770s, and the 1900s, as governed by incompatible presuppositions about causality. Newton, Kant, and Einstein may perhaps have organised their respective explanations around different theoretical principles, but certainly physics as a whole developed continuously over the period from 1680 to 1910. All three men, that is to say, participated in a single, continuing rational debate, and the differences between their presuppositions were not (as Collingwood implied) rationally unbridgeable.

So, we can rescue Collingwood's genuine philosophical insights only by denying his stronger claim: viz. that each separate epoch organizes its thought around a different self-contained and self-sufficient 'constellation of presuppositions'. We can avoid being driven all the way to historical relativism (that is) only by weakening Collingwood's distinction between absolute and relative presuppositions, and matching it against the historical process of conceptual change in a more complex manner. The one thing we cannot do is pay the full price that Collingwood's actual argument exacts: i.e., give up all hope of finding rational procedures for comparing the concepts and beliefs current in different historical epochs, and restrict ourselves to discussing conceptual changes in terms of causes and effects alone.

Having looked independently at Frege and Collingwood, our representative absolutist and relativist, we may now set them alongside one another and identify the shared question to which they give opposite answers. As we have seen, the central question in each man's discussion is one about the rational authority of a certain conceptual system or systems. According to the mathematically-minded Frege, ultimate authority attaches to a single, ideal conceptual system, which is to be arrived at by stripping away 'irrelevances' or 'accretions' from all actually existing systems. The conceptions current in any historico-cultural context will then carry genuine authority, only as approximating to this unique system of 'pure' concepts. According to the historically-minded Collingwood, by contrast, idealized formal systems have no intrinsic sovereignty over the actual products of conceptual history, and every existing system of absolute presuppositions can claim sovereign authority within its own historico-cultural context. Intellectual history is thus divided into a sequence of distinct phases, and in each of these ultimate rational authority belongs to a different constellation of concepts.

Behind this direct opposition, however, Frege and Collingwood agree completely over one more fundamental point. Both absolutist and relativist treat the problem of rationality as requiring us to give final intellectual authority to one or another logical system: either an axiomatic system of propositions or a presuppositional system of concepts. Both men (that is) assume that our concepts and propositions are—and should be—linked in logically systematic ways; they disagree only over the question, which particular system is rationally authoritative. In Frege's philosophy, the importance of systematicity is evident enough: and, seeing that Frege was interested above all in arithmetic, it is also understandable. What is less immediately obvious, and less obviously necessary, is Collingwood's concurrence in the same traditional view. Yet, as we saw, Collingwood presents his whole argument in terms not of 'aggregates' or 'populations' of concepts and presuppositions, but of presuppositional 'systems'. He describes these systems as hierarchies whose internal relations are, in their own way, as tightly 'logical' as the deductive connections of an axiomatic system. Instead of challenging this anti-historical assumption— that concepts and propositions can carry rational authority, only if they form a 'logical system'—he merely offers an alternative account of the logical relations by which the elements of conceptual systems are connected together.

This fact had consequences for Collingwood of two different kinds. On the one hand, it saved him from lapsing into the sort of private jargon typical of the idealist philosophers who influenced him so much in his early years. Faced with the transition from eighteenth-century empiricism to nineteenth-century historicism, Collingwood—like F. H. Bradley, his immediate predecessor at Oxford—baulked at the final hurdle. Whereas the classical German idealists, such as Hegel, blurred the differences between synchronic logical relationships (holding within the conceptual systems of any one era) and diachronic/ dialectical relationships (holding between successive systems) to the extent of describing History as the process by which

Reason progressively unfolds its own Logic, Collingwood could not go along with this style of rhetoric. His long acquaintance with Locke, Berkeley and Hume had its effect:

> In spite of all temptations
> To belong to other nations,
> He remained an Englishman.

Whatever his feelings as a historian, as a philosopher he recoiled from this invitation to telescope logic and history into a unified 'historical dialectic'. Instead, he agreed with his empiricist predecessors that logic must be confined to the internal structure of conceptual systems. For him, as a result, logical or rational relationships held good only synchronically, within the theories of a given epoch, and relative to its particular constellation of presuppositions. This ruled out discussing conceptual change in the quasi-logical Hegelian jargon about the Inner Rationality of History. Unfortunately, it ruled out at the same time all diachronic discussion of conceptual systems at the level of 'absolute presuppositions' in terms of 'reasons' and 'rationality'—even with a small r—and it left him with no alternative theory of conceptual change, other than a vague historical materialism which accounted for it in terms of quasi-causal metaphors—'processes of unconscious thought', socio-cultural 'strains', and the like.

On the other hand, the assumption of systematicity had a second significance for Collingwood. Without it, he had no way of justifying his division of intellectual history into a succession of distinct 'phases'. Unless successive 'constellations of absolute presuppositions' were clearly different, there would be no procedure for marking off one historical 'period' from the next, or for telling just where the proper limits of rational discussion must be drawn. And, indeed, failing clear criteria for distinguishing the Newtonian, Kantian, and Einsteinian 'systems' of physics (say) as absolute and self-sufficient, the historical relativist has no clear grounds for telling his historical epochs apart. If he had dropped the assumption of logical systematicity, Collingwood would thus have destroyed his only criterion for marking off one phase of conceptual history—together with its characteristic concepts and doctrines—from another later phase, having its own distinct ideas and beliefs incommensurable with those of the earlier phase. That done, he would no longer even have had words in which to express the relativist conclusion for which he was arguing. Given distinct historical 'time-slices', each with its own 'presuppositional system', Collingwood could at least concede sovereign authority to those transient systems, even while denying the claims of any universal or 'absolute' system. If the divisions between successive phases were seen as arbitrary, there were no longer any clearly separate epochs for intellectual standards and concepts to be 'relative to'. Historical relativism then lost its objects of reference. Instead, the succession of intellectual phases, each defined by its own specific constellation, collapsed into a historical flux, with concepts and propositions changing not systematically but kaleidoscopically; and the problem of finding a source of rational authority was once again thrown wide open.

That, in fact, is what our own conclusion must be. The problem of conceptual change is intractable for both Frege and Collingwood, just because they both subscribe to the philosophical cult of systematicity, i.e. to the belief that concepts must form 'logical systems', and to the consequent equation of the 'rational' with the 'logical'. Likewise for so many other philosophical writers: whenever rational questions arise about conceptual change on the fundamental level, anyone who accepts these 'systematic' assumptions will be unable to tackle them. Whatever type of systematicity he favours—whether he interprets the 'logico-rational' relations between coexisting concepts and propositions as taxonomic or axiomatic, as hypothetico-deductive or presuppositional—in each case, his insoluble problem will be the same. He can admit questions about justification, intellectual merit, and rationality, only so long as they arise *within* the scope of some one particular 'logical system', so that 'giving a reason for p' involves relating p to the rest of the same system. At points of transition between self-contained systems, he will be forced to suspend all questions about justification and rationality. The intellectual steps involved in such transitions cannot be rationally justified in terms drawn from either system singly; and this—given the equation of rationality with logicality—will imply for him that they cannot be justified *at all*.

* * *

At a deeper level, then, Frege's absolutism and Collingwood's relativism both construe the demand for a universal impartial standpoint of rational judgement as calling for a system of objective or absolute standards of rational criticism. The absolutist asserts that, on a sufficiently abstract, quasi-mathematical level, such standards can still be formulated as 'eternal principles'; while the relativist simply denies that any such standpoint can have any universal validity. And this same common assumption prevents both men from coming to terms with the rationality of conceptual change.

How, then, are we to avoid the difficulties of both these reactions? The first step is to reject the commitment to logical systematicity which makes absolutism and relativism appear the only alternatives available. This decision brings us to the heart of the matter. For it was, in fact, always a mistake to identify rationality and logicality—to suppose, that is, that the rational ambitions of any historically developing intellectual activity can be understood entirely in terms of the propositional or conceptual systems in which its intellectual content may be expressed at one or another time. Questions of 'rationality' are concerned, precisely, not with the particular intellectual doctrines that a man—or professional group—adopts at any given time, but rather with *the conditions on which, and the manner in which, he is prepared to criticize and change those doctrines as time goes on*. The rationality of a science (for instance) is embodied, not in the theoretical systems current in it at particular times, but in its procedures for discovery and conceptual change through time. Formal logic—as Quine and Collingwood agree —is concerned simply with the inner articulation of intellectual systems whose basic concepts are not currently in doubt; such logical relations can be considered either at some particular time, or apart from time. In this sense, there is of course no 'logic' in the discovery of new concepts. But that in no way entails that conceptual changes in science do not take place in a 'rational' manner, i.e. for good or for bad reasons. It entails only that the 'rationality' of scientific discovery—of the intellectual procedures by which scientists agree on well-founded conceptual changes—necessarily eludes analysis and judgement in 'logical' terms alone.

From this point on, accordingly, we must set aside the

traditional cult of systematicity, and carry our analysis of concepts—in science and elsewhere—back to its proper starting-point. The intellectual content of any rational activity forms neither a single logical system, nor a temporal sequence of such systems. Rather, it is an *intellectual enterprise* whose 'rationality' lies in the procedures governing its historical development and evolution. For certain limited purposes, we may find it useful to represent the provisional outcome of such an enterprise in the form of a 'propositional system', but this will remain an abstraction. The system so arrived at is not the primary reality; like the notion of a geometrical point, it will be a fiction or artefact of our own making. In all our subsequent enquiries, therefore, our starting-point will be the living, historically-developing intellectual enterprises within which concepts find their collective use; and our results must be referred back for validation to our experience in those historical enterprises.

This change of approach obliges us to abandon all those static, 'snapshot' analyses in which philosophers have for so long discussed the concepts current in the natural sciences, and other intellectual activities. Instead, we must give a more historical, 'moving picture' account of our intellectual enterprises and procedures, in which we can finally hope to understand the historical dynamics of conceptual change, and so recognize the nature and sources of its 'rationality'.[1] From this new point of view, no system of concepts and/or propositions can be 'intrinsically' rational, or claim a sovereign and necessary authority over our intellectual allegiance. From now on, we must attempt instead to understand the historical processes by which new families of concepts and beliefs are generated, applied, and modified in the evolution of our intellectual enterprises; and recognise how the grounds for comparing the adequacy of different concepts of beliefs reflect the respective parts they play in the intellectual enterprises concerned.

1.3 : *Rationality and its Jurisdictions*
The problems of human understanding and rationality can be discussed in either of two contexts: theoretical or practical.

[1] See R. Causey in the collection *The Structure of Scientific Theories*, ed. F. Suppe (forthcoming).

By concentrating exclusively on theoretical difficulties, we turn these problems into questions of 'pure' philosophy, calling for analyses, definitions, and formal distinctions. ('Is the term "rational" descriptive or prescriptive?', 'How can the "same" concept operate, properly speaking, within two "different" conceptual systems?') By focusing rather on their practical implications, we treat them as methodological issues, requiring indices, criteria, and/or pragmatic measures. ('Just how far from the statistical mean does an experimental reading become a "significant" deviation from expectation?', 'By what tests will an English law-court decide that some Soviet legal concept has the same standing as, say, the concept of "negligence" in common law?') Here, we are aiming to discuss our problems, as far as possible, with an eye to both contexts at once: relying on the practical experience of substantive enterprises to suggest how our philosophical analyses might be improved, and using these improved analyses in turn to sharpen up our practical understanding of the substantive issues. In reformulating our theoretical problems about the collective roles of concepts within historically developing enterprises, we can therefore begin by asking what is to be learnt about them from the practical manner in which our rational enterprises are conducted, and how the difficulties we have here posed in purely theoretical terms are dealt with in actual practical situations. Selecting from a great variety of possible fields, let us consider examples taken from law, physics, and anthropology.

(1) To begin with the law: in the course of legal history, lawyers, judges, and professors of jurisprudence have repeatedly dealt with questions of practical procedure which came up for formal philosophical analysis only later. Solon, the practical architect of Athenian jurisprudence, preceded Socrates, the abstract philosopher of justice. Men like Bodin and Selden developed, within a legal context, those techniques of comparative analysis on which subsequent academic historians relied in developing the methods of modern historiography. Similarly, our own central theoretical problem—of making rational comparisons between the concepts and standards of different historico-cultural milieus—has a long history, on a practical level, in the legal procedures of common-law jurisdictions. In each case, the exigencies of practical affairs have

provided the material for subsequent theoretical analysis, rather than the other way around. So let us take a brief look at the term, 'jurisdiction', which corresponds most nearly in legal practice to our own problematic notion of an historical or cultural 'milieu'. Between the idea of jurisdictions and that of milieus, there are two important differences. Jurisdictions have well-defined boundaries, in a way that historico-cultural milieus do not; and clear rules exist for invoking in one jurisdiction arguments originating in another. Somehow, lawyers and judges have managed to work their way in practice through problems for which philosophers have not stated any coherent or satisfactory theoretical solution. Thus, if we demand exact criteria for distinguishing successive and supposedly distinct 'periods' of intellectual history, the answers of philosophers will tend to be either arbitrary or question-begging, or both. Phrases like 'the Periclean Era', 'the High Middle Ages' and 'the Enlightenment' are notoriously vague; yet any methodical attempt to sharpen these boundaries soon removes us to a level of generality whose relevance to actual examples is unclear. ('When does a "presupposition" become truly "absolute"?') And it is the same if we insist on exact anthropological criteria for distinguishing one 'culture' from another. The culture of Spain is, no doubt, different from the culture of Finland, but what about Galicia and Castile, Catalonia and Roussillon? Are we there concerned with so many different cultures, or rather with so many variants of a single culture? Evidently, neither 'cultures' nor 'periods' can be divided off with complete theoretical precision, and 'historico-cultural milieus' are doubly ill-defined, combining the vaguenesses of both.

By contrast, the boundaries between jurisdictions are normally clear, sharp and agreed, for the sufficient—though pragmatic—reason that we cannot afford to leave them vague. Whether geographical or topical, any such uncertainty would quickly lead to a situation in which rival courts were claiming judicial authority over the same territory or case, thus damaging the effective authority of both courts. So, however much Scottish Nationalists or advocates of States' Rights might welcome a legislative redistribution of judicial powers, the question whether in America a particular crime is a Federal or a State offence, or whether a particular matrimonial cause falls under Scots or English law, is one to which, as a matter of practical necessity, we need to give an unambiguous answer.

Courts of law have, accordingly, developed practical procedures for separating jurisdictions, and comparing their respective concepts, arguments, and standards of judgement, which circumvent the quandaries of philosophical theory. Within the common-law tradition, for instance, earlier decisions may always be called up for reconsideration, and precedents from one jurisdiction may be cited as bearing on a case now before the courts of another. Thus, the highest courts of the United Kingdom have a standing right to reconsider in the 1970s (say) decisions arrived at in any earlier decade; and they can even declare the judges responsible for some previous decision 'mistaken', however well that ruling was supported by arguments accepted at the time. Where no relevant precedent exists from the United Kingdom itself, they likewise have a further standing option: to adopt, or set aside, precedents from other common-law jurisdictions, such as New York State or the Commonwealth of Australia. In exercising these options, of course, judges always respect the relativity of judicial norms and decisions. Ancient rulings and decisions from remote jurisdictions are always scrutinized with particular care, before being accepted as bearing on a case to be decided here and now. Yet the existence of jurisdictional boundaries never by itself eliminates from consideration judicial precedents from older cases or other countries. A decent respect for judicial relativity (that is to say) never plunges the courts into mere judicial relativism. On the contrary, the judicial experience of all mankind is kept available as a reserve, on which the courts can call—with due attention to historical and cultural differences—in arriving at a just resolution of current cases.

The task of finding impartial rational procedures for comparing judgements from different milieus is a similar one; and this parallel with the law may help us to recognize our own middle way between absolutism and relativism, by focusing our attention on the practical manner in which concepts are employed in the actual conduct of intellectual enterprises. The important thing will be not to mistake genuine difficulties for outright impossibilities, or authentic problems for irresolvable paradoxes. While the relativity of legal standards to particular jurisdictions can make it genuinely difficult to compare cases from different jurisdictions, and while those difficulties are greater the more remote the jurisdictions, this difficulty in no way amounts to an outright impossibility. Given care, discretion, and sufficient attention to relevant differences, we can make all the appropriate allowances. Given a complete legal relativism, on the other hand, such comparisons would be ruled out from the start. Every jurisdiction would then be the sole and sovereign guardian of its own legal concepts, criteria, and standards of judgement, and these would have only an historical or anthropological interest to onlookers from other jurisdictions.[1] Our own task, too, will be—more generally— to make proper allowance for the relativity of rational standards to different milieus, without being misled into treating comparisons between milieus as downright impossibilities.

(2) Physical scientists are accustomed, in practice, to drawing a similar subtle yet fundamental distinction, between relativity and relativism. They too, are compelled to acknowledge the relativity of their judgements and measurements, not just as an abstract conundrum, but as a substantive difficulty arising in their own professional work. And they, too, have developed 'relativistic' procedures, which do justice to this relativity, without falling into the trap of outright relativism. The exact nature of their solution has sometimes been misunderstood. By a historical coincidence, men finally admitted the thoroughgoing relativity of moral and intellectual standards, at the very time when physicists were developing their own so-called Theory of Relativity; and for some years the slogan, 'Everything is relative!'—meaning, 'Cross-contextual judgements are impossible!'—was popularly attributed to Albert Einstein. As the general reader supposed, Einstein had somehow demonstrated that the human relativities of history and anthropology had roots in the fundamental structure of the physical world itself. Physical theory was thus believed to justify broader attitudes of subjectivism; and the title of Edward Westermarck's

[1] On this whole subject, see the excellent monograph of E. H. Levi, 'An Introduction to Legal Reasoning', *University of Chicago Law Review*, 15, No. 3 (1948), 501–74.

widely read defence of cultural relativism (*Ethical Relativity*) was read as a deliberate echo of Einstein.[1]

Yet this appeal to Einstein's authority was entirely misconceived. The intellectual strategy of relativistic physics is quite contrary to relativism. Its methods are, in fact, much more like those of the common-law tradition. Einstein began, of course, by recognizing that measurements of spatial and temporal magnitudes (e.g. position and duration, velocity and acceleration) involve a hitherto unsuspected relativity to the choice of 'reference-frame' or 'reference-objects'. What interpretation can be placed on such measurements depends on whether they are made and considered relative to, say, the surface of the Earth or a freely-falling elevator, a high-energy particle or a distant galaxy. Yet what exactly followed from this discovery? If we understood it as involving an outright relativism, we should then have to dismiss physical comparisons between different frames of reference as meaningless. Yet that was not Einstein's conclusion. On the contrary: he set out, instead, to establish general impartial procedures for making just these comparisons.[2] Einstein was, after all, a physicist. He could never have agreed to treat spatio-temporal measurements as comparable only within a single reference-system, for that would have meant abandoning his fundamental obligations as a natural philosopher. Instead, he worked out new equations for converting to the needs of one 'frame' measurements originally made relative to a different reference-frame. As a result, spatio-temporal judgements made in different frames could be safely 'transformed', without running into the theoretical difficulties that had begun to afflict the earlier 'Galilean' transformations. While the actual content of our spatio-temporal judgements depended upon the particular frame to which they

[1] E. Westermarck, *Ethical Relativity* (London, 1932). As Gerald Holton has pointed out to me recently, Einstein himself did not originally use the term 'relativity' in describing his early work on the electrodynamics of moving bodies. He spoke of his approach, rather, as an *Invariantentheorie*; and it is interesting to reflect on the different impact his ideas would have had, if he had stuck to this description throughout. For this makes it clear that his own approach was fundamentally concerned with discovering *invariants* as between different 'frames of reference', and so was *opposed* to any equation of 'relativity' with 'relativism'.

[2] See A. Einstein and L. Infeld, *The Evolution of Physics* (Cambridge, Eng., 1938), Part III.

were referred, rational procedures were thus devised for comparing measurements relative to different frames.

Our own problem has essentially the same form. While all concepts and judgements have to be considered 'relative to' the problems and traditions of the relevant milieus, that is no reason for lapsing into relativism, or rejecting cross-contextual judgements as meaningless. Once again, we must not mistake a substantive difficulty for a formal impossibility; since, by abandoning our problem as insoluble in principle, we should be ignoring our own obligations as philosophers. The practical example of Einstein's work, like the everyday procedures of the law, should encourage us rather to consider what rational procedures will circumvent the very real practical difficulties posed by differences in cultural and historical milieu. We must not, of course, pitch our expectations too high. Some of the judgements of different milieus will no doubt prove genuinely incommensurable, while other comparisons can be brushed aside as far-fetched or misconceived. ('Was the character of Brutus more or less noble than Kepler's scientific insights were profound?') It will be enough to show how, at some times and in appropriate cases, rational cross-comparisons can be meaningful. Certainly—on the face of it—'rational enterprises' do exist, within which the ideas of different epochs and cultures can be quite properly described as more or less 'well-established', 'adequate', 'discriminating', 'true', or 'rationally based'; and can be described in these terms, without merely rationalizing a parochial preference for our own current standards. If we can understand—in general terms—how such comparisons are to be made, even in a single case, that will be enough to demonstrate that the relativity of our concepts and judgements is a source of substantial but soluble problems, not of abstract and inescapable paradoxes.

(3) The attractions of relativism remain strong, even today, in one substantive field of intellectual inquiry: namely, anthropology itself. There, the problem of comparing different standards of rationality has become a serious and active methodological issue. For the working anthropologist must decide, in actual cases, how far he should pay attention to his own ideas of rationality, and how far to the considerations regarded as 'rational' within the tribe he is studying, in passing judgement

on the rational adequacy of their activities and customs. If a tribe with a long tradition of sympathetic magic insists on using homeopathic medicines in preference to antidotes, must the anthropologist necessarily accept this as 'rational' behaviour? No doubt, the members of the tribe will give their own reasons for doing so—reasons which seem to them good and sufficient—yet, in judging the adequacy of those procedures, what attitude should the anthropologist himself adopt? Confronted by this question, anthropologists were for a long time tempted to change the subject. It was easier to take the relativist way out: of considering only what was regarded as rational by any particular tribe, and avoiding the question whether that attitude was sound or unsound, well-founded or groundless. For a long time, therefore, this latter question was dismissed as illegitimate—or, at any rate, as entirely irrelevant to anthropology. Only recently, in fact, have professional anthropologists begun to discuss this problem explicitly and analytically, distinguishing the various contexts in which, and purposes for which, the question of 'rationality' can be raised.[1]

Once again, an elementary philosophical move presents itself which, at first glance, promises a straightforward escape from our problem. Surely, it may be argued, we must distinguish two different approaches to questions about rationality, and so two distinct senses in which the term 'rational' can be used. We may adopt a strictly factual approach, in which case our use of the term will be purely descriptive; we shall then simply report how, within different societies or tribes, intellectual and practical judgements are or have been arrived at —i.e. what is the case. Alternatively, we may adopt an evaluative approach, in which case our use of the term will be correspondingly prescriptive; we shall then be laying down, from our own standpoint, rules for deciding how other peoples should arrive at their own judgements—i.e. what ought to be the case. If we only respect the difference between these two approaches, this argument concludes, we can keep out of trouble. As empirical anthropologists, we can confine ourselves to the facts about rationality, in the descriptive sense. If we

---

[1] See *Rationality*, ed. B. R. Wilson (Oxford, 1970): especially Chapter 9, 'Some Problems about Rationality', by Stephen Lukes, originally published in *Archives Européennes de Sociologie*, VIII (1967), 247–64.

choose to raise further questions about rationality, in its other prescriptive sense, we shall then be raising questions of value, and acting as moralists or logicians not anthropologists.

Yet how far can anthropologists really afford to remain purely factual in their approach and dismiss all evaluations as irrelevant or unprofessional? Arguably, this self-denial is both unhelpful and unnecessary. In the first place, it distracts attention from a whole range of perfectly proper questions: e.g., questions about the actual degree of therapeutic success achieved by sympathetic magic or homeopathic medicines, and so about a tribe's motives in retaining such methods, despite their comparative ineffectiveness. Indeed, unless the anthropologist is occasionally prepared to stand back and evaluate the customs of a tribe in his own terms, he can hardly hope to understand the full significance of those customs even for the tribe itself. An exclusive preoccupation with 'factual' questions is also unnecessary. The possibility of distinguishing between 'facts' and 'values' does not oblige us to consider them in entirely separate terms. On the contrary, as the example of the common law reminds us, there are perfectly good ways of reaching evaluative conclusions 'in the light of' the factual record.

The common-law tradition is, in fact, based squarely on the method of 'precedents', in which 'factual statements' are repeatedly invoked as the sole but sufficient basis for 'judgements of value'. On a formal or functional level, it is true, common-law argumentation acknowledges the theoretical distinction between values and facts. When first promulgated, the force of verdicts and rulings, decisions and findings, is unambiguously prescriptive: on such occasions, their very purpose is to ascribe guilt or liability, and to prescribe penalties or punishments. Yet those same decisions subsequently become items in the judicial record, to be cited as historical 'matters of fact'. As such, they are reported and criticized, quoted and glossed, using a descriptive idiom. The distinction between prescriptive and descriptive usages is, thus, more easily stated in philosophical theory than applied in judicial practice. For, whether judicial findings are promulgated normatively or described retrospectively, their substance is the same in either case. So, when legal precedents are cited as bearing on some current case, they are commonly presented in a declarative idiom—'The Court of Criminal Appeal, in 1935, held that so-and-so'—even though the immediate function of these 'factual' statements is the 'normative' one of directing the Court towards a particular decision.[1]

Nor is this duality mysterious. What, after all, is the central purpose of 'appealing to precedents'? It just is to marshal the judicial experience of the past in such a way as to direct the judicial practice of the present. Within common-law arguments, of course, factual statements are never equated with normative statements. It would never be self-contradictory to say, 'It was decided thus in a parallel case, but it ought not to be so decided in the present case.' Rather, the historical record of past decisions creates standing presumption for the future. Critical comparisons with past rulings establish presumptive rules for future decisions in similar cases; and those rules are set aside, only when specific grounds can be advanced to rebut the presumptions so created. Within the forum of the courts, indeed, that is how common-law arguments acquire their binding force, how factual considerations become legally relevant, and how the historical decisions of earlier courts come to serve as judicial precedents at all.

Let me here summarize a point to which we shall have to return later. The foundations of comparative legal reasoning were analysed with particular depth by Oliver Wendell Holmes, in his book, *The Common Law*. Up to a certain level, Holmes argued, legal reasoning quite properly proceeds in a routine manner, by reapplying established rules and precedents. At a deeper level, however, its character radically changes. When it encounters the limits of the accepted rules, judicial reasoning acquires a new and more 'functional' character. At the risk of anachronism, one might even describe Holmes's account of the deepest foundations of judicial argument as expounding a kind of 'legal ecology'.[2] Once we pass beyond the scope of current

[1] Cf. G. Gottlieb, *The Logic of Choice* (London, 1968); and also H. L. A. Hart *The Concept of Law* (Oxford, 1961).
[2] See H. S. Commager, *The American Mind* (New Haven, 1950), Chapter 18, especially pp. 376, 378, and 386–8: referring particularly to O. W. Holmes, *The Common Law*. Notice in addition Mr. Justice Holmes's profound remark, 'The present has a right to govern itself, so far as it can . . . Historical continuity with the past is not a duty, only a necessity', in *Collected Legal Papers* (New York, 1920),

principles and methods, he asks, what else is there to do except to consider how far, in a changing social and historical situation, one or another extension of current procedures will best serve the proper purposes of the law? And the nature of the resulting historical process—the manner in which legal concepts emerge, develop and are eventually abandoned, through the reapplication and extension of precedents to novel, unforeseen situations —is documented in detail in such books as Edward H. Levi's *Introduction to Legal Reasoning.*[1]

Our own general analysis of rational argument will follow a similar direction. Rationality, we shall argue, has its own 'courts' in which all clear-headed men with suitable experience are qualified to act as judges or jurors. Within different cultures and epochs, reasoning may operate according to different methods and principles, so that different milieus represent (so to say) the parallel 'jurisdictions' of rationality. But they do so out of a shared concern with common 'rational enterprises', just as parallel legal jurisdictions do with their common judicial enterprises. So, if we look and see how, within the rational enterprises which are the loci of conceptual criticism and change, new concepts are introduced, develop historically and prove their worth, we may hope to identify the deeper considerations from which such conceptual changes derive their 'rationality'. As in Holmes's jurisprudence, our analysis of conceptual development will concentrate on the 'ecological' relationships between Men's collective concepts and the changing situations in which those concepts have been put to work; for this purpose, we shall be characterizing the general processes by which conceptual populations develop historically, in the same kind of way that common-law historians have characterized the historical development of legal concepts. Given a full enough analysis of these processes, we shall try to show

---

p. 139, as quoted in *American Constitutional Law: Historical Essays*, ed. L. W. Levy (New York, 1966), p. 3. If we were to read the word 'necessity' in this quotation in the traditional philosophical sense, it would give the appearance of a quite irrelevant irony, like the famous description of the situation in the disintegrating Austro-Hungarian Empire, as 'desperate but not serious'. We shall be taking up the implications of Holmes's notion of 'necessity' again much later in our enquiries, in Part III.

[1] See above, p. 89, n. 1; On the relation between anthropology and law, see also the symposium, *Law in Culture and Society*, ed. Laura Nader (Chicago, 1969).

both how the practical criteria of judgement accepted in different enterprises are in fact arrived at, and at the same time how these criteria acquire the binding force on which their authority depends.

1.4: *The Revolutionary Illusion*

By abandoning the philosophical presupposition that the human understanding should operate necessarily and universally in accordance with some system of fixed principles, we have now shifted the burden of proof. Through most of intellectual history, the stability and universality of our fundamental forms of thought has been regarded as proper and natural: intellectual change has been the 'phenomenon' needing to be explained, or explained away. Our present stance reverses the situation. Intellectual flux, not intellectual immutability, is now something to be expected: any continuous, stable or universal features to be found in men's actual patterns of thought now become the 'phenomena' that call for explanation. The burden of justification having shifted so far, the really puzzling thing about our collective human understanding becomes the possibility that certain intellectual forms, structures and procedures might in fact prove to be universal.

For instance, Noam Chomsky has claimed that all human thought and language display certain universal patterns of grammatical structures, and this suggestion can no longer strike us as natural and self-explanatory.[1] On the contrary, it must be highly surprising. Whereas philosophers in earlier centuries might see fixed and necessary 'forms of judgement' in all rational thought, for us any such constancy is no less mysterious than change. As well as asking what factors explain the actual changes in those concepts and categories that are historico-cultural variables, we must therefore consider how those concepts and categories that turn out to be historico-cultural universals can end up with the universality they do. What can exempt those universals from the variability that is now the general rule? From now on, any supposition of a universal and invariant structure of thought or grammar calls for a further scientific or historical explanation: an explanation which may

---

[1] N. Chomsky, *Aspects of the Theory of Syntax* (Cambridge, Mass., 1965); see also C. Lévi-Strauss, *Structural Anthropology* (Paris, 1958, Eng. tr., New York, 1963).

perhaps be given in neuro-anatomical terms, perhaps by appeal to our evolutionary ancestry, perhaps as reflecting the common exigencies of all human life—or a combination of all three.

This demand, that conceptual invariance and conceptual variability should be treated on a par, will impose on us a break with the whole Kantian tradition as drastic as that which Kant himself made with the ideas of his predecessors. And it is one which will oblige us, in due course, to dissociate ourselves from all those Kantian assumptions which remain influential today, in the form of Claude Lévi-Strauss's 'structuralism', in the 'necessary operations' of Piaget's psychology, and in a dozen other guises. Kant's own restatement of the problem of human understanding had, for him, the same radical simplicity and charm as the transition to Copernican astronomy. The irresoluble conundrums of traditional metaphysics had, after all, been side-effects of an ill-chosen intellectual standpoint: so many Ptolemaic epicycles, forced on philosophers as a consequence of misidentifying the true 'locus' or 'centre' of rationality, and capable of being dispelled if men only reorganized their analysis around a new centre. Yet, for all its originality and fruitfulness, Kant's new 'critical method' left philosophers as committed as ever to 'fixed and universal' principles of understanding. He simply looked for the origin of those principles in a new direction.

It is as well that, in this respect, Kant compared himself to Copernicus, rather than to Galileo or Newton. For, to twentieth-century eyes, his position has something of the same antique air as Copernicus's system. His continued assumption that the existence of 'invariant forms of reason' is entirely natural and self-evident, for instance, recalls Copernicus's continued assumption that the traditional circular orbits in planetary astronomy were perfect and self-explanatory. Today, by contrast, the inescapable necessities of Euclidean geometry and the categorical imperative are no more *self-evident* than the natural perfection of rest on the Earth or circular motion in the Heavens. And the parallel between the two men goes further. For there is reason to think that Kant's position was no more final than that of Copernicus. Where Galileo went beyond Copernicus by demonstrating that the same general kinematic rules govern both motion and rest, and Newton explained them equally as alternative effects of similar dynamical causes, Kant's position now needs to be supplemented and fulfilled by corresponding theories of conceptual stability and change, kinematics and dynamics.

From the seventeenth century on, an adequate theory of mechanics had to explain both motion and rest in the same terms; so too, an adequate theory of conceptual development should now account for both conceptual stability and conceptual change in the same terms. We have no more reason to take immutability as self-explanatory in mental philosophy (epistemics) than we have in natural philosophy (physics), or to regard stability as more 'natural' or 'intelligible' than change. Rather, we must set out to show how a single set of factors and considerations, interacting in different ways, can be used to explain both why our 'forms of thought and perception'—concepts, standards of rational judgement, *a priori* principles and the rest—vary rapidly in some cases, situations, and circumstances, and also how, in other cases, situations, and circumstances, they can remain effectively unchanged.

For the purposes of our present argument, this test is crucial; and it is one which the most widely favoured account of conceptual development entirely fails to meet. That account is constructed around a distinction between historical phases of two contrasted kinds, which are referred to respectively as 'normal' and 'revolutionary', and it was expounded most persuasively in T. S. Kuhn's book, *The Structure of Scientific Revolutions*, published in 1962.[1] This theory of 'intellectual revolutions' accounts for the processes involved in these two kinds of phases in quite different terms: so much so, that the contrast between normal and revolutionary change has acquired something of the same spurious absoluteness as the medieval contrast between rest and motion.

In Kuhn's form, the theory of scientific revolutions can be read as a response to the question which had been posed by R. G. Collingwood more than twenty years earlier, but left unanswered: viz., 'On what occasions, and by what processess is one set of fundamental concepts, or constellation of absolute

[1] Thomas S. Kuhn, *The Structure of Scientific Revolutions* (Chicago, 1962).

presuppositions, displaced by another?' During the 1950s this question had been taken up again and widely discussed, by historians and sociologists as well as philosophers. N. R. Hanson tackled them in explicitly philosophical terms in his book *Patterns of Discovery* (1957);[1] it was the topic of my own Mahlon Powell Lectures for 1960 at Indiana University, as published in *Foresight and Understanding*;[2] and it has been implicit in much recent sociology of science, e.g. in the work of Thomas Merton,[3] Bernard Barber,[4] and Joseph Ben-David.[5] However, it will be convenient here to take Kuhn's position as an object of study, and the progressive changes in this position during the 1960s will then be highly revealing. For these changes serve to underline the importance of our earlier conclusion: namely, that the choice between absolutism and relativism appears inescapable, only if we adopt an over-systematic view of the conceptual relations within the current intellectual content of a science.

There is no direct evidence that Kuhn's theory was intended as an explicit successor to Collingwood's; yet, in each case, similar assumptions led by very reasonable steps to similar conclusions. In its most familiar form, indeed, Kuhn's position displays such close parallels to Collingwood's that a glossary can be established for translating between them. To begin with Collingwood: he makes a fundamental distinction between the synchronic (or logical) relations holding between the presuppositions of any one particular culture, phase, or epoch, and the diachronic (or historical) relations holding between the presuppositions of successive cultures, phases or epochs. Within a given milieu, men normally share a particular constellation of presuppositions, and operate at the fundamental level within a common conceptual system; so they can discuss all their disagreements in straightforward, rational terms. Their joint task is then to compare the subsidiary assumptions on which those disagreements turn; and, just because their more fundamental

---

[1] N. R. Hanson, *Patterns of Discovery* (Cambridge, England, 1958).

[2] Stephen Toulmin, *Foresight and Understanding* (London and Bloomington, Indiana, 1961). Referred to below as *Foresight*.

[3] See, e.g., Robert K. Merton, *Social Theory and Social Structure*, rev. edn (New York, 1957), Part V.

[4] See, e.g., B. Barber, *Science and the Social Order* (New York, 1952).

[5] See, e.g., Joseph Ben-David, 'The Scientific Role: Conditions of its Establishment', *Minerva*, 4 (1965), 15–54.

concepts are shared, they will have a common vocabulary of debate and common procedures for resolving their differences. By contrast, at moments of transition from one intellectual epoch to another, the strains within a given system of thought become insupportable and need to be removed. When this happens, the absolute presuppositions of the era are themselves called in question, and normal rational debate ceases to be possible. There are no longer agreed procedures for settling differences, or a shared vocabulary in which to discuss them. Until a new constellation of presuppositions has established its authority, and the basic strains have been eliminated, the normal procedures of rational debate are held in suspense. At this fundamental level, conceptual changes can be discussed only in terms of unconscious thoughts, socio-economic influences, and other such causal processes.

All these conclusions can be found—though in other words —in Kuhn's *Structure of Scientific Revolutions*. Kuhn places as much weight as Collingwood on the contrast between two alternative modes of conceptual change. During long periods of 'normal' science, current ideas in (say) physics are dominated and shaped by the authority of an overall master-theory or 'paradigm'. In accepting such a paradigm, the scientists concerned determine for the time being the intellectual preoccupations and rational standards of their particular field of inquiry: what questions will arise, what forms of explanation will be acceptable, what interpretations will be recognized as legitimate. In this respect, a paradigm has the same logical role as a constellation of absolute presuppositions. The scientists working under the intellectual authority of any paradigm then form a school, of much the same kind as a school of artists. Within a school of scientists, this shared framework of ideas provides an acceptable vocabulary for expressing theoretical disagreements, and agreed procedures for resolving them—for just so long as those disagreements are not so deep as to challenge the authority of the paradigm itself. In consequence, most scientific change proceeds in a consolidatory way, controlled by the ultimate authority of the accepted paradigm. But very occasionally this normal development is interrupted by a period of crisis. At such times, the current paradigms are first challenged, and then overthrown. As a result, sovereign intellectual authority passes from one overall master-theory, e.g. the system of mechanics established by Galileo and Newton, to a successor, e.g. the mechanics of Einstein and Heisenberg; and the outcome of such a crisis is, in Kuhn's terminology, a scientific 'revolution'.

Only some of the examples discussed by Kuhn in his 1962 book, under the heading of 'paradigm-switches', conform fully to this Collingwoodian pattern; others are less drastic in their implications. Those that do so, however, are both all-embracing and rationally catastrophic in their effects; and Kuhn devotes a whole chapter to a discussion of rival paradigms as 'alternative world-views'.[1] On this fundamental level, a scientific revolution involves a complete change of intellectual clothes. Its effects are so profound that a scientist working under the authority of the new paradigm shares no theoretical concepts with one whose intellectual loyalties are still committed to its predecessor. Lacking a common vocabulary, they can neither communicate with one another about their disagreements, nor formulate common theoretical topics for discussion and research. Each man will even end up by 'seeing' the world in ways organized according to his own schema or *Gestalt*. For what he 'sees' when he looks down (say) a microscope will be governed not only by the structure of his eyes and his instruments, but also by his particular theoretical paradigm; this will determine what any particular specimen is 'seen as'—whether the scientist concerned will view it as (say) a tissue or globule, as a vesicular sac or nucleated cell.[2] In this way, tacit reliance on an idealist theory of knowledge encourages Kuhn to accept an idealist theory of perception also.

On the classical revolutionary view, therefore, 'normal' scientific growth and 'revolutionary' change are in total contrast. Normal science involves no unavoidable incomprehension between rival scientists; nor does it lead to radical changes in the *Gestalt* of our sensory experiences. Its tasks are essentially consolidatory, with all the scientists of the particular school concerned operating according to a common framework of rational ground-rules. In an outright scientific revolution, on the other hand, the displacement of one fundamental paradigm

[1] Kuhn, op. cit., Chapter 10, pp. 110–34.
[2] See *Architecture*, esp. pp. 342 ff., and the associated film, *The Perception of Life*.

by another represents an absolute and complete change. New-think then sweeps aside Oldthink entirely; so much so that, in the nature of the case, the reasons for replacing Oldthink by Newthink can be explained in the language of neither system. Like men committed to different constellations of absolute presuppositions, a Newthinker and an Oldthinker have no common vocabulary for comparing the rational claims of their respective theoretical positions. A Newtonian physicist committed to the ideal of constructing mechanical explanations for all physical phe-nomena, for example, will then be separated from a typical twentieth-century physicist by something more than the passage of time and the normal sequence of new empirical discoveries. They will be divided also by conceptual barriers which—if the full definition of a 'revolutionary' change is applicable—are rationally insurmountable. Their commitment to incompatible 'paradigms' is then reflected in unavoidable incomprehension.

On this literal-minded interpretation, Kuhn's account has the same relativist implications as Collingwood's theory. The merits of intellectual 'revolutions' cannot be discussed or justi-fied in rational terms—since no common set of procedures for judging this rationality are acceptable, or even intelligible, to both sides in the dispute. So the considerations operative within a revolutionary change must apparently be interpreted as causes or motives, rather than as reasons or justifications. Only after the victorious new paradigm is securely enthroned in acknowledged power can the rule of rationality be restored. As for Collingwood earlier, the 'rational procedures' of a science are secondary authorities, whose writ carries weight thanks to the sovereignty of the primary intellectual powers—whether paradigms or absolute presuppositions. Transfers of sovereign power and authority between one paradigm and its successor thus take place on the very frontiers of rationality. Rather than conforming to established canons or procedures, they establish novel canons of scientific rationality. New frameworks of funda-mental theory cannot themselves be arrived at in a 'rational', or 'rule-following' manner. Paradigms are sovereign; they make their own laws.

These conclusions will apply to an actual 'paradigm-switch' or conceptual change—we have said—in those cases, and only those cases, which answer to the fully-fledged definition of a 'scientific revolution', as involving the adoption of an entirely new world-view. And the first question we must raise is whether any theoretical change within a given scientific discipline has ever in fact produced so radical a discontinuity; or whether this fully-fledged definition does not exaggerate the depth of the conceptual changes actually occurring within the natural sciences. If we press this question, we shall at once be led to the same reservations about the doctrine of intellectual revolutions, regarded as a theory of conceptual change, as we entered earlier against Collingwood's doctrine of absolute presuppositions, regarded as an account of the structure of conceptual systems.

What examples might we pick on as possible illustrations of such total changes in scientific world-view? Two promising candidates are the changeover from pre-Copernican astronomy to the new science of Galileo and Newton, which was the topic of Kuhn's first, historical book,[1] and the more recent changeover from the classical physics of Newton and Maxwell to the relativistic and quantum physics of Einstein, Heisenberg and their successors, on which Kuhn has also done extensive historical research.[2] Yet in neither case does the full revolutionary schema fit the facts. It is a caricature, for instance, to depict the changeover from Newtonian to Einsteinian physics as a complete rational discontinuity. Even a cursory consideration of Einstein's influence on physics will show how little his achievement exemplifies a full-scale scientific revolution. In a highly organized science like physics, every proposed modification— however profoundly it threatens to change the conceptual structure of the subject—is discussed, argued over, reasoned about, and criticized at great length, before being accredited and incorporated into the established body of the discipline. Indeed, the more radical and comprehensive the theoretical changes proposed, the more elaborate and prolonged this discussion will commonly be. So far from physicists committed to rival paradigms reaching a state of complete mutual incom-

[1] Thomas S. Kuhn, *The Copernican Revolution* (Cambridge, Mass., 1957), esp. Chapters 5, 6.
[2] See Kuhn's recorded conversations with a number of leading twentieth-century physicists, deposited in the archives of the American Institute of Physics.

prehension—the state of affairs supposedly characteristic of a fully-fledged 'revolutionary crisis' in science—changes as profound as Einstein's must be justified by stronger reasons than are needed where less is intellectually at stake.

This point is the crux. For recall: on the fully-fledged account of a genuine scientific revolution the two parties involved (e.g. Newtonian and Einsteinian physicists) will share no common language for theoretical discussion, and no agreed procedure for comparing results. Yet what does the actual record show? The professional careers of many theoretical physicists spanned the years from 1890 to 1930, and these men lived through the changeover in question. If there had in fact been any breakdown in communication, of the sort to be expected in an authentic scientific revolution, we should be able to document it from the testimony of these physicists. What do we find? If there was such a revolution, the men directly involved were curiously unaware of it.[1] After the event, many of them explained very articulately the considerations that prompted their decision to switch from a classical to a relativistic position; and they reported these considerations as being the reasons which justified their change, not merely the motives which caused it. They did not see the switch, in retrospect, merely as an intellectual conversion, to be described by a shoulder-shrug and a disclaimer: 'I can no longer see Nature as I did before . . .' Nor did they treat it as the outcome of nonrational or causal influences: 'Einstein was so very persuasive . . .', or 'I found myself changing without knowing why . . .', or 'It was as much as my job was worth . . .' Rather, they presented the arguments that sanctioned their change of theoretical standpoint.

On the face of it, then, the switch from Newtonian to Einsteinian physics was more than a process of unconscious thought, and more than a conversion to a radically novel world-view. So the example of a scientific revolution on which Kuhn places greatest weight fails to illustrate his definition. Nor does the actual historical record about the changeover

[1] See, e.g., Max Planck, *A Scientific Autobiography* (London, 1950); Max Born, *Physics in my Generation* (London and New York, 1956); and also Einstein's intellectual autobiography in the volume, *Albert Einstein, Philosopher-Scientist*, ed. P. A. Schilpp (Evanston, Ill., 1949).

from pre-Copernican to Newtonian physics illustrate the fully-fledged revolutionary specification any better. True: a four-teenth-century scholar like Bradwardine would, initially, find it very difficult to make intellectual contact with an eighteenth-century Newtonian astronomer like Laplace. The total effect of 450 years of conceptual change in physics and astronomy would, at first, make it hard for them to understand each other's questions—to say nothing of their respective answers. Yet does this imply that the gulf between their theoretical positions was rationally unbridgeable? Were Copernicus and Galileo, Kepler and Newton the authors of a totally new and all-embracing paradigm, whose novel world-picture snapped all intellectual connections with the physics of earlier times? Once again, this is a caricature of the actual facts.

Over this point, the writings of T. S. Kuhn the historian are the best commentary on the theories of T. S. Kuhn the philo-sophical sociologist. As his historical analysis makes clear, the so-called 'Copernican Revolution' took a century and a half to complete, and was argued out every step of the way. The world-view that emerged at the end of this debate had—it is true—little in common with earlier pre-Copernican concep-tions. Yet, however radical the resulting change in physical and astronomical *ideas and theories*, it was the outcome of a continuing rational discussion and it implied no comparable break in the intellectual *methods* of physics and astronomy. If the men of the sixteenth and seventeenth centuries changed their minds about the structure of the planetary system, they were not forced, motivated, or cajoled into doing so; they were given reasons to do so. In a word, they did not have to be converted to Copernican astronomy; the arguments were there to convince them.

Taken at its face value, then, the fully-fledged definition of an 'intellectual revolution' in science is plagued by paradoxes of the same kind as afflicted Collingwood's theory of absolute presuppositions. If we are to make the theory of paradigms and revolutions fit the actual historical evidence, accordingly, we can do so only on one condition. We must face the fact that paradigm-switches are never as complete as the fully-fledged definition implies; that rival paradigms never really amount to entire alternative world-views; and that intellectual discontinuities on the theoretical level of science conceal underlying continuities at a deeper, methodological level. This done, we must ask ourselves whether the use of the term 'revolution' for such conceptual changes is not itself a rhetorical exaggeration.

The two key notions in Kuhn's account—'paradigm' and 'revolution'—are in fact separate and independent, both in their implications and in their historical origins. Originally, those who supported the doctrine of paradigms were in no way committed to a revolutionary view of paradigm-switches; and the full theory of intellectual revolutions is forced on us, only if we construe the term 'paradigm' as equivalent to the phrase 'conceptual system', as understood in the 'logical' sense of traditional philosophy.

The idea of analysing the network of explanations in physical science as built around certain fundamental patterns of explanation, or *paradeigmata*, is in fact an old one. Georg Christoph Lichtenberg, who was Professor of Natural Philosophy at Göttingen in the mid-eighteenth century, introduced the term *paradeigma* for just this purpose. (This was a period at which the foundations of modern grammatical analysis were also being laid, and the term found a parallel use in linguistics, to refer to the standard forms for the conjugation of verbs, the declension of nouns, and so on.) In physics, Lichtenberg argued, we explain puzzling phenomena by relating them to some standard form of process, or paradigm, which we are prepared to accept for the moment as self-explanatory. During the heyday of Kantian and Hegelian philosophy, this idea was temporarily eclipsed, but it was resurrected at the end of the nineteenth century, when Lichtenberg's work had the same liberating influence on German-speaking philosophers that David Hume's had on English-speaking ones. Ernst Mach, for instance, described Lichtenberg as the major influence on his own empiricist theories of perception;[1] while the term 'paradigm' was picked up by Ludwig Wittgenstein, who applied it both for its original purpose in the philosophy of science and also, more generally,

---

[1] See Mach's intellectual autobiography, in the essay 'Die Leitgedanken meiner naturwissenschaftlicher Erkenntnislehre', in *Physikalischer Zeitschrift*, 11 (1910) 599–606, reprinted in *Physical Reality*, ed. Stephen Toulmin (New York, 1970), esp. p. 38, n. 6.

as a key to understanding how philosophical models or stereo-
types act as moulds or—in the parlance of engineers—as
'cramps', shaping and directing our thought in predetermined,
and sometimes quite inappropriate directions.[1] In this form,
the term entered the general philosophical debate, first in
Britain and subsequently in the United States, where it arrived
in the early 1950s.

Among philosophers of science, the theory of paradigms was
explored in the 1930s by Wittgenstein's student, W. H. Watson,
in his book *On Understanding Physics*,[2] and subsequently by
Hanson and myself. (The term figured largely in *Foresight and
Understanding*, for instance, where I connected it up with the
related notion of 'ideals of natural order', to which we shall be
returning in later essays.) In none of those contexts, however,
was it implied that changes of paradigm necessarily take place
in an abrupt, discontinuous, or 'revolutionary' manner. On
the contrary, all the philosophers concerned would have
rejected this implication; clearly, reasons can be given for such
changes, and nothing in the theory of paradigms need imply
anything else. The theory of scientific revolutions is, thus, quite
independent of the theory of paradigms. This, rather than the
term 'paradigms', is the distinctive feature of Kuhn's analysis,
and we must study rather closely the changing manner in
which Kuhn has applied the term 'revolution' in his successive
accounts of scientific change.

His use of the term has gone through five distinguishable
phases. These are represented, chronologically, by (1) his historical work on *The Copernican Revolution*;[3] (2) his first public
unveiling of the explanatory theory of revolutions at Worcester
College, Oxford, in the summer of 1961;[4] (3) the first edition
of *The Structure of Scientific Revolutions*, largely written earlier but
published only in 1962;[5] (4) a series of transitional papers
written in response to criticisms of his theory, and delivered

---

[1] The term recurred repeatedly in his Cambridge lectures from 1938 to 1947:
see, e.g., the notes published by R. Rhees in *Philosophical Review*, 77 (1968), 271–
320, esp. p. 274.

[2] W. H. Watson, *On Understanding Physics* (Cambridge, England, 1938).

[3] *The Copernican Revolution* (Cambridge, Mass., 1957).

[4] 'The Function of Dogma in Scientific Research', in *Scientific Change*, ed. A. C.
Crombie (London, 1963), pp. 347ff.

[5] *The Structure of Scientific Revolutions*, 1st edn. (Chicago, 1962).

between 1965 and 1969;[1] and (5) his most recent, revised and amended views, as presented in his appendix to the second edition of *Structure*[2] and in his 'Reflections on my Critics', printed in the symposium, *Criticism and the Growth of Knowledge*,[3] both of which appeared in 1970.

(1) In his first, historical phase, Kuhn used the word 'revolution' in a purely descriptive manner, and its implications were, at most, prophylactic. By calling the changeover from pre-Copernican to Newtonian science a revolution, he did nothing to explain how that switch took place, or what kind of intellectual processes were involved. At this stage, indeed, the term claimed no explanatory significance. It simply marked a profound redirection of men's intellectual loyalties on the theoretical level, and cast doubt on any 'uniformitarian' suggestion that intellectual progress in science always—and properly—depends on the application of a routine Scientific Method. As Herbert Butterfield had insisted earlier, empirical observation alone could never have led to the adoption of the Copernican theory: a 'new thinking-cap' was needed.[4] Kuhn's choice of title echoed this claim.

Initially, then, the description of (say) Copernicus or Galileo, Newton or Lavoisier, Dalton or Maxwell, Darwin or Einstein, as producing a 'revolution' in his particular science, implied only that the effect of his work was to discredit an older, established set of theoretical concepts, and to replace it by another one, based on novel principles. It made no pretence at answering Collingwood's residual questions about the manner in which—the occasions on which, and the processes by which—such a transition was brought about.

(2) In the Worcester College paper and the first edition of *The Structure of Scientific Revolutions*, Kuhn employed the term 'revolution' in a very different manner. At this stage, he

---

[1] See, e.g., the paper, 'Logic of Discovery or Psychology of Research', originally prepared for the collection, *The Philosophy of Karl Popper* (forthcoming) and delivered at Bedford College, London, in the Summer of 1965; printed in a revised form in *Criticism and the Growth of Knowledge*, ed. I. Lakatos and A. Musgrave (Cambridge, England, 1970), pp. 1–23. See also the paper delivered by Kuhn at the symposium on *The Structure of Scientific Theories*, held at the University of Illinois in 1969 (ed. F. Suppe, forthcoming).

[2] *The Structure of Scientific Revolutions*, 2nd edn. (Chicago, 1970), Appendix.

[3] Lakatos and Musgrave, op cit., pp. 231–78.

[4] H. Butterfield, *The Origins of Modern Science* (London, 1957), p. 35.

treated the description of conceptual changes as 'revolutionary' phenomena as entailing the need to give a correspondingly 'revolutionary' explanation of their manner of occurrence. When our scientific ideas have to be reconstructed from the ground up, he argued, the actual processes of change take quite a different form from that found in the course of 'normal' scientific change. And he emphasized this difference, on the occasion of its first public presentation, by speaking of paradigms as 'dogmas'.[1] The role of a paradigm, as dictating the unquestioned principles around which the intellectual loyalties of a school of scientists are organized, was comparable to the role of the system of theological dogmas around which the loyalties of a religious order were built; and a paradigm-switch involved a transfer of loyalty, or intellectual conversion, comparable to that involved in the abandonment or modification of a theological dogma.

This comparison was provocative, and it succeeded in provoking. But it could be made appealing only at the price of ambiguities, both in the status of Kuhn's whole theory, and in his account of paradigms. Was he putting forward a philosophical thesis about the intellectual role of paradigms in justifying the rational establishment of new scientific insights; or was he advancing an historico-sociological hypothesis about the persuasive role of paradigms in promoting their acceptance? Correspondingly, did his use of the term 'paradigm' refer to the rational pattern of some form of theoretical explanation, whose intellectual authority was derived from its own proved explanatory merits; or did it refer, rather, to some standard presentation of that theory, in a classic book or monograph, whose authority derived at second hand from the personal authority of the man who composed it? Fuzziness over this point made Kuhn's Worcester College paper needlessly paradoxical, yet his rhetorical comparison of paradigms in science to dogmas in theology was made plausible only by blurring over these distinctions.

If we take sufficient care to distinguish the intrinsic authority of an established theory from the magisterial authority of an individual scientific master, on the other hand, the comparison between paradigms and dogmas at once loses its charm. In des-

[1] See Crombie, op. cit., pp. 347ff.

cribing the role of Newton's ideas in shaping the preoccupations of eighteenth-century physicists, for instance, Kuhn's presentation drew no distinction between the influence of (i) Newton's *Principia*, which was the founding document of classical mechanics, and (ii) Newton's *Opticks*, which had a dominant influence on eighteenth-century physical thought more generally. Taking the *Principia* first, we can use it to illustrate a worthwhile philosophical point: namely, that one function of an established conceptual scheme is to determine what patterns of theory are available, what questions are meaningful, and what interpretations are admissible, for a physicist working within (say) the Newtonian tradition of mechanics—so that, for as long as Newton's theory retained intellectual authority, its principles could serve as the theoretical Court of Last Appeal, i.e. were 'paradigmatic'. Yet this (to repeat) was a philosophical point, reflecting the methodical character of scientific procedures; and it did nothing whatever to support any claim that 'dogma' plays any part in science. On the contrary, during the years between 1700 and 1880, physicists acted quite reasonably—and undogmatically—in taking Newtonian dynamics as their provisional starting-point; and then, as now, it was always open to them to challenge the intellectual authority of the fundamental Newtonian concepts. (This permanent right of challenge, as Karl Popper insists,[1] is one of the features marking off an intellectual procedure as genuinely 'scientific'.)

Alternatively, if we take Newton's *Opticks* as our example, we can use it to illustrate a different, sociological point: namely, that in historical fact secondary or derivative scientists, like the eighteenth-century Newtonians, tend to see less of the whole picture than the primary, original workers who were their mentors, and who provided their inspiration. Notoriously, disciples tend to narrow their minds, admitting as meaningful questions, legitimate interpretations, and acceptable patterns of thought only those which they regard—rightly or wrongly—as sanctioned by the example of the master within whose 'school' they are working. Historically speaking, this failing can be advantageous, since it enables a major scientist like Newton to exert his magisterial authority, and so provide guidelines

[1] See, e.g., K. R. Popper, *Conjectures and Refutations* (London, 1963), Chapter 3.

within which lesser men are expediently confined. Yet this (to repeat) remains a point of sociology, rather than philosophy; and we can usefully discuss the function of dogma in scientific research only at this sociological level. Thus, while Newton's dynamical theories retained a legitimate intellectual authority of their own until the year 1880 or later, the influence of the *Opticks* was already having a narrowing effect before the end of the eighteenth century. By 1800, in fact, the continued authority of the *Opticks* represented little more than the magisterial dominance of a great mind over lesser ones, and the ways in which Newtonian scientists invoked this authority were beginning to lapse into dogmatism. (One need only recall the manner in which respectable Newtonians reacted to Young's wave theory of light, which they rejected out of mere respect for Newton's supposed corpuscularian views, and persisted in rejecting even after it had substantial experimental support.)[1]

By itself, then, the practice of basing theories on 'paradigms' has nothing dogmatic about it. If we cite both the *Principia* and the *Opticks* as illustrating a single theory of scientific change, we must recognize that they served as paradigms in significantly different senses of the term; and we must take care to respect the distinction between the two corresponding kinds of authority—the intrinsic intellectual authority attaching to a well-established conceptual scheme, and the magisterial or institutional authority exercised by a dominant individual or school. If we do so, we shall at once see that, taken in its philosophical sense—as defining the intellectual role of a 'paradigm' in the rational development of scientific theory—Kuhn's suggestion that scientists necessarily adhere to their 'paradigms' in a dogmatic spirit was, at best, a rhetorical exaggeration. And this, in turn, inevitably raises the question whether the description of paradigm-switches as revolutionary was not itself, from the start, a rhetorical exaggeration also.

(3) Significantly, when Kuhn's full-scale presentation of his theory appeared in 1962, with the publication of *The Structure of Scientific Revolutions*, it made no reference to the comparison between paradigms and dogmas. Kuhn had picked on this analogy (it seems) solely for effect, when writing the Worcester

[1] Cf. E. Whittaker, *A History of Theories of Aether and Electricity*, revised edn., Vol. I (Edinburgh, 1951), pp. 100ff.

College paper as a 'trailer' for the fuller-scale account of his theory. In every other respect, the book took the same directions as the paper, and was exposed to the same ambiguities. At this stage, Kuhn entered no reservations about his contrast between 'normal' and 'revolutionary' changes in science. The differences between the two types of historical processes involved were correspondingly clean, sharp, and well defined. During periods of 'normal' change, all the scientists of a particular school were working under the authority of an accepted paradigm, and were engaged in theoretical consolidation: extending the application of the same general schema of explanation ever more widely across their field of study, in accordance with procedures and patterns of theorizing whose rationality was common form. During a 'revolutionary' period, by contrast, no single paradigm was accepted as authoritative, and the science in question was temporarily in crisis: waiting for comprehensive new procedures and patterns to provide the focus around which a new scientific school could form.

The former, consolidatory state was normal, the latter revolutionary; and a well-established science like physics went through a full-scale scientific revolution only quite exceptionally. The changeover from pre-Copernican to the new science of Galileo and Newton—or so Kuhn's readers inferred—was presumably one such change, and that from classical physics to relativistic quantum mechanics was another. Aside from these rare and revolutionarily deep transformations some two or three centuries apart, conceptual change in physics had presumably taken place in a 'normal' manner, controlled by a single paradigm—whether Aristotelian, Newtonian, or Einsteinian. As a result, scientific change which was 'normal', in Kuhn's technical sense of the term, was presumably also normal, in the more familiar sense of 'usual' or 'customary'. The smooth sequence of intellectual progress in physical science (it seemed) had been interrupted by 'revolutions' only on a few, rare occasions.

(4) Between 1962 and 1965, Kuhn's critics pressed one major objection against his theory. This called in question the applicability of his central distinction. For had scientific change, in fact, ever been as revolutionary as Kuhn made out? If his definitions were applied strictly, could genuine examples

be found of a 'scientific revolution' at all? This question was potentially embarrassing; if no actual theoretical change in science qualified fully for Kuhn's title of 'scientific revolution', this fact put him in a quandary. Certainly, in course of time, the accumulation of gradual changes in scientific theory might have collective effects so profound as to be described in retrospect as 'revolutionary'—so there was no need to conclude that, in the absence of drastic, clear-cut 'revolutions', all scientific change was therefore 'normal' in Kuhn's technical sense. Yet how could such revolutionary results come about, without 'revolutions' to produce them? Such an outcome faced Kuhn with a hard choice. For in this case, he must either withdraw his account of full-scale scientific revolutions completely, or else modify it piecemeal, so that the original sharp distinction between 'normal' and 'revolutionary' change became progressively blurred.

The consequent debate has lasted through most of the 1960s, and by 1965 the main emphasis in Kuhn's account had already shifted greatly. Instead of focusing attention on the rare occasions when physical theory has been reconstructed comprehensively, from the ground up, Kuhn now began to concentrate his attention, rather, on cases involving less drastic conceptual changes. For, on closer examination, it turned out that even the Copernican and Einsteinian revolutions were something less than 'revolutionary', in Kuhn's technical sense of the word. His theoretical account of a full-scale scientific revolution thus failed to reflect the actual experience of the sciences involved, even in those phases of scientific change whose effects were most 'revolutionary', in the familiar, descriptive sense.

Faced with this discrepancy, Kuhn might have chosen to withdraw his original claims about the explanation of scientific change, and have substituted a less drastic theory, in which the role of intellectual revolutions was played down. Instead, he took the opposite course. By 1965, he was conceding that his first distinction between 'normal' and 'revolutionary' change in science might have been too sharply drawn; but he was arguing, in reply, that scientific revolutions were in fact, not less frequent, but more frequent than he had previously recognized.[1] His critics' objections had convinced him that the

[1] See Kuhn's Bedford College paper; p. 108, n. 1 above.

sciences are exposed to profound conceptual changes, not just every 200 years or so, but continually. So he went on to redescribe theoretical change in science as comprising an unending sequence of smaller revolutions or (to coin a term) 'micro-revolutions'. Every serious theoretical change in science, even if less than a complete 'paradigm-switch', now committed us to refashioning our concepts in a 'revolutionary' way.

This emendation at first looked innocent enough, yet its consequences went to the very heart of Kuhn's theory. For it transformed the historical development of scientific theory into a 'revolution in perpetuity', even in the cases hitherto labelled as 'normal'; and, in the process, it quietly abandoned the central distinction around which his whole theory had been built in the first place—viz., that between conceptual changes taking place within the limits of an overall paradigm, and those involving the replacement of an entire paradigm. From now on, there were no longer two contrasted types of theoretical change in science, one 'normal', the other 'revolutionary'. Instead, there was only a continuous spectrum of cases, all of them revolutionary, but some more so than others. So, the original explanatory contrast between alternative, alternating phases in the historical development of scientific theory was transformed, step by step, into a logical distinction: between (i) those scientific arguments which involve no conceptual or theoretical changes, and so can be presented in terms drawn from formal logic, and (ii) those which involve conceptual or theoretical novelties, and cannot be presented in this way. And, since every genuinely theoretical change in science involves a conceptual innovation of some kind—however minor—all genuinely theoretical changes clearly involved, in some measure, arguments of the second, 'revolutionary' kind.

(5) Confirmation that this was the destination towards which the development of Kuhn's theory was headed finally came in 1970, with the publication of the second edition of *The Structure of Scientific Revolutions*. For, in the appendix of this new edition, Kuhn insisted on retaining the original terminology of 'normal' and 'revolutionary' changes, but introduced qualifications designed to meet the objections of the intervening years; and, in so doing, he revealed clearly the underlying logical basis of the terminology.

Many of his readers (Kuhn now complained) had taken too seriously and literal-mindedly the description of paradigm-switches in his first edition as 'changes of world-view', whose effect was to introduce absolute rational discontinuities into the scientific argument. He had never intended (he protested) to suggest that the mutual incomprehension between scientists of successive generations is ever more than partial; nor had he ever meant to deny that scientists have good reasons for adopting some new conceptual scheme, or paradigm, in place of an older one. The point of calling paradigm-changes 'revolutionary' was simply to underline the fact that the good reasons advanced in support of conceptual changes 'cannot be cast in a form that fully resembles logical or mathematical proof'.[1]

Now, any author is at liberty to reinterpret the claims he made ten years before, in the light of his readers' misunderstandings. But, in this case, the revolutionary theory of scientific change which Kuhn was earlier 'misunderstood' to be advocating was a great deal more interesting than his own reinterpretation now allows; and, if this reinterpretation had been the whole truth, his original choice of the term 'revolution' was not merely a rhetorical exaggeration, but something worse. For this final account reduces the difference between 'normal' and 'revolutionary' changes to the distinction between *propositional* changes which involve no conceptual novelties, and so lend themselves to some kind of a deductive or quasi-deductive justification, and *conceptual* changes which go beyond the scope of all merely formal or deductive procedures. So interpreted, of course, any scientific change whatever will normally have both something 'normal' and something 'revolutionary' about it. And, if this were indeed all that Kuhn had ever meant by his use of the phrase 'scientific revolutions', that choice of phrase was grossly misleading; for it simply disguised a familiar (but atemporal) logical distinction in an irrelevantly historical fancy dress. Rather than distinguishing two historical kinds of scientific change, it merely indicated two logically distinct aspects of any theoretical change in science.

Over the years, then, Kuhn's account of scientific revolutions has become not less ambiguous, but more so. He began, like Collingwood, by recognizing the defects of the traditional

[1] *The Structure of Scientific Revolutions*, 2nd edn., Appendix.

inductive logic, which can never be stretched to cover profound theoretical transformations such as those involved in establishing the Copernican theory; and, at that stage, the term 'revolution' was merely a descriptive historical label. He went on to offer an historico-sociological hypothesis to explain the processes of scientific change, which rested on a fundamental contrast between drastic (or 'revolutionary') processes and 'normal' (or consolidatory) processes; and at this stage the term 'revolution' was raised to the level of an explanatory category. Under the pressure of counter-examples, he has qualified this account, bit by bit, in so many ways that all genuinely theoretical changes have ended by qualifying as revolutionary, so emptying the term of its distinctive explanatory value. And, by the time he was finished, he has been explaining the contrast between the 'normal' and the 'revolutionary' as implying no more than the logical distinction between conclusions which are 'deductively justifiable' and those which are not.

For some readers, these ambiguities have only made Kuhn's argument more attractive, since they have compelled us to attend more closely than was earlier customary to the relations between the ideas of a natural science, and the men who conceive/argue for/hold/reject those ideas. As Kuhn shewed, a comprehensive account of conceptual development must not merely consider concepts in the abstract, and in isolation from the men who conceive and use them, but also relate the history of ideas to the history of people, so placing the development of our conceptual traditions within the evolution of the activities by which those traditions are carried. At the time Kuhn first wrote, most philosophers of science were excessively wary of the genetic and psychologistic fallacies. Writing as a historian of science, he was able to do a useful service, by re-emphasizing the close connections between the socio-historical development of scientific schools, professions, and institutions, and the intellectual development of scientific theories themselves.

Still, it was one thing to re-establish the neglected links between conceptual changes and their socio-historical contexts; it was another matter entirely to identify intellectual, and still more logical, issues with socio-historical ones. Indeed, the more keenly one is aware of the interdependence of concepts and their contexts, the more indispensable certain distinctions become: for instance, that between the intrinsic authority of ideas and the magisterial authority of books, men and institutions, or that between the methodical acceptance of concepts whose merits have been demonstrated and the dogmatic acceptance of concepts whose merits are unproved. Unfortunately, the effect of Kuhn's proper emphasis on the sociology of science was, first, to distract him from these necessary distinctions, and finally to land the whole notion of 'intellectual revolutions' in complete incoherence. To put the central confusion bluntly: if we take Kuhn's own final statements seriously, and construe the distinction between 'normal' and 'revolutionary' scientific arguments in logical terms, this has one highly embarrassing consequence. On his latest reinterpretation, Kuhn's account of 'scientific revolutions' rests on a logical truism and—as such—is *no longer a theory of conceptual change at all.*

These difficulties might perhaps have been foreseen. For, in other fields of study, the idea of 'revolutions' has had an unfortunate backhistory, and a glance at those earlier difficulties throws light on the deeper significance of Kuhn's own moves: first, from a revolutionary to a micro-revolutionary, and eventually to a logical account of scientific change. In other historical sciences—e.g. political theory, geology, and palaeontology—the considerations that forced Kuhn to change his position have long been familiar. Political theorists, for instance, learned by bitter experience many years ago that the term 'revolution' must be handled with great circumspection, and by now they have recognised that it can safely be employed only as a classificatory label devoid of explanatory power. Initially, liberal democratic thinkers were tempted to treat the term as something more. In their eyes, steady constitutional change represented a 'rationally intelligible' political continuity; by contrast, political revolutions were disruptions of 'normality', which introduced historical discontinuities unanalysable in normal rational terms. Nowadays, however, political scientists try, rather, to avoid exaggerating the contrasts between 'normal change' and 'revolution'. Even the most unconstitutional change does not involve absolute and comprehensive breaches of political continuity. The most dramatic revolutions never lead to an absolute break with the past. Continuities of law, custom, and administration always survive, and often outweigh revolutions in the pattern of political allegiance and authority. Whether we consider the French system of public administration, the American legal tradition, or the Russian practice of escorting foreign tourists: all of these survived a political revolution with very little change. In each case, a revolution in ultimate political loyalties affected many other institutions only marginally. As a result, the post-revolutionary state of each country resembled its own pre-revolutionary state, in these respects, much more closely than it did the post-revolutionary states of the other two countries.

So the occurrence of a 'revolution' does nothing to excuse the political historian from giving a more profound and explicit historical analysis. On the contrary: to say merely '. . . and then there was a revolution'—as though this were some kind of divine intervention—would be to shirk the historian's proper intellectual task. (Such an attitude belongs to the world of *1066 and All That*, and has little to do with the complex facts of real historical change.) Statements about political revolutions, therefore, are acceptable to historians nowadays only as posing deeper, Collingwoodian questions about 'the occasions on which' and 'the processes by which' supreme authority changes hands in a revolutionary way. In a late-twentieth-century historiography, the term 'revolution' can still be helpful as a descriptive label, indicating a transfer of political authority that takes place in a more than usually profound and precipitate way. But, at an explanatory level, the differences between normal and revolutionary change turn out to have little real theoretical significance.

The same is true in the historiography of science. Any suggestion that a complete paradigm-switch involves conceptual changes of a totally different kind from those that take place within the limits of a single overall paradigm—that they represent some sort of a 'rational discontinuity', and lead to inescapable incomprehension—is quite misleading. At most, the two sorts of conceptual changes differ only in degree; and certainly they must, in the last resort, be accounted for in terms of the same set of factors and considerations. Assuming that 'normal' scientific changes can be explained in historical terms which are somehow suspended in the rare event of a scientific 'revolution' would, once again, be to fall back into an entirely naïve historiography.

We can recognize the source of Kuhn's difficulties even more clearly, if we recall the development of geology and palaeontology between 1820 and 1860.[1] By the early years of the nineteenth century, a few men were already convinced that profound geological and zoological changes had marked the Earth's history. All the same, the accepted theories of geological and palaeontological development, culminating in the early editions of Charles Lyell's *Principles of Geology*, remained 'uniformitarian'. According to these theories, the same natural agencies—both inorganic (wind, water, etc.) and organic (plants, animals, etc.)—were active at every stage of the Earth's history, in exactly the same manner as we can still observe today. As against these over-simplified assumptions, George Cuvier advanced the rival theory of geological 'catastrophes'. The starting-point of this theory was an authentic fact of real geological importance: namely, the discontinuities between adjacent strata apparent in the Earth's crust today, which are not at first sight explicable as products of smooth, continuous sedimentation or compression. Cuvier's explanation of these stratigraphical discontinuities was a naïve 'revolutionary' one. In his view, they eluded the categories of normal geological explanation: such discontinuities of effect must, he argued, be the products of equally discontinuous causes.[2] So, if they were to be explicable at all, they must be interpreted as the product of Divine Intervention. The 'normal' processes of natural change were punctuated by rare supernatural 'revolutions' or 'catastrophes'.

For some thirty years from 1820 on, the 'uniformitarian' and 'catastrophist' theories remained in deadlock. The uniformitarians dismissed Cuvier's appeals to theology as scientifically illegitimate, and redoubled their efforts to explain all features

---

[1] For a fuller discussion of this topic see C. C. Gillispie, *Genesis and Geology* (Cambridge, Mass., 1951), Chapter IV; also *Discovery*, Chapter 7, pp. 159ff.

[2] The evidence on which Cuvier based his belief in drastic historical discontinuities in the history of the Earth is surveyed (e.g.) in G. Cuvier, *Discours sur les Révolutions de la Surface du Globe*, 3rd edn. (Paris. 1825).

of the Earth's crust in terms of natural causes operating in familiar ways. Yet the difficulties facing both parties remained formidable, and the deadlock was eventually resolved only through the slow erosion of both extremes. On the one hand, the uniformitarians were compelled to admit natural changes of progressively greater magnitudes. Thus in 1835 Lyell's admirer, Charles Darwin, personally observed an earthquake which dislocated adjacent strata on the Chilean coastline by as much as twenty feet in a single movement.[1] As a result of such observations, the natural causes admitted by uniformitarians became ever more 'catastrophic' in their effects. On the other hand, catastrophists like Louis Agassiz at Harvard found themselves obliged to introduce a progressively larger number of hypothetical 'catastrophes' into their account of the Earth's history—for instance, the successive Ice Ages— and, in the process, these 'drastic and inexplicable' events became correspondingly smaller in scale.[2] Eventually, the resulting geological micro-catastrophes ceased to be 'world-historical events', comparable in theological significance to Noah's Flood. Instead, they began to rank as new types of natural phenomena, displaying recognizable uniformities of their own. Though still divided in doctrine, the rival geological systems gradually converged in practice. Once Lyell's uniform causes had become sufficiently drastic, and Agassiz's catastrophes had become sufficiently uniform, the original criterion for telling normal (or 'natural') from catastrophic (or 'supernatural') changes had disappeared.

Notice how in this sequence, from Cuvier through Agassiz to the eventual reconciliation with the uniformitarians, the term 'catastrophe' went through a series of reinterpretations similar to that which has affected Kuhn's term 'revolution'. To begin with, the catastrophist geologists were merely interested in drawing attention to a certain class of phenomena —viz. stratigraphical discontinuities—which had no obvious explanation in terms of 'normal' or 'natural' geological processes; at this stage, the term had no use beyond that of describing these discontinuities as 'catastrophic' in their effect.

---

[1] See Darwin, *The Voyage of the Beagle* (London, 1839), Chapter XIV, Feb. 20– Mar. 4, 1835.

[2] See, e.g., E. Lurie, *Louis Agassiz, A Life in Science* (Chicago, 1960), pp. 94–106.

At the second stage, Cuvier claimed an absolute difference between the historical causes of stratigraphical continuities and discontinuities: the continuities were effects of normal natural processes, the discontinuities of abnormal and supernatural 'catastrophes'. Under pressure of criticism, Cuvier's insistence on this absolute contrast was then abandoned, in favour of a less drastic thesis: every major geological change now had something 'catastrophic' about it, and the assumption that every such change was an 'Act of God' was tacitly abandoned.

So with the changing theories of conceptual development since the 1930s: Collingwood's rigid division of intellectual history into separate phases, controlled by incompatible constellations of absolute presuppositions, had the same virtues as Cuvier's theory in geology. It drew attention to the profoundity of actual conceptual changes, and challenged the orthodox assumption that intellectual progress results, always and everywhere, from a single, universal and uniform 'method'. If Collingwood was the Cuvier of intellectual history, Kuhn then becomes its Agassiz. Although his original emphasis put him very close to Cuvier and Collingwood, the micro-revolutions of his more recent accounts have become as minor and innumerable as Agassiz's final micro-catastrophes. By now, indeed, any conceptual change whatever counts as a micro-revolution; instead of intellectual revolutions being the antitheses of normal conceptual changes, the new micro-revolutions are the units of normal and revolutionary change alike. So the hyperrational aspects of Kuhn's 'revolutions', like the supernatural features of Cuvier's 'catastrophes', have vanished; and, in the process, the original criterion for distinguishing normal from revolutionary change has disappeared. We are left with nothing more drastic or revolutionary than a sequence of conceptual changes, differing from one another in speed and in depth, whose underlying processes, procedures, and/or mechanisms remain as unexplained as ever.

After chasing the revolutionary hare as far as it can lead us, we are therefore back where we started. An adequate theory of conceptual change must answer the questions left over by Collingwood: viz., 'On what occasions, and by what processes and procedures, does one basic set of collective concepts—in science or elsewhere—come to displace another?' It must answer these questions in a way that explains, in one and the same set of terms, both why our ways of thinking in some fields remain effectively unchanged over long periods, and also why in other fields they sometimes change rapidly and drastically. And it must answer them, finally, in a way that makes clear the respective roles, in this historical development, of 'rational procedures' on the one hand, and 'causal processes' on the other. What we need, therefore, is an account of conceptual development which can accommodate changes of any profundity, but which explains gradual and drastic change alike as alternative outcomes of the same factors working together in different ways. Instead of a *revolutionary* account of intellectual change, which sets out to show how entire 'conceptual systems' succeed one another, we therefore need to construct an *evolutionary* account, which explains how 'conceptual populations' come to be progressively transformed.

In taking this further step, we must keep in mind one distinction which Kuhn disregards. The 'micro-revolutions' characteristic of his fourth account of scientific change are in one respect ambiguous: he leaves it unclear whether he is referring (i) to the class of novel theoretical suggestions that are current at all times within a given science—novelties that may circulate for weeks, months, or years, before being definitely accepted or rejected—or (ii) to the smaller sub-class of novelties which actually win a place in the accepted body of theory, and thereby modify the conceptual tradition. From now on, we ourselves must take care to distinguish, as Darwin did in his account of organic evolution, between

(i)   the units of variation, i.e. the tentative conceptual variants circulating within a discipline at any particular time; and

(ii)   the units of effective modification, i.e. the conceptual changes that are actually incorporated into the collective tradition of a discipline.

Correspondingly, we must discuss the development of our collective concepts under two distinct heads:

(i)   innovation—asking what factors and/or considerations lead the bearers of an intellectual tradition to propose certain ways of moving ahead from the currently accepted position: and

(ii) selection—asking what factors and/or considerations lead them to accept certain of these innovations as established in preference to others, and so to modify the collective conceptual tradition.

In each case, the question 'What factors and/or considerations lead the bearers of an intellectual tradition to propose (or to accept) . . .?' will mean, 'What kinds of reasons and/or causes are relevant to our understanding of their proposals and acceptances?' Once again, that is to say, our question is as much a rational or historical question as a sociological or psychological one. And the development of our concepts can then be properly understood, only when we come to see how both socio-historical (i.e. causal) processes and intellectual or disciplinary (i.e. rational) procedures play a part in the historical sequences by which intellectual variants in a science are first proposed, and then selectively perpetuated.

Before we leave this discussion of the theory of intellectual revolutions, it is worth adding two further remarks. The first has to do with the difference between two distinct kinds of principle, or presupposition, which the Kuhnian theory of paradigms can too easily lead us to confuse. The second connects the outcome of our present analysis with the conclusions we arrived at earlier: showing that the idea of paradigms has 'revolutionary' implications, only if the structure of our paradigms is itself thought of in over-systematic terms.

(1) The central paradox in the classical theory of scientific revolutions is the implication that, between supporters of different paradigms, mutual incomprehension is unavoidable. This conclusion is inescapable so long as we regard paradigms, or constellations of absolute presuppositions, as unitary and indivisible. In practice, however, we need to draw a further distinction, which neither Collingwood nor Kuhn makes sufficiently explicit. Within any actual science, we can distinguish concepts and principles of two very different sorts. On the one hand, there are the basic 'theoretical' principles of the science, such as Newton's Principle of Universal Gravitation, and Mendel's genetic Principles of Segregation and

Recombination. On the other hand, there are the 'disciplinary' principles—e.g., that all physiological functions are to be explained in chemical terms—which define the basic intellectual goals of a science, and give it a recognizable unity and continuity. And the central paradox looks very differently when a scientific 'paradigm' is defined in terms of the accepted theoretical principles only, and when it refers to a set of theoretical and disciplinary principles taken together, as a single package.

If we confine our attention to theoretical principles, paradigm-switches certainly need not lead to inescapable incomprehension, and new concepts can perfectly well be introduced into a science without any radical discontinuity. Thus, the theoretical concepts of Einstein's relativistic physics (say) may perhaps be incompatible with those of Newton's classical theories, in this first sense; yet supporters of the two positions shared enough disciplinary aims for them to be able to discuss, in a vocabulary intelligible to both sides, which of the two theories 'did the better explanatory job' for theoretical physics. To say this is not, however, to imply that complete incomprehension never occurs at all. It is to say, rather, that it becomes a real problem, only when we consider disagreements which involve the boundaries between different scientific disciplines, and which therefore involve cross-purposes at both the theoretical and the disciplinary level. The most familiar real-life example is, perhaps, the case of Goethe's *Farbenlehre*. In writing about the theory of colour, Goethe denounced Newton's optical theories on the grounds that they grossly misrepresented a whole range of familiar facts evident in our direct experience of colour. Instead of accepting the supposed relationship between the visible colours of things and the wavelengths of the light coming from them, Goethe set aside Newton's whole approach to optics, with a sigh of regret that a passion for mathematics had seduced so talented a man into such a wrong-headed direction. In place of Newtonian optics, he set out to reconstruct the theory of colour around quite different principles and concepts, which disregarded the Newtonian theories as a temporary aberration of the human intellect.[1]

[1] See, e.g., C. Sherrington, *Goethe on Nature and on Science* (Cambridge, England, 1949) pp. 13ff.

At a quick glance, Goethe's argument looks like one more attempted scientific takeover, which differed from those of Copernicus and Einstein only in that it aborted. The *Farbenlehre*, we may be tempted to conclude, offered a new 'paradigm' for optics which was to take over supreme intellectual authority from the 'spectrum' theory advanced by Newton a century earlier. On closer inspection, however, the cases turn out not to be genuine parallels. Goethe's ambition was to explain not physical phenomena but psychological ones: the theory he dreamed of was not a physical theory of light-rays, or light-waves, regarded as the 'material bearers' of colour, so much as a psychological theory of visible light, regarded as the 'immediate object' of colour-perception. He rejected Newton's theory, that is to say, not because it gave inadequate solutions to the problems he himself was facing, but because for his purposes Newton's solutions were largely irrelevant.

Goethe's difference with Newton was, thus, not a straightforward disagreement within a single common discipline, since the interests of the two men in fact overlapped only at the margin. Whereas there is a direct intellectual conflict between the electromagnetic theories of (say) Maxwell and Heisenberg, the colour theories of Goethe and Newton were largely at cross-purposes. Newton was primarily concerned with the physics of light, and his excursions into perception (e.g. the Newton-Young three-colour theory) were merely incidental to that primary purpose; but Goethe was concerned with the psychology of perception and never seriously faced the physical issues around which Newton's *Opticks* had been organized. (The true successors of Goethe were not physicists like Maxwell and Heisenberg, but sensory physiologists and psychologists, such as Helmholtz and Lettvin.[1]) There was accordingly no real hope of Goethe and Newton sharing a common disciplinary view about what a 'theory of colour' should be, and should aim at achieving. Their intellectual goals were so disparate that their theories of colour inevitably lacked any common theoretical concepts. In this respect, Goethe's attack on Newton reflected a basic divergence of purposes.

[1] H. von Helmholtz, *Physiological Optics*, Vol. II (Leipzig, 1860), esp. Chapters 1 and 20: also, the lectures on the theory of colour-perception given by J. V. Lettvin at the American Museum of Natural History in 1969 (forthcoming).

We can see, now, how the distinction between theoretical and disciplinary considerations enables us to escape the paradoxes of the classic revolutionary view. It may well be that no proposition within Einstein's theoretical physics can be strictly translated into Newtonian terms, or *vice versa*; yet this fact by itself does not impose any 'rational discontinuity' on the science. On the contrary: when two scientific positions share similar intellectual aims and fall within the scope of the same discipline, the historical transition between them can always be discussed in 'rational' terms, even though their respective supporters have no theoretical concepts in common. Radical incomprehension is inescapable, only when the parties to a dispute have nothing in common even in their disciplinary ambitions. Given the very minimum continuity of disciplinary aims, scientists with totally incongruous theoretical ideas will still, in general, have a basis for comparing the explanatory merits of their respective explanations, and rival paradigms or presuppositions—even though incompatible on the 'theoretical' level—will remain rationally commensurable as alternative ways of tackling a common set of 'disciplinary' tasks.

(2) The other distinction we must here remark on is that between 'propositional systems' and 'conceptual populations'. The development of Kuhn's arguments is illuminating, not least, because it demonstrates the connection between over-systematic accounts of scientific theory and over-revolutionary accounts of conceptual change in science. When Kuhn first wrote on this subject, American philosophy of science had been dominated for some forty years by a self-sufficient, anti-historical logical empiricism inherited from the Vienna Circle; and one real virtue in his work was to emphasize the need for a more historical and less formalistic approach to science.[1] Evidently, the arguments involved in changes of paradigm could not be described in the cut-and-dried terms of current inductive logic; one continuing strand in Kuhn's arguments has, accordingly, been an anti-formalistic one. A 'revolutionary' change in the concepts of science is one too profound to be analysed in terms drawn from formal logic alone.

At the outset, however, Kuhn challenged the accepted

[1] See my review of C. G. Hempel, *Aspects of Scientific Explanation*, in *Scientific American* (Feb. 1966).

analysis of scientific argument only in cases where entire paradigms, or 'systems of theory', were up for review. In 'normal' science, by contrast, scientists were presumably still bound by the formal requirements of the particular 'system' in question; and these requirements could be evaded only at the price of overthrowing the 'system' in its entirety. During 'normal' science, conceptual changes might be governed by formal inductive principles of some sort, but 'revolutionary' or non-logical changes were confined to the overthrow of entire paradigms. So understood, the very title of his book should have been, not *The Structure of Scientific Revolutions*, but *Revolutions in Scientific Structure*. Comprehensive and thoroughgoing revolutions involved the replacement of one entire systematic theoretical 'structure' by another, but these revolutions themselves could have no structure.

Kuhn's later accounts were significantly different, for they recognized that a complete body of scientific theory (e.g. Newtonian physics) is not a single, coherent logical system, which must be accepted or rejected in its entirety, but rather something in which we can make radical changes piecemeal. So long as the switch between alternative paradigms was seen as a change between complete 'systematic structures' of concepts and propositions, the classical distinction between revolutionary and normal phases in scientific change was forced on us, with all its paradoxical consequences. Once that assumption is given up, however, we need no longer hang on either to a strictly 'systematic' account of the conceptual structure of the natural sciences, or to a strictly 'revolutionary' account of the changes between successive paradigms.

Once again, the moral of this story is the importance of distinguishing the 'logicality' of propositional systems in (e.g.) pure mathematics from the 'rationality' of conceptual changes in the natural sciences or elsewhere. The intellectual content of an entire science normally lacks the unitary structure characteristic of any single theoretical calculus employed within that science; and discussing them both in the same terms can lead only to philosophical confusion. In suitable cases, perhaps, we can represent the content of some single theory, or 'conceptual system', in the form of a 'logical system', as Hertz did when he presented classical mechanics in an axiomatic form. But it is a mistake, even here, to treat the form of the resulting logical system as capturing the scientific content of the conceptual system which it is used to represent. (The systematicity of a conceptual system is not formal or syntactic; rather, it is that of a segmented semantic domain.) The intellectual content of an entire science can thus be represented in a strictly 'logical' form only in quite exceptional circumstances. More typically, a science will comprise numerous coexisting and logically independent theories or conceptual systems; and it will be none the less 'scientific' for doing so.

Rather than treating the content of a natural science as a tight and coherent logical system, we shall therefore have to consider it as a conceptual aggregate, or 'population', within which there are—at most—localized pockets of logical systematicity. Seen in this light, the problem of scientific rationality can be restated in new terms. Many independent explanatory procedures, concepts, and methods of representation are normally current as means of fulfilling the proper disciplinary missions of any particular science. Between a few of these concepts and procedures there will be formal or 'logical' links—as there are, for instance, between the Newtonian concepts of *force*, *mass* and *momentum*. Alongside these systematically related concepts and procedures, there will normally be others which are logically independent of, and even at variance with, one another. An adequate analysis of the 'reasonableness' of scientists and the 'rationality' of scientific procedures will then require us to consider

   (i)   the various non-formal relations between the coexisting concepts, explanatory procedures, and methods of representation current in different sciences;
   (ii) the ways in which conceptual problems arise in any particular field of science, and are recognized as such; and
   (iii) the nature of the rational considerations, in the light of which concepts and explanatory methods are modified and/or replaced, in the development of the science.

The extreme 'revolutionary' view of conceptual change remains attractive, therefore, only if we make the twin mistakes of equating 'rationality' and 'logicality', and supposing that an entire science has the same logical coherence as (say) Euclid's geometry or Newton's mechanics. So the idea of systematicity and the idea of intellectual revolutions are cognate and correlative ideas. Those who assume that an entire science necessarily forms a single, coherent intellectual system will correctly infer that 'radical' changes in its intellectual content must also be 'revolutionary'. In this respect, the problems arising over Kuhn's account of scientific change have significant parallels in sociology and elsewhere. The belief that society as a whole forms a single coherent and functional 'social system', for instance, is a direct counterpart to the belief that physics as a whole forms a coherent 'logical system'. An over-systematic analysis of social structure has, in fact, dominated a great part of sociological theory for almost as long as its logical counterpart has dominated the philosophy of science; and, in each case, the revolutionary view is an understandable over-reaction to that domination.

In both cases, however, this revolutionary reaction has the same defects. We can view an entire society as forming a single functional 'system', only if we fail to distinguish the looser social and political relations between the different institutions of a society from the tighter and more formal relations existing within any single institution. There may well be strict, functional relationships between the different roles in any single institution; between the Chairman of the Board and the General Manager, or between the Commander-in-Chief and the Chief of Staff. But construing the relations between (say) Army and Industry, Church and Legislature, on this same systematic model, would be as much of a mistake as treating the relations between mechanics and optics (say) as strict logical relations, comparable to those within mechanics itself. In each case, this mistake invites the same response. If the component elements of an entire society—or an entire science— are as tightly interrelated as this view implies, there will then be no way of modifying them piecemeal or one at a time: the only chance of radical change will lie in rejecting the entire 'system' as a whole, and starting afresh.

Conversely, the only sure way of avoiding this conclusion is to undercut both the over-systematic and the over-revolutionary positions at a single stroke. In the case of science, this means recognizing that the different concepts of a scientific discipline are related more loosely than philosophers have assumed. Instead of being introduced at one and the same time, and all of a piece, as a single logical system with a single scientific purpose, different concepts and theories are introduced into a science independently, at different times and for different purposes. If they still survive today, this may be because they are still serving their original intellectual functions, or else because they have since acquired other, different functions; and we are free to replace, modify, or supplement those concepts independently in the future, as our legitimate scientific occasions require. This means recognizing that an entire science comprises an 'historical population' of logically independent concepts and theories, each with its own separate history, structure, and implications.

In sociology, likewise, the different institutions of a society are related more loosely than has recently been assumed. Instead of being intelligible only if considered all of a piece, as a single integrated 'social system', they need to be thought of in more historical terms, since they were originally established at different times and with different ends in view. If they still survive today, this may—once again—be because they are still serving the functions for which they were first established, or else because they have since acquired other, different functions; and we are again free to replace, modify, or supplement our institutions independently in the future, as our legitimate social occasions require. This, too, means moving beyond the systematic theory of social structure and the revolutionary theory of social change which is its antithesis, and allowing that entire societies comprise 'historical populations' of institutions, each with its own history and internal structure. Later in our inquiries we shall return and see what questions such a 'populational' theory of social changes would raise. For the moment, our immediate task is to develop the corresponding 'populational' account of conceptual change in intellectual disciplines, and this is the task to which we must now turn.

SECTION B   Rational
Enterprises and
Their Evolution

Introduction

We are now ready to embark on the constructive part of these inquiries. To summarize, first, the steps by which we have reached this point, the argument has run as follows:

(1) The historical and cultural diversity of men's ideas underlines the need for an impartial standpoint of rational judgement; and it appears at first sight, that this can be provided only in abstract logical terms, claiming absolute and universal rational authority over the concepts and judgements of all milieus.

(2) The moment we go beyond pure mathematics and formal logic, however, the attempt to identify such an absolute standpoint runs into difficulties of historical relevance, leaving us with no evident alternative except to abandon our search, and lapse into historical or cultural relativism.

(3) But both the absolutist and the relativist positions turn out to rest on the common erroneous assumption, that 'rationality' must be an attribute of some particular logical or conceptual 'system'; they differ only in locating our rational standpoint, in the one case, in an idealized, abstract system, and in the other case in some actual but arbitrarily chosen system.

(4) We must begin, therefore, by recognizing that rationality is an attribute, not of logical or conceptual systems as such, but of the human activities or enterprises of which particular sets of concepts are the temporary cross-sections: specifically, of the procedures by which the concepts, judgements, and formal systems currently accepted in those enterprises are criticized and changed.

(5) When we turn to consider the process of conceptual change, however, a similar dilemma reappears on the historical level: now, we seem forced to choose between a uniformitarian account, which assumes the universal relevance of a

single set of rational methods, and a revolutionary account, which treats conceptual change as a sequence of radical switches between rationally incommensurable positions.

(6) But this second dilemma can also be avoided, provided that we recognize two further distinctions: (i) between the theoretical concepts and principles of a discipline—which can, and may, change discontinuously—and the disciplinary concepts and principles which are for the time being constitutive of the discipline, and change more gradually; (ii) between the specific theories of a discipline—each with its own particular family and/or system of concepts—and the intellectual content of the entire subject, which comprises a changing 'population' of concepts, and families of concepts, that are in general logically independent of one another.

Our analysis of the communal aspects of concept-use will accordingly focus on 'rational enterprises' and their historical development. It must account, at the same time and in the same terms, for both the continuities and the changes in such enterprises, by treating their intellectual content as forming 'conceptual populations'. The development of these populations will be characterized—here, as elsewhere—as reflecting a balance between factors of two kinds: innovative factors, responsible for the appearance of variations in the population concerned, and selective ones, which modify it by perpetuating certain favoured variants. And the 'rationality' of these enterprises will be vindicated, in turn, by identifying the specific loci within them at which conceptual variants are exposed to critical selection, and appeals to 'rational considerations' play an effective part in their development.

Two key phrases here clearly indicate the form of our present problem. The phrases 'conceptual populations' and 'selective perpetuation of variants' imply that our analysis should be an 'evolutionary' one, not just in the broad sense of being non-revolutionary, but in a quite precise and strict sense of the term. For the nature of populational change, regarded as a general type of historical process, is already well understood on one special case: viz., that of organic species. And we shall greatly simplify our own arguments here if we are prepared to take the populational analysis of organic evolution as a template or standard of comparison, in identifying the questions to be tackled in our analysis of conceptual evolution.

To do this, it will not be necessary to assume—as Ernst Mach unfortunately supposed—that intellectual evolution has something 'biological' about it, or even that the process of conceptual change in the sciences displays any substantial resemblance to the process of organic change. We shall be committed only to a more modest hypothesis, namely, that Darwin's populational theory of 'variation and natural selection' is one illustration of a more general form of historical explanation;[1] and that this same pattern is applicable also, on appropriate conditions, to historical entities and populations of other kinds. And we shall return, at the end of this Section, to reconsider this hypothesis, and show more explicitly the conditions which must be borne in mind, if the populational mode of explanation is to be applied without disaster to historical entities other than organic species.

The general pattern of historical explanation implicit in evolutionary zoology can be condensed into four basic theses, or commonplaces, each of which has a counterpart in the case of conceptual evolution.[2]

(1) To begin with, it is as much of a problem for zoology to understand why there are recognizable organic species at all as it is to explain why these species change as they do. Darwin's theory was (as he said) an account of the origin of species, quite as much as of their evolution. Until well into the nineteenth century, organic species had been thought of in two different ways, neither of them historical. Most naturalists were zoological 'realists', believing that the total population of living creatures on our globe—or presumably, anywhere else—was divided into distinct and separate kinds, each of which maintained itself unchanged by producing offspring of the same kind.

---

[1] Cf. M. T. Ghiselin, *The Triumph of the Darwinian Method* (Berkeley, 1969). I have had fruitful conversations on this point with Mark Adams, who has been exploring a similar line of thought independently.

[2] See, e.g., E. Mayr, *Population, Species and Evolution* (Cambridge, Mass., 1970); T. H. Hamilton, *Process and Pattern in Evolution* (New York, 1967); G. L. Stebbins, *Processes of Organic Evolution* (Englewood Cliffs, 1968); J. Maynard Smith, *The Theory of Evolution* (London, 1958).

More radical zoologists tended, by contrast, to be 'nominalists,' arguing that living things are classified into species only by our own arbitrary intellectual decisions. (As Buffon remarked, in an early essay, 'In reality, there exists nothing but individuals.')[1] Either way, no question arose about the historical mutability of organic species; if taxonomic classification itself was a human artefact, one might ask about the history of the 'species-concept', but hardly about the history of organic species themselves.[2]

The contemporary position is an intermediate one. Biologists nowadays are neither consistent realists nor downright nominalists. Rather, they believe that, while the forms of living things have changed repeatedly and are still changing, the actual range of forms existing at any particular time or place has been neither continuous nor unlimited. Any local environment normally contains definite and discrete populations. From the entire conceivable spectrum of animals and plants, only some have come into existence, and these commonly fall into distinct 'kinds'. We therefore have to explain both why, within continually varying populations of living things, any such definite and discrete species are found at all; and also how the species existing at one epoch, instead of losing their initial distinctness, can be transformed into other, equally distinct forms, or can divide into separate successor-populations having all the distinctness of different species.

(2) The central Darwinist insight was the recognition that both these facts—both the continuity of organic species, and their manner of change—can be explained in terms of a single dual process of variation and selective perpetuation. Every generation of living things comprises more individuals that survive to reproduce themselves; and every generation comprises individuals having variant forms or features, only some of which transmit these novel features to the subsequent population. Darwin's phrase 'natural selection' merely summarized those processes which put most novel variants at a disadvantage in the competition for reproduction and so maintain the gener-

---

[1] Buffon, *Histoire Naturelle*, Vol. I (Paris, 1749), p. 38.
[2] Cf. A. O. Lovejoy, *The Great Chain of Being* (Cambridge, Mass., 1936), Chapter 8; and 'Buffon and the Problem of Species', in *Forerunners of Darwin, 1745–1859*, ed. B. Glass *et al.* (Baltimore, 1959), pp. 84ff.

ally stable character of the species; but which occasionally enable advantageous novelties to become established in an organic population, so producing slow changes in its overall character.

(3) The combined action of variation and natural selection gives rise to authentic new species, only where several further conditions are satisfied. Darwin himself simply assumed the regular occurrence of inheritable variations: by now, we have genetical reasons for regarding this assumption as well-founded, of a kind that Darwin lacked. The problematic questions now have to do with the circumstances in which 'advantageous' variants can be selectively perpetuated, and so become dominant within a population. For instance, a novel variation can demonstrate its 'advantages', only in a situation involving enough 'selective pressure'. Failing serious competition, variant individuals have no chance to out-reproduce their rivals and the species will gradually lose its distinct character. Again, natural selection can be effective, only where the 'forum of competition' is not too extensive. If animals and birds interbreed freely over large areas, variants 'advantageous' within one particular locality will be swamped by cross-breeding over the larger area, and so fail to establish themselves permanently even within that favourable locality. In fact, Darwin and Wallace hit on the idea of natural selection independently, as a result of being presented with ready-made 'laboratories of evolution': Darwin in the Galápagos Islands, Wallace in the Malay Archipelago. In both cases, the different islands were close enough to ensure occasional cross-colonization, so that their varied populations of finches, tortoises, or orchids presumably shared common original ancestors. Yet they were also sufficiently isolated to provide separate 'forums of competition', in which the balance of advantage had favoured different variants—here a seed-eating finch, there an insect-eating one. Given appropriate environmental conditions, one could therefore appeal to 'variation and natural selection', as explaining both how the members of a population on any one island had come to form a single species, and also how related populations on nearby islands had grown sufficiently apart to form distinct species.[1]

[1] See, e.g., D. Lack, *Darwin's Finches* (Cambridge, Eng., 1947).

(4) Finally, the zoological use of terms like 'selection' and 'advantage' involves an implicit metaphor of choice. One key to the whole Darwinian theory of speciation lies in recognizing the implications of this metaphor. Variants are selectively perpetuated if and only if they are sufficiently 'well-adapted'. Here, the word 'adaptation' simply refers to the effectiveness with which different variants cope with the 'ecological demands' of the particular environment, and the term 'demands' itself embraces both physical conditions of life—e.g. climate, soil and terrain—and other coexisting populations of living creatures—predators or prey, shade-plants or camouflage, parasites or intestinal flora. The whole theory is in fact built around an interlocking set of notions. Competition and ecological demands are correlative notions; wherever individuals 'compete', some comparative measure of 'success' is implied, by which the 'winner' succeeds more completely than the 'loser'. In the Darwinian competition, this measure is the breeding test; the 'successful' forms have most representatives in subsequent generations. Correspondingly, the ecological demands of an environment determine the local requirements for evolutionary 'success': the term 'demands' focuses attention on those factors which, within that 'niche', influence the opportunities for any novel variant to contribute offspring to later generations. A novel variation may make its possessors more or less visible to predators, more or less prone to endemic disease, more or less dependent on regular water-supply, more or less available for mating . . .; and any of these factors may accelerate or inhibit the spread of the variation through the relevant population. All these factors thus imply 'demands', which are most fully 'met' by the 'successful' forms that outbreed their competitors.

Considered as a contribution to the historiography of nature, the Darwinian mode of explanation had some remarkable features. It explained both the continuities and the changes in the development of species in terms of the same familiar 'pressures', or 'demands', from the modes of life to which living populations are naturally exposed. It showed how, in some situations, these 'demands' serve to maintain existing species, in others to encourage the establishment of new ones. It indicated what conditions must be satisfied, if the prevailing ecological demands and pressures were to result in lasting changes. And in this way it provided a framework of theory which linked three groups of hitherto unrelated factors into a single coherent account: viz., (i) the long-term historical patterns of organic change, (ii) the short-term pressures exerted on any population by the immediate environment, and (iii) the medium-term standing conditions which provide the 'leverage' required, if those immediate pressures are to have any long-term effect. And, for our present purposes, these historiographical features of the Darwinian method are more important than any specifically biological details, about (say) mutation or natural selection. In seeing how populational categories can be extended to historical processes of other kinds, e.g. to the processes of conceptual change, we must not be drawn into discussions about genetics or predators or water-supply. Rather, we are concerned with the general relationships to be found in these historical processes: e.g., between (i) the long-term patterns of conceptual change, (ii) the day-to-day activities of concept-users, and (iii) the standing conditions on which the immediate decisions of concept-users depend for their longer-term effect.

Now let us restate these four basic commonplaces in terms applicable to conceptual development: (1) Within any particular culture and epoch, men's intellectual enterprises do not form an unordered continuum. Instead, they fall into more-or-less separate and well-defined 'disciplines', each characterized by its own body of concepts, methods, and fundamental aims. Considered over a long enough period, the intellectual content of such a discipline can change quite drastically; and so also, though more slowly, may its intellectual methods and aims. Yet each discipline, though mutable, normally displays a recognizable continuity, particularly in the selective factors that govern changes in its content. An evolutionary account of conceptual development accordingly has two separate features to explain: on the one hand, the coherence and continuity by which we identify disciplines as distinct and, on the other hand, the profound long-term changes by which they are transformed or superseded.

(2) These continuities and changes both involve the same dual process. In any live discipline, intellectual novelties are

always entering the current pool of ideas and techniques up for discussion, but only a few of these novelties win an established place in the relevant discipline, and are transmitted to the next generation of workers. The continuing emergence of intellectual innovations is thus balanced against a continuing process of critical selection. Some conceptual variants are picked out for incorporation, others are weeded out or ignored; yet, in suitable circumstances, this same process can account either for the continued stability of a well-defined discipline, or for its rapid transformation into something new and different.

(3) This dual process can produce a marked conceptual change, only given certain further conditions. We assume that, at any given time, enough men of natural inventiveness and curiosity exist to maintain a flow of intellectual innovations or 'variants'. The problematic questions then have to do with the conditions on which such novelties can prove their 'advantages', and so win a place in the relevant body of ideas. Once again, there must exist suitable 'forums of competition', within which intellectual novelties can survive for long enough to show their merits or defects; but in which they are also criticized and weeded out with enough severity to maintain the coherence of the discipline. So Karl Popper's capsule description of scientific method, as a dialectical succession of 'conjectures' and 'refutations', can at once be reinterpreted in evolutionary terms: it lays down the ecological conditions on which alone variation and selection can lead to effective scientific change.

(4) Finally, an evolutionary analysis of intellectual development once again involves a set of interdependent notions, which between them define the 'intellectual ecology' of any particular historical and cultural situation. In any problem situation, the disciplinary selection process picks out for 'accreditation' those of the 'competing' novelties which best meet the specific 'demands' of the local 'intellectual environment'. These 'demands' comprise both the immediate issues that each conceptual variant is designed to deal with, and also the other entrenched concepts with which it must coexist. And, once again, terms like 'competition' and 'merits', 'demands' and 'success' express correlative notions, which can be properly understood only by seeing them as so many aspects of the entire historical process of conceptual variation and

disciplinary selection. In both the zoological and the intellectual case, accordingly, historical continuity and change can be seen as alternative results of variation and selective perpetuation, reflecting the comparative success with which different variants meet the current demands to which they are exposed. What links the historical development of intellectual disciplines to populational processes of other kinds is thus no specifically biological analogy, but simply the general pattern of development by innovation and selection.

If intellectual disciplines comprise historically developing populations of concepts, as organic species do of organisms, we may then consider how the interplay of innovative and selective factors maintains their characteristic unity and continuity. Just as organic populations form distinct species rather than unstructured aggregates because the available 'ecological niches' impose a sufficient unity and continuity on the population, despite the continual diversification of individual organisms, so here the balance between intellectual innovation and critical selection divides up our entire repertory of concepts into recognizable 'sets' representative of distinct disciplines, despite the continual appearance of intellectual novelties within any particular set. Yet a populational approach debars us from giving permanent definitions of the resulting disciplines, marking off different fields of enquiry by immovable boundaries in terms of supposedly unchanging 'essential properties'—whether methods or problems, theories or concepts, techniques or subject-matter. The same features are not absolutely and eternally constitutive of 'physics' or 'biochemistry', at every stage in its development; scientific disciplines, like organic species, are evolving 'historical entities', rather than 'eternal beings'. In intellectual history as in natural history, the old philosophical ideal of 'permanent entities', which preserve an essential identity through a continuing sequence of 'accidental' historical changes, can now be superseded by a more lifelike, and less mysterious notion: viz., that of 'historical entities' which, though possessing no absolutely unchanging characteristics, preserve enough unity and continuity to remain distinct and recognizable from one epoch to another. What makes (say) the work of Buridan and Galileo, Maxwell and Feynman so many successive contributions to the same discipline has not been a common commitment to any single, permanent and unchanging, or essential 'physics'. It has simply been the characteristic unity and continuity which their common intellectual enterprise has preserved, despite all its changes, throughout the last 600 years.

We shall be developing our account of conceptual evolution in two instalments. Regarded as an historically developing human activity, any well-structured rational enterprise has two faces. We can think of it as a discipline, comprising a communal tradition of procedures and techniques for dealing with theoretical or practical problems; or we can think of it as a profession, comprising the organized set of institutions, roles, and men whose business it is to apply or improve those procedures and techniques. And these two faces represent alternative views of the same historical changes, as seen from different directions. If we consider a rational enterprise in disciplinary terms, its temporal development is a topic for the history of ideas. The historian's task is then to ask how the concepts of a natural science (say) come to have the life-histories they do: how, for lack of a sufficient basis in our experience of nature, they begin by being entirely speculative, how they subsequently acquire such a basis and so become well-founded, and how they eventually lose all intellectual authority, falling into the category of mere approximations or even of superstitions. Alternatively, if we consider the same enterprise in professional terms, its temporal development becomes a matter for the history of scientific organizations, institutions, and procedures. The historian's task is then to demonstrate how the activities of individual scientists and scientific groups have changed: how certain professional groups or technical procedures which initially had no standing or authority were subsequently accepted as authoritative or well-established among the members of the profession, only to be relegated to a secondary role, or even discredited, at a later stage.

So long as the intellectual content of a science (say) was construed as a 'logical system', the history of scientific ideas could still be separated from the history of scientific institutions and activities, and the resulting histories could be written independently. For there then seemed to be no common element linking the sequence of conceptual or propositional 'systems', which formed the 'intrinsic' substance of a scientific discipline, to the social, economic, and political structures which provided their 'extrinsic' professional framework. At most, the success of novel ideas might be the occasion for creating new institutions, while the organizations of science served as a human expression of its ideas. The development of scientific concepts thus came to appear an autonomous, self-directed process, which could at most be facilitated or handicapped by institutional and socio-political factors. Once we begin treating disciplinary and professional development as alternative aspects of the same populational process, this autonomy must be called in question, and the two parallel histories of the enterprise can no longer be wholly independent. In demonstrating the established place of any new concept in a scientific discipline, for instance, we must now pay attention to the selection-procedures actually used in evaluating the intellectual merits of each new concept, and these procedures must themselves be related to the activities of the men who form, for the time being, the authoritative 'reference-group' of the profession concerned. To this extent, we shall find the disciplinary or intellectual history of the enterprise interacting with its professional or sociological history, and we can separate the 'internal' life-story of ideas from the 'external' life-stories of the men whose ideas they are, only at the price of over-simplification.

Approaching our present problems from this populational point of view, we may distinguish six main groups of questions:

(i)   What defines the limits of an intellectual discipline, and why are there distinct disciplines at all?

(ii) What is the nature of conceptual variation, and how does the current pool of conceptual variants provide the material for disciplinary change?

(iii) To what processes and procedures of intellectual selection is such a pool of variants exposed?

(iv) By what channels of transmission and perpetuation are selected variants incorporated into a discipline, so as to modify its established content?

(v) How do differences in the degree of isolation and competition affect the influence of intellectual selection, and so react on the unity, character, and development of intellectual disciplines themselves?

(vi) Within what sorts of environment do intellectual disciplines operate, and how do the standing demands of those environments affect the processes and procedures by which conceptual variants are judged?

In Chapters 2 and 3 we shall be looking at rational enterprises from the disciplinary point of view, with an eye to their characteristic goals and patterns of historical development. In Chapter 4, we shall be looking at them from the professional point of view, with an eye to their sociological organization, patterns of authority, and changing reference-groups. In each case, our task will be to recognize how the human activities characteristic of a rational enterprise define 'intellectual niches', within which questions of both disciplinary and professional kinds arise, and to demonstrate how both kinds of issues can be related to the 'ecological demands' of their respective 'niches'.

CHAPTER 2

Intellectual
Disciplines: Their
Goals and Problems

2.1: *Scientific Disciplines and their Explanatory Ideals*

THE division of men's intellectual lives and activities into distinct disciplines is easy to recognize as fact. But it is less easy to explain, and our first task is to decide in what terms this subdivision is to be understood. How, for instance, are such disciplines to be classified and defined? Here verbal definitions will hardly help. Premature attempts at capturing the meaning of our basic notions tend to be quite unenlightening—as with Euclid's unhappy phrase, 'A *point* is that which has no part.' So, here, we may know that atomic physics, molecular biology, and jurisprudence constitute recognizably distinct disciplines, but we may still be uncertain how the very term 'discipline' itself should be defined. Evidently, professional scientists and jurists have effective practical criteria for deciding what properly belongs to their respective disciplines, and for recognizing their continued existence through time. Indeed, these criteria can be very exacting; scholars and scientists have a sharp nose for impostors, and are quick to reject one man's arguments as being 'not real physics at all', another's as 'romantic nature poetry masquerading as zoology', a third's as 'left-wing politics disguised as common law'. The first step in analysing the character of intellectual disciplines is to make the nature and source of these criteria explicit. What, then, is the crucial mark that an intellectual activity has the character of a discipline? By what indices should we determine when to apply and when to withhold that term?

We might tackle that question in any of a number of ways. We might start by looking for these criteria in the specific content of the disciplines concerned: perhaps in the theories, concepts, or conceptual systems of the different natural sciences. But this would again be to mistake the part for the whole. Specific theories, concepts, and conceptual systems are transitory products, or cross-sections, of entire historically developing sciences; the unity and continuity of those sciences must reflect, not just the formal relationships within any such cross-section, but also the substantive relations embracing the entire succession of developing ideas. The operative question is, thus, 'What makes the later phases of a science the "legitimate heirs" of the earlier?' These different phases are linked, neither by identities nor by logical entailments, but by relationships of legitimate ancestry and affiliation; and our problem is to discover how their legitimacy can be explained.[1]

Certainly, the continuous affiliations of the sciences make themselves felt in several different ways. The groups of men who work as atomic physicists, or cell biologists, or neuro-anatomists, are linked as masters and pupils in scholastic genealogies; the learned societies and research centres of each science are linked in similar institutional genealogies; while, within a given science, other genealogies link the experimental apparatus, the explanatory models, the terminologies, mathematical techniques, and subject-matters of earlier phases to those of later phases. In one context or another, it can be illuminating to describe the historical development of a science in terms of any one of these sequences: following out the succession of influential teachers or the refinement of equipment the improvement of computation techniques or the enlargement of empirical scope. The difficulty is to know which of these various strands has the deepest significance; or whether, like the fibres of a rope, they are collectively definitive of the science, without any one of them being indispensable.

Take as an example the science of atomic physics, as it developed from its beginning around 1890 through its heyday in the 1920s, up to its fragmentation into several successor-sciences during the 1950s. Suppose that, in the year 1930, we had described atomic physics as an intellectual discipline; what would this label have implied? What was it that gave the atomic physics of 1930 its unity as a self-contained subject?

---

[1] Cf. Ludwig Wittgenstein's reply to critics who complained that what he was doing was 'not philosophy', which was to answer, 'Maybe not, but what I am doing is the legitimate heir to that which has previously been called "philosophy".'

Faced with this question, one might have been tempted to point to some standard textbook, replying simply, 'By *atomic physics* I mean the subject expounded in this book.' Such a response, however, would trap us in historical relativism of an unintended kind. The intellectual content of atomic physics in 1910 was very different from what it became twenty years later, but we certainly would not agree, on that account alone, to deny J. J. Thomson and Rutherford the title of 'atomic physicists'. A textbook written in 1930 gives us, in fact, only one time-slice of the entire subject we know as atomic physics, viz., the ideas and arguments representative of the science at the time in question; and this 'representative set' qualifies for the name of 'atomic physics', only because it was the legitimate successor of Thompson's and Rutherford's earlier ideas and arguments. In order to characterize the science adequately, therefore, we must give a definition (i.e., criteria of unity, coherence, and continuity) in terms that cover the ideas current among atomic physicists in 1910, 1930, and 1950 equally; and one which indicates how the subject is related to its precursors in the broader physics of the 1890s, as well as to the more specialized sub-disciplines which have succeeded it in the 1960s and 1970s.

Let us quickly run through a number of possible indices. One way of arriving at such criteria might be in terms of the human continuity provided by atomic physicists themselves. For a start, then, we might say: 'Atomic physics is the subject invented by J. J. Thomson, Ernest Rutherford, and Niels Bohr, and developed by their pupils and successors.' Yet, though this might prove to be a successful practical index, it is no basis for a proper definition. We recognize Thomson and Rutherford, Bohr and Schrödinger, Fermi and Lawrence as members of a single professional guild, not merely because they learned from one another, but because of what they learned from one another. Their common membership of a professional guild helped to sustain the unity of atomic physics, but it did not give atomic physics its disciplinary unity and continuity. What united these men into their common profession was their shared commitment to the proper concerns of atomic physics—as identified by some other test.

How, then, are these 'proper concerns' to be identified?

That question is already significantly broader than the one from which we began. For, although we might define the concepts and theories characteristic of a particular science in impersonal terms, 'concerns' are always the concerns of people. Following up this clue, we may now ask: 'What continuing elements did the intellectual concerns of atomic physicists display, over the whole period from 1900 to 1950?' Notice that we are here asking about continuing elements, not about invariant ones. The terminologies, theoretical models, and fundamental equations of atomic physics underwent several drastic changes during this period, so that by 1930 Rutherford himself could no longer think about atoms in the way that his pupils' pupils took for granted. And, though a few theoretical terms remained in use throughout the entire fifty years, these were hardly the crucial element responsible alone for the continuing existence of atomic physics. In any case, scarcely more than the words survived; the concepts 'electron' and 'nucleus', as discussed by Heisenberg and Dirac in the 1930s, were far removed from what they had been in the theories of Thomson and Rutherford thirty years before. We shall do better, then, to look for the continuity of atomic physics in the *problems* by which successive generations of atomic physicists were faced; and these must be specified, not so much in terms of any single unchanging question or group of questions, as of a continuing *genealogy* of problems.

Thus, the ideas which Bohr advanced, to deal with his own problems about atomic structure, are best related to the points left over by his teacher, Rutherford; while the difficulties which he himself left unresolved posed, in turn, the new problems on which his own pupils had to work . . . In this way, the problems around which successive generations of scientists focus their work form a kind of dialectical sequence; despite all the changes in their actual concepts and techniques, their problems are linked together in a continuous family tree. Analysed in these terms, the continuity of a scientific discipline then rests as much on the considerations governing the changes between successive 'generations' of affiliated problems—with all their associated theories and ideas—as on any considerations leading to the survival of unchanging problems, or accepted theories and concepts. And, after all, if the sciences are to be the genuinely 'rational' and historically developing enterprises we take them for, this is just what we should expect.

It is this 'genealogy of problems' which accordingly underlies the other genealogies by which the development of a science can be characterized. In the sequence of theories, later models and concepts owe their legitimacy to having resolved problems for which earlier models and concepts were inadequate. In the sequence of experimental instruments, later designs permitted measurements which threw light on questions unanswerable using earlier designs. Even the developing subject-matter—or empirical scope—of a scientific discipline is controlled by this underlying genealogy of problems. If we mark sciences off from one another (using Shapere's term) by their respective 'domains', even these 'domains' have to be identified, not by the types of objects with which they deal, but rather by the questions which arise about them.[1] Any particular type of object will fall in the domain of (say) 'biochemistry', only in so far as it is a topic for correspondingly 'biochemical' questions; and the same type of object will fall within the domains of several different sciences, depending on what questions are raised about it. The behaviour of a muscle fibre, for instance, can fall within the domains of biochemistry, electrophysiology, pathology, and thermodynamics, since questions can be asked about it from all four points of view: and, in principle, the same fibre could be brought within the scope of still other sciences, by making it a topic for (say) quantum-mechanical or psychological questions.

A genealogy of scientific problems is, of course, something far less impersonal than a sequence of theoretical terms, mathematical equations, or axiomatic systems. To press this point home, we must now carry our questions one stage further, and ask what it is that makes a problem 'problematic', as seen from the point of view of a particular scientific discipline. Only if we dig down to this deeper level can we make it clear why an adequate definition of a 'discipline' must refer not only to the subject-matter on which its disciplinary activities are

[1] On this subject, see Dudley Shapere's essay in *The Legacy of Logical Positivism* (p. 64, n. 2, above), and also his contribution to the symposium, *The Structure of Scientific Theories*, ed. F. Suppe (forthcoming).

focused, but also to the professional attitudes by which those activities are guided.

To put the point in general terms: the problems of science have never been determined by the nature of the world alone, but have arisen always from the fact that, in the field concerned, our ideas about the world are at variance either with Nature or with one another. Galileo and Descartes liked to talk of the scientist's task as being to decipher, once and for all, the secret cipher in which the Book of Nature was written, and so to arrive at the 'one true structure' of the natural world. Yet that was always a Platonist ideal or aspiration, rather than a plain description of the task facing scientific research, as it is in fact done. Speaking less portentously, we might rather say that the task of a science is to improve our ideas about the natural world step by step, by identifying problem areas in which something can now be done to lessen the gap between the capacities of our current concepts and our reasonable intellectual ideals.

Long-term large-scale changes, in science as elsewhere, result not from sudden 'saltations', but from the accumulation of smaller modifications, each of which has been selectively perpetuated in some local and immediate problem-situation. In attempts to understand conceptual change in science, therefore, we must be on the lookout for the local and immediate demands of each intellectual situation, and for the advantages attaching to different conceptual novelties. These demands are rarely simple, and always highly specific. So the precise succession of problems facing a science reflects not the external timeless dictates of logic, but the transitory historical facts about each particular problem-situation. Necessarily, then, a 'rational method' in science is responsive to the specificity of every intellectual situation; and the quality of a scientist's judgement is shown less in his commitment to a general 'method' than in his sensitivity to differences in the requirements of his problem. The source of scientific problems lies, therefore, in a delicate historical relationship between the attitudes of professional scientists and the world of Nature which they study. Problems arise (I say) where our ideas about the world are at variance either with Nature, or with one another: i.e., where our current ideas fall short, in some remediable respects, of our intellectual ideals. This way of formulating the scientific task brings back to the surface an element in scientific inquiry which formal treatises of inductive logic frequently submerge. Conceptual problems in science emerge from the comparison, not of 'propositions' with 'observations', but of 'ideas' with 'experience'. Our present explanatory capacities must be judged in the light of the relevant intellectual ambitions and ideals. And the nature of scientific 'problems' cannot be properly defined, without considering also the character of those ideals.

What does it mean, then, to talk of the 'intellectual ideals' or 'explanatory ambitions' of a science? And what part do such ideals play in the development of scientific thought? To resume our earlier illustration: throughout the first half of the twentieth century, 'atomic physicists' shared a certain very general conception of the general form which a complete account of the structure of matter should take, if it were perfectly to explain the behaviour of material bodies on the atomic level; this conception was based on the notion of 'sub-atomic structures', as advanced by J. J. Thomson in the 1890s. They shared also a certain collective ideal, or communal goal: namely, to find ways of accounting for the relevant properties of actual objects and substances in detail, and so to put flesh on that intellectual skeleton. This collective ambition may have varied in part, as between individuals or sub-groups within the guild of atomic physicists; and with the passage of time, notably after 1925, it changed in some fairly substantial respects.[1] Yet, without being universal or immutable, this disciplinary ambition did develop continuously; and its continuous development, from Thomson and Rutherford up to Feynman and Bohm, is the crucial strand in the fundamental genealogy of 'atomic physics'.

True: at every stage, the atomic physicists' intellectual

[1] Cf. J. B. Conant, *Scientific Principles and Moral Conduct* (Cambridge, England, 1967), p. 8: 'In order to meet the challenge of relating these ways of seeking truth, I have to substitute for the words "scientific methods" some such phrase as "the way science has advanced". Such a change is necessary because the success of natural scientists, which is the point of reference, is not due primarily to their methods, but to the aim of their efforts. And curiously enough the aim is determined every few years by what has been the outcome of the experiments and observations of the preceding years. I appeal to the record of experimental science to justify this statement.'

reach exceeded their grasp in some fundamental respects. But that is true of all science. Taken at their face value, ideals of 'completeness' or 'perfection' are as unrealizable in science as anywhere else. Yet to say this is not to diminish their importance; it is simply to recognize the purpose they are there to serve. Throughout this whole period the general conception of a 'complete explanation' of the physical properties of matter, given in terms of 'atomic sub-structure', imposed an intellectual unity on the questions of atomic physics and provided the profession of atomic physicists with its collective goal; while the associated models of intelligibility—what in *Foresight and Understanding* I called 'ideals of natural order'—were the means of identifying the atomic physicists' unsolved problems and unanswered questions.[1] To generalize this point: scientists locate and specify the intellectual shortcomings of their current concepts by recognizing the shortfall between their current ability to 'account for' the relevant features of the natural world and the explanatory ambitions defined by their current ideals of natural order, or models of complete intelligibility. In short,

Scientific Problems = Explanatory Ideals — Current Capacities.

So, when Thomson and Rutherford invented 'atomic physics', their first achievement was not one of empirical observation or mathematical calculation; it was one of intellectual imagination. Lacking any clearly relevant evidence, their predecessors had stopped short at a theoretical subdivision of matter into permanent, indivisible 'atoms' combining into 'molecules'. Thomson and Rutherford saw how novel conceptions of 'atomic sub-structure' would provide new scope for physical explanation even on the macroscopic level; and, in doing so, they enlarged both the reasonable expectations of physicists, and the rational demands by which their explanatory ambitions were directed. In saying this, of course, we must notice that their programme was a severely practical one, directed at explaining some quite definite and identifiable phenomena. It was not conceived as a bare 'theoretical' possibility in an intellectual void, like Prout's Hypothesis some

[1] *Foresight*, Chapters 3 and 4

eighty years before; still less was it advanced in a Humean spirit, as a merely 'formal' possibility. Yet, however specific their interests, Thomson's and Rutherford's programme was none the less based on a new intellectual ideal, and other physicists could accept their programme only by a corresponding effort of the imagination. Indeed, J. J. Thomson's more conservative colleagues, who could not make this imaginative leap, began by treating his suggestion that the electron was a material object of 'sub-atomic' dimensions as some kind of practical joke.[1]

From this time on, accordingly, one indispensable step in any atomic physicist's apprenticeship has been to enter imaginatively into Thomson's and Rutherford's intellectual ideals. (Here, of course, the word 'imaginatively' does not mean the same as 'using images', or 'thinking of them as imaginary, or fictitious'.) Only by matching natural phenomena against the intellectual template of these ideals can one in fact identify the characteristic problems of 'atomic physics'. So, from 1900 to 1950 the history of the subject has been the story of a three-way interplay, involving (i) the fundamental 'ideals of natural order' which have determined its explanatory ambitions, (ii) the specific concepts and theories put forward in the hope of realizing those ideals, and (iii) the practical experience physicists have accumulated of the phenomena that 'atomic physics' gives us the means of explaining.

Certain aspects of any science can thus be described in impersonal terms: their historical development can be discussed as an episode in the history of ideas. Other aspects can be discussed in human terms alone: their historical development forms an episode in the history of human activities. At a more fundamental level, however, the intellectual activities of a scientific profession reflect both the stance with which the men concerned approach their experience of Nature, and the concepts in terms of which they interpret that experience. At this level, we can no longer separate the activities of scientists sharply from the concepts and theories which are the outcome of those activities. In this respect, the central problems of an

---

[1] See I. B. Cohen, 'Conservation and the Concept of Electric Charge', in *Critical Problems in the History of Science*, ed. M. Clagett (Madison, 1959), esp. p. 358; and also J. J. Thomson, *Recollections and Reflections* (London, 1936), p. 341.

intellectual discipline are, at the same time, the central pre-occupations of the corresponding profession. To reconstruct the historical evolution of 'atomic physics' is, therefore, to trace the affiliations between the changing problems of successive decades and to show how the rational continuity of the subject was preserved throughout those changes.

To sum up: the historical transformation by which the content of a scientific discipline evolves is intelligible only in terms of the current explanatory ambitions of the relevant professional guild. Yet, in its turn, the character of those ambitions can itself be explained only by using terms drawn from the vocabulary of the discipline. The subject-matter of a science is in fact 'problematic' at all, only when considered in the light of those intellectual ambitions of the scientists concerned; yet those very ambitions can be formulated realistically, only in the light of experience of the relevant subject-matter . . . In this dialectical manner the task of defining the current ideals of a science with all the necessary precision implicitly mobilizes the whole of its historical experience.

In the last resort, then, the nature of an 'intellectual discipline' always involves both its concepts and also the men who conceived them, both its subject-matter or 'domain' and also the over-arching intellectual ambitions uniting the men who work within it. If the notions of an 'intellectual discipline' and a 'learned profession' are correlative, so also are the factors which maintain the intellectual coherence of the discipline and the self-identity of the profession. In either case, the experience which men have accumulated in a particular domain leads them to adopt certain explanatory ideals. Those ideals determine the collective ambitions to which a man commits himself when he enrols in the corresponding profession, and works *qua* biochemist or *qua* atomic physicist; and these same ideals maintain the coherence of the discipline itself, by setting boundaries within which hypothesis and speculation are confined and improving the selection-criteria for judging conceptual innovations.

An account of scientific disciplines given in these populational terms provides the means of recognizing the rational continuity lying behind the superficial appearance of 'revolutionary' change. For the continuing intellectual ambitions of a discipline, while not immmutable, change both more gradually and more continuously than the concepts and theories which are their transitory end-products. In short, the existence and unity of an intellectual discipline, regarded as a specific 'historical entity', reflects the continuity imposed on its problems by the development of its intellectual ideals and ambitions.

2.2: *Scientific Concepts and Explanatory Procedures*

In the remainder of this chapter, I shall try to show how the intellectual ideals characteristic of a scientific discipline act as the link between its explanatory techniques, its concepts, its theoretical problems, and empirical applications. Two preliminary remarks are needed. Firstly: in its early stages, a science is marked not by sheer ignorance about the relevant phenomena but rather by uncertainty about its proper intellectual goals or explanatory tasks. Frequently, we have a profusion of information at our disposal—about human behaviour, or the weather, or the movements of the planets—without yet knowing 'what to make of it'. (Notice the constructive implications of that idiom.) Correspondingly the eventual creation, or 'speciation', of a new scientific discipline is associated with the acceptance of an equally specific research programme. Within a well-established field of scientific inquiry, therefore, we shall normally find an agreed division of labour between coexisting sub-disciplines having different explanatory goals, between which there are—at worst—marginal territorial disputes (e.g. those between molecular biochemistry and cell biology).

Secondly: our approach invites a new kind of history of 'natural philosophy'. The chief explanatory patterns, forms of theory, or 'themata' of science were all of them worked out in advance of any clear recognition of their empirical scope.[1] John Dalton's 'discovery' of chemical atoms (for example) had a complex back-history. Strictly speaking, what Dalton discovered was a new way of matching ideas to facts: viz. how the atomistic picture of matter conceived by Democritus and Epicurus, and later elaborated by Newton and his successors,

[1] The term 'themata' is used by G. Holton, in 'The Sources of Scientific Hypotheses', *Eranos-Jahrbuch*, 31 (1963), 351–5.

could be used to account in detail for the chemical laws of constant combination and multiple proportion. Natural philosophy thus blends into scientific theory at the point where some mode of explanation whose implications had been previously explored only 'in the abstract' is given an actual explanatory use, and the precise range of its empirical application begins to be mapped. Seen in these terms, the historical development of scientific ideas has involved two complementary enterprises: (i) the analytical refinement of intellectual patterns, and (ii) the progressive recognition of their empirical scope. The Presocratic philosophers were concerned, almost entirely, with the first of these two enterprises, and they left behind them a rich legacy of possible explanatory forms or patterns, whose actual empirical relevance was not yet clear. With Plato and Aristotle, the task of giving these explanatory patterns an empirical application, in planetary astronomy or marine biology, was at last faced in general terms. And we could represent the developing relationship between philosophy and science since classical times, not as a sequence of sharp transitions—as hard-headed, empirically-minded scientists in one field after another learned to disregard *a priori* speculations and face their empirical subject-matter without philosophical preconceptions—but as a more gradual process, by which men progressively learned how to give their pre-existing theoretical modes of explanation an application in specific empirical situations.

This symbiosis between natural philosophy and empirical science—between the abstract analysis of possible explanatory forms, and their application to actual classes of natural phenomena—is closely related to our central topic here: viz. the key relationship, between the intellectual ideals of a scientific discipline and its explanatory procedures, concepts, and theoretical problems. The heart of recent arguments about conceptual change in science is the insight that no single ideal of 'explanation', or rational justification—such as Plato and Descartes found in formal geometry—is applicable universally in all sciences at all times. Each effective discipline has had specific goals and ideals, which have determined its specific methods and structures, and one central strand in its historical development has been the progressive refinement and clarification of those goals and ideals. This refinement is the central activity which has created the occasions for suggesting, testing out, and adopting new intellectual methods, procedures, and structures; and we must analyse the 'rational' use of concepts within collective intellectual disciplines with an eye to that activity.

This insight has one immediate consequence. Hitherto, philosophers have taken it for granted that the term 'explanation' refers, first and foremost, to an *argument*—preferably a strictly formal or demonstrative argument. The actual *activity* of 'explaining' has been for them something secondary: consisting merely in expounding the arguments which, for the time being, represent the accepted 'explanations' of the phenomena concerned. Correspondingly, theoretical discovery has consisted, in their eyes, in bringing to light the patterns of deductive relations captured in 'explanatory arguments'. We must now reverse this philosophical relationship, between explanatory arguments and the activity of explaining. As we use them here, the terms 'explain', 'explaining', and 'explanation' will refer primarily to a range of human activities, which may or may not include the setting-out of formal, demonstrative arguments; the terms will apply only secondarily to the arguments which enter into these explanatory activities. In our primary sense, for instance, it is physicists who 'explain' physical phenomena, not physics. A physical law or theory can then be spoken of as 'explaining' phenomena only in a derivative sense, meaning that the law or theory is one which physicists successfully invoke when 'explaining' the phenomena in the primary sense. On this interpretation, the argument itself will no longer be the 'explanation' of a phenomenon; at best, it can 'serve as' an explanation, when produced in the appropriate context and applied correctly.

This verbal amendment has significant implications. For it encourages us to pay more serious attention than philosophers have usually done to explanatory activities and procedures ('explanations', in our primary sense) other than those which involve appeal to formal, demonstrative arguments. In the practical business of giving explanations—after all—scientists often enough rely not on the presentation of explicit deductive arguments, but on such alternative activities as the drawing of graphs or ray-diagrams, the construction of intellectual models, or the programming of computers. And, while logicians may wish to claim that these alternative explanatory activities must rely implicitly on formal arguments which might have been set out explicitly, this philosophical claim finds no reflection in the actual practice of working scientists. On the contrary, the explanatory procedures that working scientists actually employ all function, in scientific practice, on a par with one another. Where strictly formal, and verbal, arguments are subsequently claimed to be 'implicit in' such alternative procedures, these represent logical idealizations or abstractions derived from the scientific explanations concerned, rather than their primary and essential content.

For our purposes, then, the proper starting point will be the general category of 'explanatory procedures'; the particular procedure of presenting a demonstrative argument, involving appeal either to a law of nature or to an axiomatic system, will be no more than one specific instance of this more general type. This starting-point has one special advantage. For a *procedural* account of 'explanation' makes it easy to understand the historical process by which scientific concepts are transmitted from one generation of scientists to the next. On such an account, for instance, the concepts around which scientists organize their theories clearly serve the collective goals of the discipline concerned. Historically developing natural sciences are essentially communal affairs, outlasting a single human generation, and cannot be characterized in terms of the thoughts and procedures of individuals alone. On the contrary: scientific concepts are, by their very nature, capable of being transmitted, handed on, and learned, in the processes by which the discipline maintains its existence beyond the lifetime of its original creators. To coin a term: the set of concepts representative of a historically developing discipline forms a *transmit*. Whatever personal associations these concepts may evoke in the minds of individual scientists, those are not what serves the purposes of the scientific discipline itself, and links the ideas of successive generations into a single conceptual genealogy. The characteristic transmit of a science consists—and *necessarily* consists—in the communal or 'public' aspect of its concepts. The mental images and neurophysiological processes of individual scientists may, in particular cases, acquire a conceptual role, but they do not thereby become 'concepts'. Nor does the fact that such images or processes can play these parts make it any clearer what exactly a 'conceptual' role is; it serves only to mark off the particular images or neural traces which do acquire this role from others which do not.

To tackle the central issue head-on: if the collective concepts of an intellectual discipline form a transmit, which can be passed on from one generation of workers to another, what can it then be—in this sense—to 'have a concept'? How does the apprentice scientist who is learning a special science, and so getting a grasp of (say) thermodynamics or optics or systematics, demonstrate that he has 'caught on to' the relevant concepts? Or, to make this question more explicit: how does he show that he has been 'enculturated' into the communal procedures of the special science in question—that he has made its intellectual values his own, and can now apply its concepts critically, including (if necessary) suggesting worthwhile changes in them? If we understand the explanatory activities of a science as the collective or 'public' procedures of an entire profession, all such questions as these are readily answered in terms that make 'communal' sense.

The content of a science is, thus, transmitted from one generation of scientists to the next by a process of *enculturation*. This process involves an apprenticeship, by which certain explanatory skills are transferred, with or without modification, from the senior generation to the junior. In that apprenticeship, the core of the transmit—the primary thing to be learned, tested, put to work, criticized, and changed—is the repertory of intellectual techniques, procedures, skills, and methods of representation, which are employed in 'giving explanations' of events and phenomena within the scope of the science concerned. In order to demonstrate publicly—and to prove—his grasp of the explanatory powers of his science the newcomer must, above all, learn how and when to apply these techniques and procedures in such a way as to explain phenomena which fall within the current scope of the science. For instance, when an apprentice physicist acquires the concept 'energy', he learns to do three things: (i) to perform the calculations which embody the arithmetic of energy conservation, (ii) to recognize the particular problems and situations to which such calculations are relevant, and (iii) to identify the empirical magnitudes that properly enter into such conservation calculations.

Described in these terms, the proof that an apprentice scientist has grasped some concept of his science is evidently tangible and 'public'. For he demonstrates his ability to apply the concept in a relevant manner, by solving problems or explaining phenomena using procedures whose 'validity' is a communal matter. This demonstration yields not so much circumstantial evidence from which we draw conclusions about the apprentice's 'private' mental grasp indirectly, by speculating inferentially about an hypothetical 'inner life' on which his publicly demonstrated skills depend. More crucially, his explanatory achievements provide the most immediate and direct confirmation possible that he has grasped the significance of the concept, i.e. its current role in the relevant discipline. The intellectual transmit of a scientific discipline—the common inheritance that all practitioners of the science collectively learn, share, apply, and criticize—thus comprises a particular *constellation of explanatory procedures*; and, in showing that he recognizes how and when to apply these procedures, a man gives all the proof needed for professional purposes that he has acquired a 'conceptual grasp' of the discipline.

In a Cartesian mood (it is true) we might be tempted to say that in acquiring the essentially mental concept, 'energy', the apprentice also learns—quite incidentally—to express his grasp of this concept, by performing the associated public calculations and observations. But that would be to miss the point. In the case of communal concepts, any 'inner' or 'mental' activities are secondary and derivative; learning to perform the relevant collective activities is the indispensable thing. However rich the apprentice's inner life may be, if he is unable to give any tangible demonstration of his conceptual grasp, we can dismiss his understanding of the term 'energy' as— at best—sketchy and impressionistic. And likewise, in other cases: optical concepts (e.g. 'light-ray') and taxonomic concepts (e.g. 'organic species') are associated with their own similar constellations of explanatory and classificatory procedures.

Behind the term 'light-ray', there lie the experimental and graphical techniques of geometrical optics; behind the term 'species', there lie the classificatory principles of taxonomy and systematics, and the procedures for identifying animals and plants. Once again, the conclusive tests for deciding whether or not a young man has yet 'grasped' the concepts in question involve public demonstrations of the relevant procedures or skills. A young man who can neither draw nor interpret optical ray-diagrams, can scarcely be credited with a grasp of the concept, 'light-ray'. One who finds the most elementary tasks of botanical or zoological classification beyond him can, likewise, scarcely claim to have caught on to the concept, 'species'.

It is, thus, the procedures and techniques of a scientific discipline which form its communal—and learnable—aspects, so defining the representative set of concepts that are the collective 'transmit' of the science. If we learn only the words and equations of a science, we may remain trapped in its linguistic superstructure; we come to understand the scientific significance of those words and equations, only when we learn their application. And this means not just when we learn what objects and situations they refer to; it means, furthermore, when we learn what kinds of associated practical procedures—the drawing of diagrams, the setting-up of apparatus, the sorting of specimens—are involved in the empirical application of those words or equations.

We may now carry our analysis one stage further. In order to do proper justice to the *complexity* of scientific concepts, we must distinguish three aspects, or elements, in the use of those concepts: namely, (i) the language, (ii) the representation techniques, and (iii) the application procedures of the science. The first two aspects or elements cover the 'symbolic' aspects of scientific explanation—i.e. the scientific activity that we call 'explaining'—while the third covers the recognition of situations to which those symbolic activities are appropriate. The 'linguistic' element embraces both nouns—technical terms or concept-names—and also sentences, whether natural laws or straightforward generalizations. The 'representation techniques' include all those varied procedures by which scientists demonstrate—i.e. exhibit, rather than prove deductively—the general relations discoverable among natural objects, events and phenomena: so comprising not only the use of mathematical formalisms, but also the drawing of graphs and diagrams, the establishment of taxonomic 'trees' and classifications, the devising of computer programmes, etc.

Such 'symbolic' elements have a genuinely explanatory use in science, however, only where suitable application procedures are available for identifying the empirical occasions and manner of their application. Recognition procedures are required, for instance, for identifying the particular objects, systems, or measurable magnitudes to which any particular technical term, or concept-name, is to be applied. Corresponding procedures are required, similarly, for marking off those situations to which any particular law or generalization is straightforwardly applicable, from those irrelevant, anomalous, or exceptional cases to which it does not apply. And criteria are needed, further, for distinguishing those situations in which any technique of representation—whether mathematical, graphical, or classificatory—has a genuine scientific use from those in which it is inapplicable. In all these ways, the ability to 'put the concepts of a science to work' involves neither a linguistic ability alone, nor even a mastery of the relevant representation techniques alone; it involves also the ability to test out, and map the boundaries of, the 'scope' or 'range of application' within which these symbols and representation techniques have a genuine empirical relevance.[1]

In the context of discussions taking place securely within a given science, references to the 'scope' of our explanatory techniques are often kept in the background, and explanations presented entirely in terms of the symbolic aspects of the science. And, provided the relevance of those techniques is not in active doubt, this practice is a great economy. For philosophical purposes, however, we must keep all three groups of elements in view, and discuss scientific 'concepts' not in abstract linguistic terms alone, but in a more concrete, procedural idiom. The effect of insisting on this idiom may sometimes appear laborious: for instance, the concise statement, 'Huygens discovered that

[1] I went into this subject at some length, originally, in my book *The Philosophy of Science* (London, 1953), esp. Chapter 2.

Iceland Spar is doubly-refracting', will now have to be re-phrased to read,

Huygens discovered that the optical properties of crystals of Iceland Spar—which had been inexplicable, using only the regular techniques of geometrical optics—could be accounted for successfully, if one represented any light-ray entering such a crystal as giving rise within the crystal not to one refracted ray only, as in normal refraction, but to two rays . . . (etc).

But the longer form of statement has significant philosophical advantages. It brings into the open the full implications of the optical concepts involved, by relating changes in the linguistic aspects of the science—e.g., the introduction of the new technical term, 'double refraction'—to changes in the associated explanatory activities of the concept-users concerned. And it helps to clarify the next important point: namely, that the function of 'conceptual innovations' is to be understood in terms of the 'reservoir of unsolved problems' in the science concerned. (We shall return to this point shortly.)

An account of 'explanation' and 'scientific concepts' given in these procedural terms is, however, open to misunderstandings. To begin with, its emphasis on public procedures and activities may appear to make the whole account excessively 'behaviouristic'. We can meet this first objection in two ways. Granted, our account of the relationship between scientific concepts and explanatory procedures has quite deliberately been given in behavioural language. Yet this has been done, not out of any 'behaviouristic' desire to restrict the activities of individual scientists to overt, public utterances and writings, but merely because we are specifically concerned here with the collective aspects of concept-use in professionalized disciplines. After a time, no doubt, any experienced scientist begins to do much of his thinking 'in his head', just as we all of us learn to do elementary arithmetic 'in the head'. But the collective nature of concept-use—whether in arithmetic or science—is such that our 'internalized thinking' conforms to the same arithmetical, zoological or physical procedures, and criteria of 'correctness', as the thinking we do overtly or out loud. If our silent thoughts are to claim any 'validity', that will be because they meet the same collective requirements as a public 'explanation'.

However 'private' or 'mental' our inner scientific thinking may be, its content has to be specified in behavioural terms, by making overt and explicit the intellectual procedures followed implicitly and covertly in our heads.[1] Still less does our present account commit us to any 'behaviouristic' view that scientific concepts are the product in individual behaviour of 'associations' in the activities of would-be scientists established by simple 'conditioning'. Any attempt to fit the explanatory procedures of an intellectual discipline into an 'associationist' theory of learning runs into the same problems as arise over 'rule-governed' procedures of other kinds. For we are here concerned, not just with establishing empirical correlations between previously independent behavioural units, but with recognizing general, rule-conforming patterns whose overall form must be grasped and followed as a whole. And we follow these same rule-conforming patterns, whether our scientific and mathematical thinking is done 'inwardly', or spelled out publicly, step by step.[2]

In the second place, a procedural account of scientific explanation may at first seem as static and unhistorical, in its own way, as the logical or 'systematic' accounts we have already criticized. This objection can be overcome, with the help of two further qualifications. In talking of concept-acquisition as a variety of 'enculturation', we need not imply that the explanatory procedures of a science are so many unvarying rituals, to be gone through with liturgical exactitude. School science teaching sometimes gives that impression, by over-emphasizing the detailed forms of the current procedures, and so concealing both the reasons why these take the forms they do and the flexibility of judgement required in their empirical application. The currently established concepts of a science, however, define with precision only certain limits, within which a professional scientist's way of tackling his problems must lie. But these limits are themselves open to dispute. Theoretical innovation often has the effect of pushing

---

[1] See L. S. Vygotsky, *Thought and Language* (Leningrad, 1934: Eng. tr., Cambridge, Mass., 1964), Chapter 4; and also L. Wittgenstein, *Philosophical Investigations* (Oxford, 1953), esp. paras. 240ff., pp. 88–9.

[2] On the theoretical difficulties raised by the concept of a 'rule', see T. Mischel, in *Human Action* (New York, 1969), Introduction, p. 20.

them back; and we may be unable to say beforehand whether some particular innovation is doing so in a highly original or an unforgivably wild manner. (Considering the work of Francis Crick in 1950, ought one to have regarded his speculations as splendidly promising, or as wildly unsound? Or was there—at that time—no safe way of deciding which they were?)

Again: if we list the procedures that an apprentice must master, before he is thoroughly 'inside' a scientific discipline, we must take care to include those crucial procedures that make the science genuinely 'rational', and so save it from freezing into scholasticism: viz., those involved in its own self-transformation. During his scientific enculturation, the apprentice physicist or biologist learns not only how to explain phenomena within the scope of his science by applying its existing concepts; he learns also what is involved in criticizing those concepts and so improving its current content. Indeed, learning one without the other—learning how to apply an existing repertory of concepts, without learning what would compel us to qualify or change them—does nothing to make a man a 'scientist' at all. A mechanical engineer, for instance, is trained to apply techniques of representation devised by the physicists of earlier generations. (This is why some branches of engineering are referred to as 'applied physics'.) What differentiates the engineer from the physicist is, precisely, the physicist's obligation to apply his explanatory techniques critically: exploring the limits of their scope rather than taking them on trust, improving them rather than putting them to practical use. (A man unshakeably committed to twentieth-century scientific concepts can display as much 'irrationality' as any Flat-Earther; he simply exercises his obstinacy on a different terrain.) Viewed as historically developing 'rational enterprises' devoted to the improvement of our explanatory procedures, rather than as sequences of logically-structured propositional systems separated by drastic transitions, scientific disciplines are committed to their own self-transformation. A propositional system is essentially self-contained and static; by contrast, a rational enterprise is, equally essentially, open to development in accordance with its own built-in procedures of self-criticism.

*   *   *

All questions of logical systematicity aside, then, we shall do well to think of the procedures for conceptual change, in the natural sciences and other rational enterprises, in terms of the modes of behaviour they involve. As with irrational fears or other irrational modes of behaviour, retaining a given scientific concept becomes 'irrational', when persisted in beyond the point at which the concept loses it explanatory fruitfulness. Thus, a scientist who *fails* to criticize and change his concepts, where the collective goals of his discipline require it, offends against the 'duties' of his scientific 'station', as much as a somnolent night-watchman or an insubordinate soldier. The procedures for conceptual change in a science, like the explanatory procedures of science, are thus 'institutionalized.' With only the slightest exaggeration, indeed, we might compress our whole analysis of collective concept-use in science into the single epigram: *Every concept is an intellectual micro-institution.*

This epigram can be used to make two points. In the first place, it underlines once again the fact that no single concept, or set of concepts, ever exhausts a scientific discipline; at best, it represents one historical cross-section of a longer-term developing enterprise. Individual concepts or families of concepts bear the relation to complete disciplines that individual roles or institutions bear to complete societies. To understand a complete 'historical entity'—whether a discipline or a society—we must consider not only the current structure of relations linking its constituent theories, institutions, or other elements, but also the procedures accepted within it for modifying those elements. And the collective transmit through which a set of scientific concepts finds its professional expression—the set of rule-governed modes of explanatory behaviour in that branch of science—is itself 'institutionalized' in ways that make conceptual learning in a science comparable to initiation into a social institution.

If philosophers of science have been able, hitherto, to ignore the actual behaviour of scientists, in favour of logical questions about their arguments, this is because the intellectual coherence and systematic application of the sciences mark them off so strikingly from the more arbitrary and unmethodical activities of much social life. Yet a strong case can be made for analysing the inner structure and empirical relevance of scientific concepts, also, as elements in continuously developing human activities; and for considering their broader significance by seeing how the specific intellectual procedures which are the 'micro-institutions' of the scientific life are related to the broader professional goals by which the enterprise of science is currently carried forward. While this institutional view of scientific concepts may at first appear paradoxical, therefore, it simply carries further the same programme on which we embarked when we rejected an analysis in terms of static 'logical systems' in favour of one in terms of 'rational enterprises'. Once we insist on a historically developing picture, we must concentrate on the changing ways in which scientific concepts enter into the broader pattern of men's explanatory *activities*.

In other fields of thought, where the content of our intellectual activities is less elaborately structured and even less formalizable, this analogy between 'concepts' and 'institutions' appears much less daring. In tracing the history of a legal concept, for instance, we would pay attention as a matter of course, not merely to the formal definitions and logical entailments embodied in successive juridical codes and decisions, but also to the succession of new situations in which this concept has been applied, and the manner in which this changing application has reflected back on its original significance. No one would imagine that a technical term of common law can be fully explained by a simple verbal definition alone; rather, its legal significance develops progressively, as the accumulation of new cases gradually creates a meaning for it. As a result legal concepts have to be defined, not merely in verbal terms, but in terms of their institutional consequences. Similarly with social and political concepts: a history of political or social thought written entirely in terms of verbal definitions and entailments would be a mockery. Divorced from institutional expression in men's lives and affairs, phrases such as 'democracy', 'freedom of assembly', or 'equality before the law' express nothing more than abstract virtues and aspirations. The true measure of men's social and political thought lies, not in the formal definitions of their political terms, but in the significance they develop as institutionalized elements in social or political practice. Indeed, the only thing that redeems a social doctrine or political concept from being a mere slogan, or dead letter, is the existence of regular procedures by which it acquires some application in social and political practice. Such procedures create the institutions for which the concepts themselves serve as symbols, and through which they find significant expression.

This comparison between social and scientific concepts can be carried one step further. The existence of regular procedures for criticising the consequences of social or political institutions, and for advocating changes in social or political practice, is what makes the conduct of political affairs a 'rational' matter, rather than a mere exercise of arbitrary authority or contest for power. In politics, as in science, the 'rationality' of our present institutions requires the existence of accepted procedures for the self-transformation of social and political institutions. To assume that an entire society (or science) has the same functional coherence as a single institution (or theory) is then to take it for granted, quite needlessly, that no effective procedures can exist for the progressive self-transformation of existing institutions and patterns of activity.

So, instead of social and political concepts being totally unlike the concepts of the natural sciences—as one might initially suppose—the relations between thought and practice in science and politics are very similar. In either case, the appearance of a significant new concept is preceded by a recognition of new problems, and is associated with the introduction of novel procedures for dealing with those problems. In both fields concepts acquire a meaning through serving the relevant human purpose in actual practical cases. In both fields, successive changes in the application of those concepts are associated with the progressive refinement or complication of their meaning. And, in both fields, the overall 'rationality' of the existing procedures or institutions depends on the scope that exists for criticizing and changing them from within the enterprise itself.

Suppose, then, that we give a procedural account of scientific explanation, concepts, and representation techniques. We can then go on immediately and make two further philosophical points. Firstly, the propositions figuring within scientific theories never—except obliquely—tell us anything 'true' or 'false' about the aspects of the empirical world to which they apply. Secondly: such propositions cannot—in any straightforward manner—be fitted into the standard logical classifications, as 'universal' or 'particular' propositions.

These points need expanding. In the first place, then: philosophers have traditionally assumed that the terms used within scientific theory refer directly to classes of natural objects, and that its general propositions either assert, or directly entail, 'universal empirical generalizations' about those natural objects. Throughout the last fifty years, in fact, the whole programme in the philosophy of science—whether concerned explicitly with verification or falsification, confirmation, corroboration, or refutation—has taken the validity of these assumptions for granted. All these different ways of stating the philosophical problems about science are addressed to questions about the empirical truth, falsity, or degree of probability of theoretical principles. Our own account, on the other hand, implies that this basic assumption has been incorrect throughout, since questions about the empirical 'truth' or 'falsity' of theoretical principles in science—as such—do not arise. Rather, theoretical terms and statements acquire an *indirect* empirical content or reference, only when their scope or range of application has been made explicit with the help of auxiliary identification-statements; and when this is done, the effect is to embed the theoretical terms and principles in question within directly-empirical 'meta-statements'.[1]

Some illustrations will spell out what this point implies. Firstly, as to the nouns of science: the more strictly 'theoretical' a term is, the more conditional, hypothetical, and indirect is its application to individual objects identified here and now. We can normally tell a 'tree' when we see one, but we may well hesitate to say—at sight—whether that tree belongs in the taxonomic category of cryptogams or in that of phanerogams, since the application criteria for these categories are much stricter and more complex. Even more so, when it comes to recognizing a 'light-ray': in this case, the authentically

---

[1] Contrast the orthodox empiricist account of 'laws of nature' and other 'theoretical propositions' in science, as equivalent to 'universal empirical generalizations' and typified by the example 'All robins' eggs are greenish blue': Cf. C. G. Hempel, *Aspects of Scientific Explanation* (New York, 1965), p. 266.

empirical question is whether, for the purposes of geometrical optics, some particular visible beam of light 'counts as' a light-ray—i.e. whether the ray-tracing techniques of geometrical optics are applicable, with a sufficient degree of accuracy, to the optical phenomena involved. Despite its straightforwardly empirical appearance the question, 'Is this a light-ray?', is shorthand for the meta-question, 'Is this beam something to which we may—given all the relevant conditions—properly apply both the term "light-ray", and the associated representation techniques of geometrical optics?'

Correspondingly, for the theoretical sentences of science: the more strictly 'theoretical' a statement is, the more its empirical relevance is a matter of applicability, rather than of truth. The directly empirical question in these cases is not, 'Is this proposition true?', but rather, 'How generally does this principle apply, and on what conditions does it hold good?' Within strictly theoretical discussions, indeed, scientists only rarely use the words 'true' and 'false' at all; the operative issue is to establish in what sorts of empirical situation, and on what conditions, any particular theory—with all its associated concepts and representation techniques—will serve the explanatory purposes for which it was introduced. (The operative question about geometrical optics, for instance, is, 'On what conditions, and with what degree of accuracy, does the principle of rectilinear propagation *apply* to actual optical phenomena?' The plain question 'Is it *true* that light travels in straight lines?' does not arise.)[1]

As to the second point: in working discussions of scientific theory, scientists find little use for the logician's distinction between 'particular' and 'universal' statements, either. The operative issue which most nearly approaches that distinction is the question whether a particular theory is applicable 'universally'—i.e. unconditionally—or only 'in a restricted class of situations'—i.e. conditionally. Once again, answering this question requires us neither to assert nor to deny the theoretical propositions concerned, but rather to make empirical meta-statements about them: i.e. statements which

[1] See Toulmin, *Philosophy of Science*, Chapter 2; and also H. P. Robertson, 'Geometry as a Branch of Physics', in *Albert Einstein, Philosopher Scientist*, ed. P. A. Schilpp (Evanston, 1949), pp. 315ff.

explain on what conditions they 'hold good'. The standard programme for the philosophy of science, accordingly, involves confusing 'universal' and 'particular' empirical truths with 'universally' and 'conditionally' applicable theories.

So, the nouns (terms) and the sentences (propositions) of scientific theories both refer to empirically identifiable objects and events obliquely rather than directly. In the course of establishing the conditions on which any explanatory technique can be successfully applied, a scientist determines (i) in what empirical situations the propositions of the corresponding theory hold good—not 'whether they are true'—and (ii) what empirical objects or systems count as—rather then 'are'— instances of the corresponding theoretical entities.[1] Correspondingly, the intellectual substance of a natural science lies neither in its own direct 'empirical truth' nor in the 'empirical truth' of its logical consequences. It lies rather in its explanatory power; and this power is measured by the range, scope, and exactitude of its representation techniques. In appropriate cases, those explanatory procedures may involve the use of formal, axiomatic theories, built around general, abstract terms—e.g. 'mass', 'momentum', and 'force'—but this is not universally the case. In appropriate cases, they may involve the classification of a varied subject-matter, using a systematic array of categories or 'taxa'; but this is not universally the case, either. Those procedures may rely on intellectual analogies or 'models', by which happenings of one kind (e.g. electrical phenomena) are compared with happenings of another kind (e.g. hydraulic processes); they may—as with the ray-diagrams of geometrical optics—provide formulae or recipes for 'setting forth' the experimental situations under study in an explanatory manner; they may require one to plot and analyse graphs; they may require one to employ statistical measures and tests. . . . In suitable situations, any—or all—of these varied explanatory procedures may have a legitimate scientific use. And in each case alike these techniques will 'hold good' just to the extent that they provide the means of fulfilling the intellectual missions of the scientific discipline concerned.

---

[1] See Plato, *Timaeus*, 48D–50A, where he insists that empirically recognizable physical objects cannot properly be described as 'this' or 'that', but only as 'suchlike'.

To sum up this section of our discussion: the acquisition of scientific understanding has two aspects. On the one hand, the apprentice scientist learns to employ the general procedures of his science. On the other hand, he learns to recognize the particular situations to which each procedure is appropriate. And, when he gives a full scientific explanation of some actual event or phenomenon, he necessarily employs both kinds of knowledge. He can deal with his problem adequately, only if he both applies the 'right' (i.e. relevant) explanatory procedure, and also applies that procedure 'aright' (i.e. faultlessly).

These two aspects of scientific understanding do not always go together in the same man. A man with a theoretical cast of mind may have the capacity to perform elaborate calculations, or to follow out the detailed implications of his models, with perfect accuracy; yet, at the same time, he may lack the ability to recognize just which of these calculations or interpretations are relevant in any particular empirical situation. By contrast, a man with a more empirical bent may have the capacity to recognize subtle differences between particular empirical situations, and understand the general significance of these differences for the theory of his subject; yet, at the same time, he may lack the theoretical grasp to pursue the detailed implications of the corresponding calculations or models. Many professional physicists, for instance, can read electronic wiring-diagrams in a theoretical way while lacking the radio amateur's feel for actual assemblages of wires, transistors, and inductances; conversely, many radio amateurs are more at home with soldering-irons than with pencil and paper, and have more of a feel for their apparatus than they can rationalize explicitly in terms of circuit theory.[1] Complete scientific knowledge once again involves knowledge both of the explanatory procedures of a science, and of their application to Nature. In and by itself, even the most fully worked out axiom-system can never constitute a 'science'; since no formal system can tell us, still less guarantee, its own empirical scope and range of application. Nor can any abstract general theory ever, in and by itself, 'explain' or 'represent' natural phenomena; rather, it is scientists who employ this theory—in the particular manner

---

[1] A familiar historical illustration of this difference is the contrast between Michael Faraday and James Clerk Maxwell.

they do, in the specific cases they do, and with the degree of success they do—to represent, and so explain, the properties of behaviour of independently identified classes of system or object.

The collective concepts of any natural science thus derive their significance from the use scientists make of them in their explanatory activities. This conclusion was, in fact, already implicit in the logical point that Kant was making, when he declared that 'All our knowledge is of representations.' The empirical knowledge that a scientific theory gives us is always the knowledge that some general procedure of explanation, description or representation (specified in abstract, theoretical terms) can be successfully applied (in a specific manner, with a particular degree of precision, discrimination, or exactitude) to some particular class of cases (as specified in concrete, empirical terms). This is also, I believe, the point being emphasized by Wittgenstein, a philosopher who shared Kant's initial training in physics, when he spoke in his *Tractatus Logico-Philosophicus* of Newtonian mechanics as 'imposing a unified form on our description' of the world rather than 'asserting anything about' the world.[1] Taken together with 'all their logical apparatus', and applied in the ways we in fact apply them—Wittgenstein took care to add—'the laws of physics do still speak, however indirectly, about the objects of the world'. But 'any description of the world by means of mechanics is always of a completely general sort, never mentioning specific point-masses, but only all point-masses whatever'. As a result, he said, 'The possibility of describing the world by means of Newtonian mechanics tells us nothing about the world. But what does tell us something about it is the precise way in which it is possible to describe it by these means.' Alternatively, we may put the same point in a single sentence, and in our own words: 'Meaning in science is shown by the character of an explanatory procedure; truth by men's success in finding applications for that procedure.'

2.3: *The Nature of Conceptual Problems in Science*

The explanatory ideals of a science (we said) represent not just the logically consistent hopes of the scientists concerned

[1] L. Wittgenstein, *Tractatus Logico-Philosophicus* (originally published in *Annalen der Naturphilosophie*, 1921), secs. 6.34–6.36.

but their reasonable expectations for the subject. The scope of their current explanatory procedures indicates, correspondingly, just how far they have gone towards realizing those reasonable expectations. The gap between these two things, i.e., the difference between explanatory ideals and actualities, is a measure of the explanatory distance this particular science still has to go. More exactly, it is a measure of the distance the science still has to go in order to fulfil its current intellectual ambitions. Meanwhile, everyone concerned may be ready to admit that further horizons—fresh explanatory hopes and possibilities—will in due course become visible, beyond the goals which form their immediate destination; and that the future intellectual hopes of the science will be both more ambitious and more specific than those which at present direct its development. For the moment, however, current ideals and ambitions define the present goals and shortcomings of the subject.

The current domain of any science—the objects, properties, or events which pose problems for the science, and so contribute to its 'phenomena'—depends, however, not just on Nature, but also on the intellectual attitudes with which men currently approach Nature: not just, as Kant might have said, on Nature as a 'thing in itself' but on the character of the current 'representation' of Nature. As a result, features of the world which the scientists of one generation find unmysterious can become puzzling and 'problematic' for the men of the next generation, simply because their intellectual ambitions have expanded. To cite one revealing example: the classical nineteenth-century theory of matter took as its ultimate level of analysis the ninety-odd chemical elements, each of which possessed properties conferred on it in the Creation, and supposedly consisted of atoms having eternally fixed shapes and sizes. Within these limits, one was free to enquire what precise colour, or electrical conductivity, any particular element possessed in point of observed fact; but no scientific opportunity existed for enquiring why, in point of theory, sodium vapour (say) should emit radiation in the yellow part of the spectrum, rather than elsewhere. The only available way of answering this question was to say—quite literally—'God alone knows.' Thomson, Rutherford, and Bohr removed this limitation. Their new explanatory ambitions carried the ultimate level of analysis below the atomic level. By conceiving of all chemical atoms as composed of common sub-atomic particles, they made it *conceptually possible* to treat as 'phenomena'—i.e. as problematic —properties of matter that had hitherto been accepted as arbitrary features of Nature. From now on, facts like the existence of twin yellow lines in the sodium vapour spectrum ceased to be arbitrary (or theological) and became an active theoretical problem to be resolved in terms of the configurations of sub-atomic components in the sodium atom.

Not all the current problems of a science are, of course, equally active. At any particular time, the recognized domain of a science will include many phenomena that we are not yet ready to come to grips with, even though they present genuine and puzzling difficulties. The active problems are those for which some promising line of attack is available: e.g. those phenomena which might be brought within the scope of the current concepts by a straightforward modification of accepted explanatory procedures. Either way an explanatory gap remains. If we leave aside those formerly empirical sciences whose theories have frozen into a definitive mathematical form (e.g. Euclidean geometry and rational mechanics) every scientific discipline aims at explaining more phenomena than it yet has the means of encompassing, as well as improving the organization of its existing explanations. That is why any empirical science still has unsolved conceptual problems. And our next task is to show what form the resulting 'conceptual problems' take: how they differ both from directly empirical or factual issues and from formal, mathematical, and semantic problems.

We have defined the content of a scientific discipline by reference to three interrelated sets of [elements: (i) the current explanatory goals of the science, (ii) its current repertory of concepts and explanatory procedures, and (iii) the accumulated experience of the scientists working in this particular discipline —i.e., the outcome of their efforts to fulfil their current explanatory ambitions, by applying the available repertory of concepts and explanatory procedures. So understood, of course, the 'experience' of scientists is not at all the sort of thing assumed, either by sensationalist philosophers like Mach, for whom the ultimate data of science were supposedly 'sense-impressions', or by physicalist philosophers such as the logical empiricists, for whom 'scientific experience' simply comprises straight-forward factual generalizations.[1] Rather, the experience of scientists resembles that of other professional men: for example, lawyers, engineers, or airline pilots. All of these men accumulate 'experience' by finding out what can or cannot be achieved, using the different items in their repertory of professional equipment. So here: a typical item in the experience of a physicist is neither a 'sense-datum' nor an 'empirical correlation', but a fresh discovery about the scope or relevance of some 'technique of representation'—e.g. that the straightforward techniques of geometrical ray-tracing can (or cannot) be extended, on such-and-such conditions, from the case of normal refraction it was developed to deal with, so as to give us a way of handling 'double refraction' also.

Recalling our earlier definition of the 'pool of outstanding problems' in a science—in terms of the formula

Problems = Explanatory Ideals — Current Capabilities

we are now ready to give a first classification of the different kinds of conceptual problem commonly encountered in the sciences. Such problems arise in several different ways, and the reservoir of unsolved problems in a developing science includes problems on as many different levels, corresponding to different classes of problem-situation. Some of these call for (a) the extension of our current procedures to fresh phenomena; others for (b) the improvement of our techniques for dealing with familiar phenomena, or (c) the intra-disciplinary integration of techniques within a single science; others again for (d) the inter-disciplinary integration of techniques from neighbouring sciences, or (e) the resolution of conflicts between scientific and extra-scientific ideas. Normally, the various explanatory procedures of a science give us effective command over only certain parts of its current domain, while none of them yields a unified treatment for the entire field. In a few cases, indeed, the explanatory procedures associated with

---

[1] See, e.g., my essay, 'From Logical Analysis to Conceptual History', in *The Legacy of Logical Positivism*, ed. Achinstein and Barker (Baltimore, 1969), esp. pp. 34–5.

different coexisting theories may even be formally inconsistent; this is the case, for instance, with the alternative assumptions about the radius of the electron which have to be made in different areas of quantum electrodynamics today. More typically, there will be a direct competition between co-existing concepts and procedures. This gives rise to pressures along the boundaries between the ranges of application of different concepts, each of which aims to extend its range to include phenomena previously dealt with using some other procedure. So, like the members of any 'population', concepts hold their places in a science only by continuously re-confirming their worth; and the boundary between neighbouring concepts is a dynamic equilibrium, liable to be upset by any change in their balance of explanatory power.

To take our five typical classes of phenomena in turn: (a) there are always certain phenomena which a natural science can reasonably be expected to explain, yet for which no available procedure yet provides a successful treatment. During the period between Snell and Huygens, for instance, there was no difficulty in describing the phenomena of double refraction in terms of 'light-rays'; the problem was to explain them. Whereas the other phenomena of geometrical optics could be accounted for as the effects of mere changes in the paths of single light-rays (at mirrors, prisms, etc) the discovery of a case in which one incoming light-ray apparently split into two separate rays called for changes, both in physicists' ideas about what light-rays were and could do, and also in the techniques for tracing them through different types of transparent material. Or to take another illustration from genetics: certain familiar inheritance-patterns have a straightforward explanation in Mendelian terms, as expressions of independently acting genes. Other patterns, by contrast, call for more complex explanations, involving 'buffering' and other interactions between neighbouring genes, and these patterns of explanation were developed only by extending and refining the original Mendelian procedures.

To turn next to class (b): there are always some phenomena which can be accounted for up to a point using current explanatory procedures, but for which scientists would welcome more complete or more precise explanations. We can give a slightly over-simplified example from the kinetic theory o. matter. The most obvious regularities in the physical behaviour of gases are explained using straightforward arguments which assume that the atoms of substances in the gaseous state have a negligible size, are in continuous high-speed motion, and bounce off one another elastically whenever they collide, without significant dissipation of energy. At a first approximation, the resulting explanations fit the behaviour of all familiar gases well enough. As we make more precise measurements, however, deviations appear whose size varies from substance to substance, and we can bring these deviations into the effective scope of kinetic theory only by enlarging our repertory of concepts and calculations. (The so-called 'van der Waals' gas equations, for instance, include terms reflecting the different sizes of the molecules of each substance.) In such a case, a simple extension of existing concepts and procedures provides the extra precision required; but in other cases the insistence on increased precision may require the development of quite new concepts and theories. One elementary illustration of this is the transition from geometrical optics to physical optics; another more elaborate one is that from classical to relativistic planetary theory.

These first two types of conceptual problem spring directly from the consideration of unexplained 'phenomena'. Alternatively, however, problems may arise at the intellectual boundaries where current concepts and procedures approach, overlap, or conflict. In these cases—as Mach put it—our ideas have to adapt to other ideas, rather than immediately to the facts.[1] They call less for a direct extension of particular theories than for a conceptual reorganization. This reorganization will often enough lead to an increase in the explanatory power and scope of the sciences concerned, if only as a by-product. So a conceptual change designed primarily to deal with the internal conceptual problems of a science may help us, at the same time, to solve external conceptual problems of our first two types as well. But it will be worth while here to distinguish these two aspects of the change and consider them separately.

---

[1] See Mach's 1910 essay, 'My Scientific Theory of Knowledge, and its reception by my Contemporaries', reprinted in *Physical Reality*, ed. S. Toulmin (New York, 1970).

With this qualification in mind, we can discuss the other three types of problem. Class (c) comprises the problems that arise when we consider the mutual relevance of different coexisting concepts within a single branch of science; class (d) comprises those concerned with the mutual relevance of concepts from different branches of science. These two types are conveniently considered together, since they shade into one another, and the differences between them can in marginal cases become merely terminological. For our criteria for differentiating between 'sciences', or 'branches of science', are neither sharp enough nor permanent enough for us to say definitively, in all cases, whether two theories or concepts come from the 'same' or 'different' sciences. The proper location of the boundaries between adjacent sciences may, indeed, be one of the conceptual issues up for decision, since our success in integrating the concepts and explanatory procedures in neighbouring fields often allows us to dismantle the barriers between branches of science—or even entire sciences—that had previously been handled separately and independently.

In clear cases of type (c), the issues involved are internal to a particular science (e.g. optics), and here either of two outcomes is possible. The outcome may be that one explanatory procedure effectively 'takes over' the entire scope of another; as when all the results of geometrical optics were shown to be explicable more exactly, though in a more complicated way, using the concepts and techniques of physical optics. Alternatively, it may become possible to delineate the boundaries between the concepts and explanatory procedures involved in a new way, as when quantum mechanics explained how it was that some optical phenomena could be explained in terms of the 'wave' aspects of light, others in terms of 'particles'. Other cases, again, display something of both these features: thus, Maxwell's electromagnetic theory both arbitrated the boundaries between electricity and magnetism, which had previously been independent branches of physical enquiry, and also led to an unexpected 'take-over' of optics by the new, integrated theory of electromagnetism.

In clear cases of type (d), the problems arise at the boundaries between separate sciences, and again there are two possibilities for dealing with them. In certain instances, the reasons for the explanatory success of the concepts of one science may prove to be explicable, at a more fundamental level, in terms drawn from another. Consider the relations between chemistry and physics. For so long as the characteristic properties of different chemical elements were left unquestioned, these remained independent sciences, but in the twentieth century their mutual relations have been transformed. By now, as a result, every concept in inorganic chemistry, from 'valency' up, can be construed on a more fundamental level as raising questions in physics. Elsewhere the conceptual task is, rather, to explain more exactly why the boundary between two sciences runs along the line it does. As J. B. S. Haldane pointed out, for instance, the field of biology embraces processes taking place on many different time-scales, from nanoseconds to millennia; and this variety of scales is of fundamental importance if we are to understand the relations between (e.g.) the experimental science of genetics and the historical science of evolutionary biology,[1] or disentangle the respective scopes and domains of biochemistry and developmental biology.

A final class of conceptual problem arises on the margins of science proper, but has nevertheless played a significant part in the historical development of the sciences. These problems, (e), arise out of conflicts between concepts and procedures current within the special sciences and ideas and attitudes current among people at large. The nature of these problems has at times been misunderstood: notably, when all the everyday ideas and attitudes in question have been thought of as like the word 'malaria'—as offering unfounded answers to scientific questions—instead of being seen as deriving their meaning from human activities of quite other sorts. Yet in many cases these extra-scientific ideas are wholly legitimate, and the conflicts that arise over them have to be resolved rather by a refinement in the scientific concepts involved; while a better analytical understanding of the relations between the intra-scientific and extra-scientific concepts in question is needed to make it clear why they have the respective scopes and meanings they do. Such analytical problems, involving arbitration between scientific and non-scientific concepts, may eventually be completely circumvented: recent debates over the boundary

[1] Cf. J. B. S. Haldane, 'Time in Biology', *Science Progress*, Vol. 44 (1956), pp.385–402.

between 'life' and 'death' precipitated by surgical techniques of organ transplantation, for example, have been surprisingly free of metaphysical confusion and directed at well-defined issues of physiology and law. Other similar disputes continue to give trouble: for instance, those concerned with the implications of scientific concepts and theories from the neurosciences for such extra-scientific ideas as 'responsibility' and 'free choice'.

In classifying the different kinds of conceptual problems, we have here been deliberately using the same 'procedural' idiom as before: talking about the scope, extension and/or modification of the associated representation techniques. Using this idiom, we can also indicate what is involved in solving theoretical problems in science by making suitable conceptual changes. For conceptual changes, like conceptual problems, can be characterized most naturally and directly in 'problem-solving' terms. To exemplify some typical problems in physics, for instance, one may ask:

(i) Given that we already possess ray-tracing techniques capable of representing the passage of light out of one homogeneous medium (e.g. air) into another (e.g. water), can these techniques be extended to cover also the passage of light through non-homogeneous media?

(ii) Given that the normal techniques of geometrical optics fail us, when the slits and obstacles in the path of the light have dimensions comparable to its wavelength, can we develop more general techniques of representation and computation covering the behaviour of light in those cases also?

(iii) Given that we can present the fundamental equations of mechanics in a form that covers all 'conservative' motions (i.e. motions in which there is no total loss of kinetic energy), can we extend or modify these equations so as to apply also to 'non-conservative' motions?

(iv) Given that, in changing from one spatio-temporal reference-system to another, Newton's mechanics and Maxwell's electromagnetism require us to use different transformation-equations, can either or both of those

theories be modified so that mechanical and electromagnetic laws conform to the same set of transformation equations?

In each of these four cases, a specific theoretical problem,
stated in procedural terms, was the occasion for a conceptual
change; and, in each case, the solution provided by that conceptual change has to be judged in similar terms. Furthermore, since
every scientific concept has its three distinguishable aspects (language, representation, and application) the conceptual novelties
proposed for dealing with such problems may involve changes
in any or all of these aspects. On the one hand, they may leave
unchanged the existing symbols of a theory—its terminology,
basic equations, methods of graphical representation, and so
on—while simply adding the new refinements needed in
order to bring recalcitrant phenomena within the scope of the
existing symbolism. Alternatively, they may involve us in
changing—sometimes in changing radically—the language and
other symbolic devices used hitherto, even to deal with straightforward and easily explicable phenomena. In point of history,
the actual solutions to problems (i) and (iii) were of the first
kind. Problem (i) called for an advance within the science of
geometrical optics; problem (iii) for an advance within
Newtonian mechanics. By contrast, the solutions to problems
(ii) and (iv) were of the second kind. Problem (ii) was solved
only by developing the brand-new approach we know as
'physical optics', the resulting discovery being paraphrased
and symbolized in the shorthand statement, 'Light consists of
waves'. As for problem (iv), this was solved by the changeover
to Einstein's special theory of relativity, which involved small
but fundamental changes in equations that had served as the
axioms of mechanics for some two hundred years.

While the different aspects of a scientific concept are frequently interrelated, they may thus vary independently and
on different occasions; and the solutions to theoretical problems
in a science may thus come in correspondingly different forms.
At one extreme, the application procedures may change
without any change in the symbolic aspects of a concept. Ways
may be found either of extending the scope of a term or
representation technique, or (conversely) of ruling out from

its scope some new class of anomalies or exceptions. Alternatively, new techniques of representation may be adopted, which leave the language and application procedures of the science largely unaffected. Finally, at the opposite extreme, the language of a science may on occasion change quite drastically, without any associated change in its techniques of representation or application procedures.

In actual practice, of course, conceptual changes rarely involve one of these three elements alone, so any example we may cite to illustrate one or the other sort of change will only approximate to the 'pure' type. All three sorts of change can, however, be illustrated approximately from the nineteenth-century development of the energy concept. In the course of the nineteenth century, the scope of the energy concept changed repeatedly. It began as a term in the generalized mechanics of Laplace and Hamilton, but was reapplied successively in the theories of heat, electricity, magnetism, gravitation, and chemistry. In each case, fresh criteria, measures, and equivalences (or exchange-rates) were established, by which new systems or phenomena were brought within the scope of the concept. Nor was this process a single, unchecked advance; Liebig's attempt to associate biochemical phenomena with a specific form of 'vital energy', for instance, had later to be abandoned.[1] And all these changes in the application of the 'energy concept' were, for the most part, independent of changes either in the formal equations for representing and calculating the course of energetic processes, or in the nomenclature of the theory.

Meanwhile, on the theoretical plane, Helmholtz was formulating and establishing the Conservation of Energy as a matter of general principle: so giving the concept a mathematical representation suitable for application, not only to familiar cases of energy exchange, but wherever new forms of 'energy' could be identified and measured. And he arrived at this general treatment from a consideration of broad theoretical arguments, independently of all questions about the empirical application of the concept. He did so, furthermore, without seriously modifying the existing terminology of physics. His

---

[1] See, e.g., G. J. Goodfield, *The Growth of Scientific Physiology* (London, 1960), Chapters 6 and 7.

initial account of the general theory of energy conservation described it, indeed, as concerned with the conservation of 'force', and it was only after some years that be began to use the term 'energy', preferred by his contemporary Clausius and used fifty years earlier by Thomas Young. And when he finally accepted this change of nomenclature, he did so without retrospectively modifying his conservation principle. His theory embodied the same computation techniques, and had the same range of application as before, but he now began to draw an explicit verbal distinction between the two terms, 'force' and 'energy'.[1]

Which of these three changes—in nomenclature, equation, or application—are we to say represented the true invention of the 'energy-concept', or the true discovery of 'energy conservation'? That question has long been the subject of a frustrating controversy among historians of science. Some eight or a dozen different scientists have been claimed, at one time or another, as responsible for 'the' discovery in question; while, at other times, the episode has been hailed as a prime instance of simultaneous discovery—as though there were one and only one thing to be 'discovered', and all of the ten-odd scientists did just that. Yet the first thing to recognize, surely, is the richness and complexity of the issues involved in any such change. Just because conceptual problems necessarily arise, and conceptual changes necessarily take place, on several interrelated levels, we cannot properly characterize them by paying attention to one level alone. Instead of talking glibly about 'the' discovery of the energy concept, or the energy conservation principle, it is more helpful to acknowledge that—in the nature of the case—any such discovery is complete, only when all the varied problems of terminology, calculation techniques and application procedures have been successfully overcome.

Up to this point, we have been considering how concepts enter into the explanatory procedures, arguments, and other activities of the sciences, and how the gaps between the disciplinary ambitions of a science and its actual capacities act as a

---

[1] See the forthcoming book by Y. Elkana, on Helmholtz's development of the generalized concept of energy.

source of conceptual problems in practice. Now we must ask how the resulting conceptual issues measure up to the distinctions found in standard theoretical treatises on the philosophy of science. How far, then, do the central distinctions around which accounts of the 'logic of science' have commonly been built truly reflect the intellectual functions of concepts and conceptual issues, as we have analysed them here?

One crucial divergence is immediately apparent. Within a static, unchanging propositional system, it is normally possible to ask which of the constituent statements are 'empirical', and which are 'formal' or 'logical'. Within a historically developing scientific enterprise, by contrast, the significance of our concepts can be adequately shown by referring neither to the relevant empirical subject-matter alone nor to the formal structure of the science alone. It can be done only by viewing all the elements of the science—subject-matter, formal entailments, explanatory procedures, and all—within a larger framework, and by demonstrating how—on what conditions, in what kinds of case, and with what degree of precision—the explanatory procedures and/or arguments within which the concept is given a meaning can successfully be used to make sense of the relevant subject-matter. If we want to pin down the factual burden of a scientific concept, or theory, therefore, we must do so not in directly empirical statements, which make straightforward *use* of the concept or theory concerned, but rather in 'meta-empirical' statements, about the explanatory powers of the concept or theory: specifically, in statements about (i) their scope, (ii) their conditions of application, and (iii) their precision of application.

The same is true of the conceptual problems which are left unsolved by our current theories and procedures; and likewise of the conceptual changes taking place between one temporal cross-section and another in the development of a science. The nature of the outstanding conceptual problems in a science, the alternative ways in which concepts and explanatory procedures can be modified in order to deal with those problems, and the pattern of historical changes characteristic of the resulting development: all of these need to be considered and described in the same obliquely empirical idiom. And the true character of those problems and changes is only obscured if we insist on applying to such scientific concepts and theories any ready-made logical dichotomy between empirical/contingent/synthetic/factual propositions about the world, on the one hand, and formal/logical/mathematical/semantic propositions about our symbolism, on the other hand. The functions of scientific concepts being what they are, all serious statements about conceptual problems and conceptual changes unavoidably combine both empirical and formal aspects.

In some quite fundamental respects, therefore, conceptual issues—as here defined—elude the central distinctions of traditional philosophy, and obliterate lines of demarcation on which most philosophers have insisted. If we remain determined to have the points at issue in conceptual problems classified either as 'empirical' or 'formal', but not both, insuperable difficulties will arise on either choice. In tackling conceptual problems, is it our business to find out simply what *is in fact the case* about the world? Or are we, alternatively, simply re-ordering and improving the mathematico-linguistic formalisms used in *talking about* the empirical phenomena concerned? It would be equally misleading to describe conceptual issues in either of those ways alone. The precise point of those problems lies in their matching features of both kinds. The conceptual reorganization of our scientific understanding requires us to pay attention to the empirical facts, surely enough; but not merely with the intention of reporting on those facts, or even of generalizing about them. Our aim is, rather, to construct a better representation, better nomenclatures, and better explanatory procedures to 'account for' the relevant aspects of Nature, and to recognize more precisely on what conditions, and with what degree of accuracy, the resulting 'representation' can be applied to explain the nature of the world as we find it.

In this respect, conceptual issues stand in clear contrast both to straightforwardly empirical issues and to merely formal issues. In a straightforward empirical problem, the status of the relevant concepts is not in question. Suppose, for instance, that we have a well-established concept (such as 'specific heat'), together with standard procedures for measuring particular values of the variable in question: having previously applied those procedures to the element rhenium (say), we may then ask, '. . . and what is the specific heat of ruthenium?', and this will be a directly empirical question, requiring us to repeat for ruthenium the procedures of measurement and calculation earlier applied to rhenium. In such a directly empirical case, nobody is questioning that the concept 'specific heat' (say) applies to ruthenium, in just the same way and with just the same meaning that it does to rhenium. All that is at issue is to discover, in one further case, the empirical values of the accepted variables. 'Granted that we know how to apply a well-established concept C to the instance x,' we are asking, 'what then is true or false of x in respect of the concept C?'

In the case of purely formal problems, at the other extreme, nothing is at issue except the inner articulation of the mathematical or linguistic symbolisms employed in handling the current concepts. Such problems may require us to tidy up the mathematical aspects of our scientific procedures: i.e. to make them more compact or elegant, or to develop more rigorous significance tests. Or they may require us to explore analytically the detailed logical consequences implicit in our current concepts, regardless of their application to actual empirical situations. (This is what Isaac Newton did, for instance, in much of the first two parts of his *Principia*.) Yet, in these cases also, the explanatory merits of the relevant concepts and procedures are not—as such—being challenged. 'Granted that we know, in general, how to work with a well-established concept C,' we are now asking, 'how can we best present the implications of applying the concept C to anything whatever?' In such cases the mathematical tidying-up involved takes place on a purely formal level alone; and this distinguishes it sharply from the more substantial steps required (e.g.) in contemporary quantum electrodynamics, where theoretical inconsistencies exist that could be removed only by positive changes in the concepts of the subject. Over neither purely empirical nor purely formal issues, accordingly, is the explanatory adequacy of our current concepts in active doubt. Empirical issues merely invite us to extend the application of those concepts to new cases; formal issues merely invite us to reorganize our symbolisms internally; while both kinds of issue leave our current explanatory procedures basically unchanged. Conceptual questions thus contrast, in one crucial respect, with questions of both other kinds.

The significance of this difference is, evidently, connected with the historically developing character of the scientific enterprise. For, within any one 'representative set' of concepts from a science, conceptual changes are not apparent; such changes occur when, and only when, the existing repertory of concepts is modified or supplemented to do a better explanatory job, and a new 'representative set' of concepts is thereby created. In such cases, therefore, the operative question no longer begins, 'Granted that we have a well-established concept *C* . . .': it now begins, 'Granted that we *do not* all in cases know how to apply the concept *C* in its present form . . .' (For example: 'Granted that we *cannot* apply the conception "refraction" to Iceland Spar in any straightforward way . . .') And the general form of question arising about conceptual issues continues, '. . . can any *alternative* procedures be developed, for applying those concepts—with suitable modifications—to recalcitrant cases also?' So understood, conceptual problems are solved only by working on both the empirical and the formal levels at once. From one point of view, the resulting conceptual changes may appear merely semantic, but the semantic or formal changes they involve are made always in the light of new empirical discoveries. And the empirical discoveries in question are concerned with facts *about* the applicability of the relevant explanatory procedures, not with facts statable *in terms of* the scientific concepts in question; they accordingly need to be formulated, not directly in statements drawn from the current theory concerned, but rather in 'meta-statements' about the adequacy or inadequacy of that theory.

As we refine our knowledge of the empirical facts, we give up operating with concepts whose meanings are historically fixed. (This particular feature of conceptual issues is shared by problems of all our five types.) The curious optical properties of Iceland Spar, for instance, faced Huygens with neither factual nor semantic problems alone: the existing usage of the terms, 'refrangibility' and 'ray', provided no satisfactory way of explaining those properties, and Huygens had therefore to decide what else those terms might be employed to mean— i.e. how their use could be modified and extended—so as to make them applicable in the anomalous cases also. At the other extreme, similarly, the conceptual problems over organ transplantation have arisen as they do because, given current physiological knowledge, the existing criteria for telling the 'living' from the 'dead' are not adequate to the new kind of case. Once again, we are faced with a question which is neither straightforwardly factual, nor straightforwardly semantic. We have to ask neither—empirically—'Are organ-donors always in fact dead?', nor—semantically—'What do we now mean by the word "dead"?' (Both these forms of question assume that the term 'dead' has a single, clear and precise existing meaning.) Rather, the questions are 'Given our physiological understanding of the bodily procedures in a dying man, what is there for us to mean by the term "dead"?', and 'What exact criteria should we be using when applying the term "dead" to possible organ-donors?'

We may restate the same point in yet another form. In scientific theory and elsewhere, conceptual issues involve us in attending neither to the facts alone, nor to the definitions of terms alone. They require us, instead, to redefine our terms *in the light of* the relevant facts. Without appeal to empirical discoveries, there would never be any basis for such conceptual redefinitions; but, without such redefinitions, nothing could ever be done to improve our explanatory powers. Theoretical scientists can be content, therefore, to leave purely factual questions to naturalists, purely formal questions to mathematicians, and purely semantic questions to philosophers and lexicographers. Their own interest in facts is always to discover what can be 'made of' them in the light of current ideas; and their interest in mathematical formalisms and the meanings of words is to discover how they can be amended and reapplied, so as to 'throw light' on currently puzzling facts. What better kinds of explanation can be given, in the light of our more refined empirical knowledge? And in what respects do these novel explanations require consequential changes in the meanings of our terms? In the solution of conceptual problems, the semantic and the empirical elements are not so much wantonly confused as unavoidably fused.

In conclusion, it is worth contrasting the conceptual problems of a living science with the formal problems of a science whose empirical life has gone out of it. The moment a set of concepts achieves unchallengeable authority in any field of enquiry, that discipline no longer faces 'scientific' problems, properly so-called, and ceases to be a field of 'scientific' enquiry. For Theaetetus and Euclid, geometry was still as much the science of spatial relations as kinematics and dynamics were sciences of motion for Galileo and Newton. Nowadays, however, students learn the mathematical techniques of Euclidean geometry and formal mechanics, not as authentic empirical sciences in their own right, but under the heading of 'mathematical methods for scientists'—i.e. not as theories, but as theorems. And this is a perfectly appropriate heading since, from a scientific point of view, the ideas and calculations involved in the use of these techniques are no longer scientifically problematic. Having reached a definitive form, the concepts concerned are—as such —past further rational criticism, challenge, or change.

In the seventeenth and eighteenth centuries, Newtonian mechanics may have formed the heart of physical science, but more recently it has become a branch of pure mathematics comparable to Euclidean geometry. In the process, its concepts and propositions have ceased to be open to criticism in the light of empirical discoveries, and the intellectual tasks of the subject have become the characteristically mathematical ones, of reordering its theorems in as compact, elegant, and powerful a manner as possible. This being so, there is a certain irony about the title frequently applied to formal Newtonian dynamics today—viz., 'rational mechanics'—for this name runs directly counter to our whole account of the features that make scientific disciplines genuinely 'rational' at all. How did so inappropriate a name become adopted in the first place? Given the historical interactions between science and philosophy the answer is clear enough. This choice is one more product of the cult of systematicity—the equation of rationality with logicality—which united philosophers from Plato to Kant. Despite his other differences from Plato, Kant still believed that geometry and mechanics provided the one and only rational and coherent mode of description suitable for an intelligible account of Nature; and the name of 'rational mechanics' simply carries into our own day the traditional philosophical presuppositions about the status of geometry and other exact sciences.

Yet the consequences of this change show us, once again, how a complete commitment to formal considerations exacts as its price not just immunity from historical change, but irrelevance to it. Having committed themselves irreversibly to a particular system of concepts, pure geometers merely surrendered the natural science of space to the cosmologists and other physicists, who accepted no such commitment. Likewise, those mathematicians whose field of research is 'rational mechanics' have merely surrendered the natural science of force and motion to relativity and quantum physicists who accept no final commitment to the classical formalism. Is the pursuit of 'rational' mechanics then based on 'rational' decisions? The answer to that question depends, no doubt, upon your disciplinary standpoint. The pursuit of better systematicity is the pure mathematician's proper disciplinary goal; for the natural scientist, by contrast, committing oneself irreversibly to one particular system of concepts and theorems is the very antithesis of a 'rational' procedure, and a denial of one's own proper intellectual goals.

For Kant, the problem of building a truly 'rational' physics still turned on the central question, 'Can we, in the theories of motion and gravitation—as in geometry—expound our arguments in the form of logical systems which both tell us about the actual character of the natural world, and are immune to intellectual challenge in the light of experience?' Only a conceptual system combining both these characteristics, in his view, had mandatory authority over rational thinkers in all epochs and cultures; and his aim was to show the 'transcendental' basis for that authority. Yet, by now, the effect of Kant's argument has been the reverse of what he intended. Considering geometry and dynamics as natural sciences, we ourselves no longer believe that they rest on permanent 'metaphysical foundations'—even of the 'transcendental' kind that Kant argued for in his later critical philosophy. Rather, we believe that, as natural scientists, men show their rationality by their readiness to *give up* the dream of a single universal, uniquely-authoritative system of thought; and in being prepared to revise any of their concepts and theories, as their experience of the world is progressively enlarged and deepened.

2.4: *A Digression on 'Representation'*

Every attempt to restate the philosophical problems of human understanding gives rise, in its turn, to new views about what concepts are and how they are formed. Here, we have rejected a view of scientific concepts either as terms in formal calculi, or as names for empirical classes of objects, in favour of a procedural analysis, linking them to the alternative 'techniques of representation' in explanatory practice. At this point, however, something more needs to be said about the particular sense in which we are using this term 'representation'.

In the first place, then, we are not using the term in any inner, mental, or private sense. From John Locke to Norbert Wiener, by way of Hume and Wundt, Mach and Meynert, cognitive empiricists have tended to approach all epistemic issues from the point of view of the individual 'knower', and with an eye to processes supposedly going on inside his head—either neurological processes going on literally in the 'interior' of his head or psychological processes going in, in some more obscure sense, 'in his mind'. Anyone who adopts this approach has naturally been drawn towards one characteristic view of concepts, which identifies them either with 'private ideas' or 'faint images' formed as the after-effects of sensory 'impressions'; or alternatively with 'engrams', electrical potentials at 'synapses', or other neural 'traces' left as the after-effects of sensory stimulation. Nor are we using the term in any formal or quasi-geometrical sense. From Descartes to Hempel, formalist philosophers have tended to approach the epistemic problem, rather, from the direction of logic or mathematics—with an eye to the internal structure of demonstrative arguments, and to the terms in which they are couched. Anyone who adopts this alternative approach has naturally been inclined to equate concepts with the terms or variables standing for them in some 'deductive system'; or, at any rate, with the formal role of that term of variable within the deductive system concerned.

Our own account is designed to avoid both these positions. To begin with, we have deliberately chosen to defer questions about the individual acquisition of concepts until after we have dealt with their collective use: and, secondly, we have challenged the assumption that formal systems of propositions provide the only legitimate forms of scientific explanation. In this respect, we are following the example of Kant who, in an attempt to escape from the irresolvable eighteenth-century opposition between rationalism and empiricism, restated the problems of epistemology in terms of 'representations' (or *Vorstellungen*). We must take care, Kant argued, to bear in mind that all our experience is concerned with 'representations', rather than with 'things in themselves'. If we do so, we shall come to see the essential barrenness of the traditional epistemological arguments; and, having done so, we can move on to more constructive, 'transcendental' questions, about the conditions on which 'representations' are the subjects of coherent 'judgements'—i.e. on which our basic intellectual categories, concepts, and forms have any application to experience, as given in those 'representations'.

Unfortunately, Kant's own statement of his position is open to two quite different readings. His choice of the particular word *Vorstellung* only reinforces this ambiguity. On the one hand, his aim was certainly in part to emphasize the formal or grammatical structure of all 'judgements', and the dependence of all 'knowledge' on the particular concepts and intellectual forms in terms of which the resulting 'judgements' are expressed. From Kant's point of view, all judgements about the nature of external objects, and all knowledge of their behaviour, must be specified in propositional form; what we 'know' or what we 'judge' is always 'that *so-and-so* is the case'. If we attempt to say (or even think) anything about a 'thing-in-itself', independently of our language and forms of judgement, we shall get ourselves into difficulties of our own creation. Whereas any judgement, thought, or proposition necessary has some conceptual or grammatical structure, there can—by contrast—be nothing conceptual or propositional about 'things-in-themselves', as he defines them. Such 'things-in-themselves' are mere abstractions, towards which we may gesture inarticulately, but about which nothing can be said or known.

This particular point has, of course, nothing to do with sense-perception, or with the part which it plays in the acquisition of knowledge: it has to do only with the form and status of 'knowledge', considered as a product. In thinking about its implications, we need concern ourselves, therefore, only with the manner in which (e.g.) the forms of knowledge reflect the structure of language. Nothing in the resulting argument compels us either to treat 'representations' as being inner mental entities, or to regard 'things-in-themselves' as external objects hidden from direct perception, beyond the sensory 'representations' formed within our cerebral mechanisms. Yet that is the sense in which Kant's position was widely understood, both by physiologists and psychologists like Müller, Helmholtz and Fechner, and also by philosophers like Schopenhauer.[1] So Kant's followers came to use the term *Vorstellung*, not just in the analysis of knowledge as a product, but also in discussions about the mechanisms of perception. In this way, Kant's attempt to escape from subjective idealism was frustrated. The term *Vorstellung* came to refer simply to the 'ideas' brought into existence as an effect of repeated sensory 'impressions' (*Empfindungen*); the critical philosophy lost the transcendental character on which Kant had set so much store; and it became possible for Mach, once again, to confuse the theories of Kant's *Critiques* with those of Berkeley and Hume, which Kant himself had been attempting to supersede.[2]

Kant himself is partly to blame for this. Although he evidently saw the need to escape from the Cartesian dilemmas over sense-perception, in his own writings he never wholly succeeded in doing so. So, while we can reformulate much of his philosophical position today in 'procedural' terms—as an attempt to show how the character of our intellectual activities imposes *a priori* forms on our judgements and so on our 'knowledge'— Kant's own exposition of those views still preserved strong traces of Cartesianism. His account of the 'forms of intuition' (for instance) sought to correlate spatial and temporal terms, first with geometrical and arithmetical relations, and then with visual and auditory experiences; and this argument greatly encouraged the nineteenth-century interpretation of *Vorstellungen* as 'sensory representations' which, in due course, drove Kant's successors back into idealism and sensationalism.

Yet, setting aside for the moment all questions of sensory

---

[1] See A. Schopenhauer, *Die Welt als Wille und Vorste ung* (Leipzig, 1819); for Müller, Helmholtz, and Fechner, see, e.g., W. N. Dember, *Visual Perception: The Nineteenth Century* (New York, 1964).

[2] E. Mach, *The Analysis of Sensations* (Jena, 1886, 1900, Eng. tr., Chicago and London, 1914).

physiology and psychology, we can equally well state Kant's epistemic point in very different terms, using an alternative German word that has very different associations. Let us substitute the word *Darstellung*—as used by Hertz, Bühler, and Wittgenstein—for the *Vorstellung* of Schopenhauer, Helmholtz, Boltzmann, and Kant himself: we can then make Kant's crucial point about the role of concepts in the expression of collective judgements and communal knowledge clearly enough, without being entangled in the other quite distinct problems about sense-perception, 'inner experiences', and 'mental entities'. For a *Darstellung* is a 'representation', in the sense in which a stage-play serves as a theatrical representation, or in which an exhibition or recital provides a public presentation or representation of works of art or music. To *darstellen* a phenomenon is then to 'demonstrate' or 'display' it, in the sense of setting it forth, or exhibiting it, so as to show in an entirely public manner what it comprises, or how it operates: as when an hydraulic system of tubes and pumps is used to provide a simplified representation, or explanatory model, of a complex electrical circuit. (By contrast, the term *Vorstellung* suggests a 'representation' as private or personal as a *Darstellung* is public. A *Vorstellung* 'stands for', or symbolizes, something 'in the mind' of an individual. The term carries the same burden as words like 'idea' and 'imagination': it is, in fact, the standard German translation for the Lockean term 'idea', and runs into all the same difficulties.)

The relationship between a *Darstellung* and the reality which it 'displays' or 'represents' is, accordingly, a relationship between two public entities. When Hertz spoke of a dynamical theory as providing a *Darstellung* of the motions that it explains, and when Wittgenstein declared, more generally, that the propositions of a language *darstellen* the facts of the world, their assertions had nothing specifically 'mental' or 'inner' about them.[1] Both men were directly concerned only with the manner in which, and the conditions on which, a scientist's

---

[1] See L. Wittgenstein, *Tractatus*, sec. 4.1, where the words *darstellen* and *Abbildung* are used in a manner that reflects the use of the term *Bild* as an 'artefact' word; and also H. Hertz, *The Principles of Mechanics* (Eng. tr. by Jones and Walley, New York, 1900), Introduction, where an entire physical theory is spoken of as a *Bild* designed to provide a *Darstellung* of the corresponding physical phenomena.

calculations 'exhibit' or 'demonstrate' the forms of phenomena, or a language-user's utterances those of the corresponding facts. And we can construe much of Kant's own central thesis that 'knowledge' is incapable of grasping 'things-in-themselves', in this same sense: to the extent that the content of knowledge can be specified only in judgemental or grammatical forms, that which is 'known' (i.e. known-to-be-the-case) is not an object independent of human thought, but a linguistically-structured fact or proposition.

In Kant's own time, this was a highly damaging point to press against both sides in the traditional epistemological debate, for that dispute had repeatedly confused percepts with concepts, and propositions with images and objects. And, if we here pick on 'techniques of representation' as a key element in the meaning (or collective use) of scientific concepts, we may emphasize that the term 'representation' is to be understood solely in the *Darstellung* sense. In doing this, indeed, we are merely generalizing Hertz's own analysis of the explanatory function of classical mechanics. Just as the axiom-system of dynamics had for Hertz the task of 'exhibiting' the general relations involved in the motion of bodies, so too the function of taxonomic classification is to 'exhibit' the general relations between different kinds of living creatures, that of ray-diagrams to 'show forth' the relations involved in the optical reflection and refraction, and so on.

In saying this, we are implying nothing whatever about processes going on in the mind of the individual scientist when he employs the relevant concepts to 'understand' the phenomena concerned. It may yet be arguable that the individual understanding relies on the use of a 'mental programme', or *Vorstellung*, arrived at by internalizing the corresponding 'mode of representation', or *Darstellung*, but that thesis is not yet our concern. In this first group of enquiries, we are deliberately restricting ourselves to the collective intellectual functions of concepts and representation techniques in the explanatory activities of science; and an account of these collective functions in terms of *Darstellungen* can sidetrack, for the moment, all Cartesian and Humean puzzles about the relationship between the 'inner' concepts and 'external' phenomena. At the same time, it effectively concedes to Frege all that his anti-psychologism legitimately claimed: namely, that 'explaining' a phenomenon requires us not just to imagine inwardly how that phenomenon might be as it is, but to demonstrate publicly the nature of the relationships which it exemplifies. If we read the term 'demonstrate' in a broad enough sense, of course, that task need no longer mean constructing the kind of quasi-Euclidean proof which Descartes and Hume would alone have accepted as 'demonstrative'. But that is all to the good, since it indicates that, in one significant respect, the implications of Frege's anti-psychologistic position may themselves have been broader than he realized. While an 'objective' account of concepts must be given in terms of their public or collective use, which is what determines their 'sense', such an account need not, in the case of all concepts, take precisely the same form as that which Frege, Peano, and Russell between them gave for the arithmetical concept of 'number'. On the contrary: concepts of other kinds have different sorts of collective use, and their 'objective sense' needs to be analysed in correspondingly different terms.

Since the time of Kant (in short) the term 'representation' has been caught up in a familiar philosophical dialectic. At one extreme, it has been equated with an 'image' or 'sense-content', so corresponding to the Lockean notion of an 'idea'. At the other extreme, it has been associated with a formal deductive system, corresponding to the Cartesian network of 'self-evident' premises and demonstrations. The consequent argument has simply re-enacted, in post-Kantian terminology, the very epistemological debate Kant himself had hoped to transcend. So an important and entirely genuine question—viz., how the personal thoughts and conceptions of an individual concept-user are related to the communal or collective uses of concepts—has been confused with intractable conundrums about the relations between the 'inner' and 'outer', or private and public, aspects of mental life. As a result it has come to appear as though all the practical procedures by which groups of men collectively demonstrate their intellectual capabilities were merely a 'behavioural accompaniment' to the 'inner thoughts' which alone have rational significance. Thus 'ideas' or 'concepts'—regarded as internal representations, or *Vorstellungen*—have become logically detached from their own 'expressions' or 'applications' through the use of classification-procedures and calculations, diagrams, models and graphs—i.e. through external representations or *Darstellungen*.

Many of the difficulties arising out of this divorce will be dealt with at greater length in our second group of enquiries. In particular, we can leave over for later all questions about the relations to be found, in individual concept-use, between concepts and their application, mental life and intelligent activity, thoughts and judgements, private images and public representations, *Vorstellungen* and *Darstellungen*. At that later stage, we can ask how far such distinctions are wholly legitimate, even as applied to the thoughts of individual concept-users: for the present, we are concerned with ideas and concepts only as possible transmits in collective disciplines, and the 'public' aspects of concept-use alone are of immediate relevance.

By analysing the collective aspects of concept-use in terms of the explanatory procedures of a science, we can in fact go beyond both earlier types of analysis of concepts—'mental ideas' versus 'primitive terms'—without wholly losing touch with either. Thus, the theoretical terms appearing in axiomatic calculi and other logical systems can now be thought of, not as *identical with* scientific concepts, but rather as linguistic or mathematical *symbols for* such concepts. A formally defined term derives its scientific significance—i.e. is put to explanatory use—only from association with the constellation of explanatory procedures for which it stands. Similarly with the private, personal images or models, in terms of which an individual scientist may think about his problems. These, too, can be given a scientific sense, as one further expression or symbol of the relevant scientific concepts. From our own point of view, indeed, images and mathematical variables end up on the same level. Although the terms of a formal calculus may be related more exactly to the concepts for which they stand than the mental images or models we form of them, their relation to those concepts is no more direct; and it is no more correct to identify 'concepts' with those 'primitive' terms than with physical models or mental images.[1] On the contrary: all such terms, images and/or models are alternative and more-or-less

---

[1] As Hertz (loc. cit.) implies, there is no reason in principle why mental images —or *innere Scheinbilder*—should not be used to give a 'representation' of phenomena;

adequate means of expressing or symbolizing the collective concepts which form the 'transmit' of a scientific discipline. And while there is no doubt that mental images, neural traces and mathematical variables all in fact do have parts to play in the thoughts and activities of scientists, we can continue to analyze the content and validity of the concepts used within collective rational enterprises here, without making particular reference to any of these things.

---

but, as such, they are only one more possible instrument of understanding among others.

CHAPTER 3

Intellectual
Disciplines: Their
Historical Development

I F we consider the historical process of conceptual change in intellectual disciplines in terms of a populational model, we can then represent it in three alternative ways. We can take successive time-slices across the intellectual content of the discipline, so analysing the process into (A) a sequence of 'representative sets', comprising the totality of concepts having a place in the discipline concerned at successive times.

Alternatively, we may consider the appearance, subsequent development and eventual fate of particular concepts over their whole life-histories, so analysing the process into (B) a bundle of genealogies of particular concepts.

Or, again, we may combine these two modes of analysis, tracing the genealogical development of all the relevant concepts through a succession of representative sets: in this way, we analyze conceptual change as (C) the outcome of a dual process of conceptual variation and intellectual selection.

Each mode of analysis has its own merits and defects. The 'transverse' mode (A) helps to focus our attention on questions about rationality that more orthodox, system-orientated philosophies of science overlook. For the questions that lend themselves naturally to discussion in systematic terms have to do with the relations between contemporaneous concepts; while questions about the relations between concepts from successive representative sets necessarily elude the formal logician's grasp. And, where conceptual changes are in discussion, the crucial question of 'rationality' arises precisely over the 'non-logical' changes between successive representative sets of concepts: i.e. over the conditions on which we may say

Conceptual Population of a Historically Developing Discipline in Three Representations

C_q^w C_q^v C_q^y C_q^x

$C_r^{w''}$ $C_r^{w'}$ C_r^y C_r^x

C_s^z C_s^y $C_s^{x''}$ $C_s^{x'}$

'Envelope' of changing content

(Representative sets)

$\langle C_q \rangle$

$\langle C_r \rangle$

$\langle C_s \rangle$

Established concepts at time t_q

t_q t_r t_s

Time

(A) Transverse or Time−Slice Representation

that a new concept was added to the repertory of the discipline, or an older one displaced, *for good reasons.*

The longitudinal, or genealogical mode (B) makes this rational continuity even more evident. Here, each branch-point or interruption in the life-line of a concept corresponds to a 'unit' of conceptual change, about which we can pose the question, 'Was this change made knowingly and for good reasons, in the light of relevant and cogent considerations, or did it happen inadvertently or irrationally, because of (say) some failure in the disciplinary transmission-process, or some irrelevant prejudice?' As it stands, however, this genealogical model still fails to differentiate between the two complementary aspects of conceptual change: (i) the introduction into current debate of conceptual variants whose merits are as yet unjudged, and (ii) the adoption of selected variants into the established repertory of concepts. For our purposes, the combined or evolutionary mode (C) has the great advantage of marking clearly the differences between innovation and selection. It registers explicitly the fact that only some of the current concepts in a discipline are, at any particular stage, active topics of debate and innovation. The 'well-established' concepts form a background, against which the currently unsolved problems are discussed, so providing an occasion for introducing innovations into the few actively questionable concepts. And it reminds us, also, that each branch-point or discontinuity in the conceptual genealogy of a discipline represents the outcome not of a simple, one-shot change, but of a more complex, trial-and-error process.

In terms of this more complex pattern of analysis, we may now attack two groups of questions. On the one hand, there are questions about conceptual innovation or variation: i.e. about the pool of conceptual variants which coexist at any time with the established concepts of a discipline—as 'novice' concepts, admitted 'on probation', as possible ways of solving outstanding problems—but which have not yet been accredited or discredited. About these variants we can ask:

(i) Under what circumstances do such conceptual inno-vations appear, either in one particular discipline, or in many?

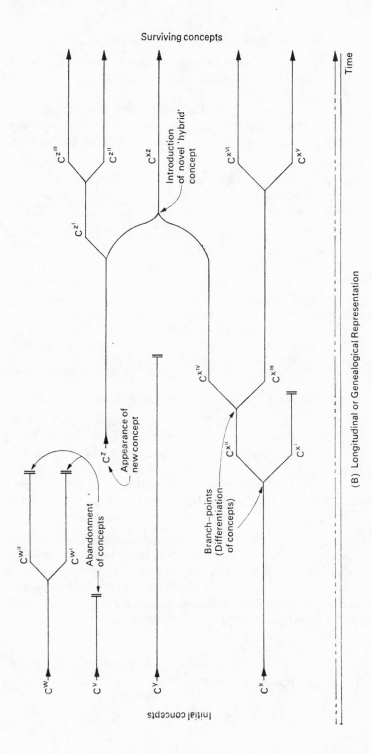

Surviving concepts

$C^{z III}$

$C^{z II}$

$C^{z I}$

C^{xz}

Introduction
of novel 'hybrid'
concept

$C^{x VI}$

$C^{x V}$

$C^{x IV}$

$C^{x III}$

Appearance of
new concept

C^z

$C^{x II}$

Branch—points
(Differentiation
of concepts)

$C^{x I}$

$C^{w II}$

$C^{w I}$

Abandonment
of concepts

Initial concepts

C^w

C^v

C^y

C^x

Time

(B) Longitudinal or Genealogical Representation

(ii) On what conditions will the input of such variants be vigorous or sluggish, or be more vigorous in one discipline than in another?

(iii) If conceptual variation takes place predominantly in certain preferred directions, what factors or considerations are responsible for the choice of those directions?

On the other hand, there are questions about the selection procedures by which some variants are accredited, others are ruled out, and others again are put on ice pending an adequate test. For instance:

(i) What kinds of factor or consideration determine which of the current conceptual variants are selected for perpetuation, and so enter the established repertory?;

(ii) To what extent does this selection rest on explicit appeals to considerations whose relevance and cogency are collectively recognized in the relevant profession?;

(iii) Can we give a satisfactory account of the criteria by which practitioners of a science distinguish well-founded, properly justified conceptual changes from ill thought out or hasty, overdue, or unintended changes?

Finally, to take these two aspects of conceptual change together:

(i) Under what circumstances will the balance between variation and selective perpetuation serve to maintain the continuity of a single, compact discipline?

(ii) Under what circumstances will it lead, rather, either to the abandonment of an earlier discipline, or to its displacement by two or more successor disciplines?

These are the specific questions into which the general problem of conceptual change in intellectual disciplines breaks down, when we approach it from a populational standpoint. And these are the questions in terms of which we may now piece together a picture of conceptual change as an historical process in which both 'rational' and 'causal' factors are operative.

3.1: *Conceptual Variation*

To begin with the innovative aspect of conceptual change: why should the appearance of conceptual variants pose any

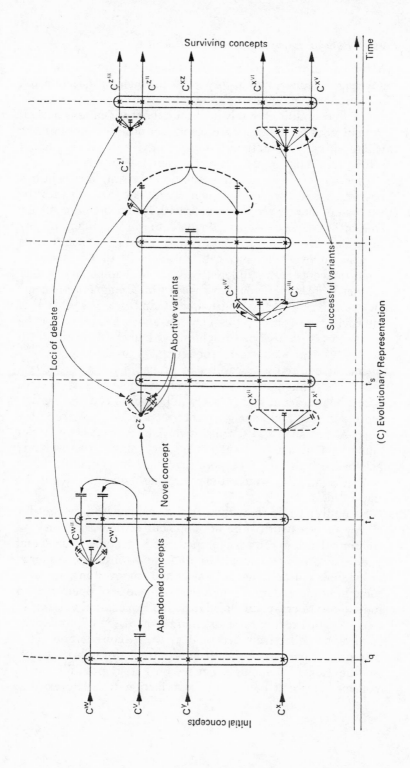

(C) Evolutionary Representation

particular problem? Can we not take it for granted that, at any stage in conceptual history, there are enough reflective and open-minded individuals to keep up a supply of novel conceptual possibilities? Surely, a certain inquisitiveness is natural to man, so that the obstinate scepticism of individual humans will eventually lead them to the truth: why, then, should intellectual innovation have any mystery about it?

This way of posing our first question oversimplifies it in two respects. Firstly: we are here concerned, not so much with the verification of new propositions, as with the introduction of new concepts. Given concepts adequate to the demands of any situation, indefatigable honesty may in due course enable men to discover whatever can truthfully be said in terms of those concepts. (Are different genera of Papilionaceae ever interfertile? Do Pi-mesons have a measurable magnetic moment? In each case, the answer is, 'Let us find out!') The task of improving the concepts themselves is another matter. It is by no means as positive or straightforward as checking the truth or falsity of an empirical proposition, or measuring the frequencies required as measures of probability: rather, it is a subtle and imaginative task, of conceiving how our concepts might be reordered so as to yield a 'better'—i.e. a more exact, more detailed, and generally more intelligible—picture of the objects, systems, and events involved. And men will not embark on this task, in the first place, unless they already acknowledge the claim of more exacting intellectual ideals, by comparison with which their existing concepts might be improved.

Secondly: while individual initiative may lead to the discovery of new truths, the development of new concepts is a communal affair. For us to speak of a genuine 'conceptual variant', it is not enough to find some obstinately honest individual entertaining a conceptual innovation; it takes something more than the personal reflections of open-minded individuals to create an effective pool of conceptual variants in a science. Before a novel suggestion becomes a real 'possibility', it must be collectively accepted as worth consideration, i.e. as worth experimenting with and putting through its paces. So the task of launching a new conceptual possibility calls not merely for a collective dissatisfaction with the existing conceptual repertory, nor for an individual proposal of some alternative explanatory procedure, but the matching of the one against the other. The individual's innovation, that is, must be seen as providing a possible way of dealing with the problems which are the sources of collective dissatisfaction.

With these preliminaries in mind, let us look at the conditions for effective conceptual innovation. To begin with, what *intellectual* conditions must be fulfilled before a conceptual innovation becomes a genuine 'possibility' at all? Certainly, when professional scientists treat some conceptual proposal as a possibility, they do not—like David Hume in his study—think of that change in isolation, or in the abstract: as being 'possible' *tout court*. They cannot take a conceptual proposal seriously merely because it avoids contradicting the certified results of previous experience, and is therefore 'possible' on merely logical grounds. Rather, they will consider any new proposal with an eye to some problem, or group of problems, and judge its status as a 'possibility' by inquiring how it might contribute to a 'possible solution' of those problems. The class of 'scientific' possibilities is thus far narrower than the philosophers' class of 'logical' possibilities. Scientific possibilities are always 'possible ways of tackling such-and-such problems'; and the context of discussion implicitly shows what specific problems the suggestions are directed at, and how the proposed innovations will help to deal with them. Unless this minimum condition is satisfied, such innovations—however formally consistent—do not even acquire the tentative status of 'possibilties'.[1]

In practice, this minimum condition can be satisfied in many different ways, depending on the character of the problems under consideration. Indeed, we might in principle list at least fifteen basic types of conceptual innovation, all of them having the form of genuine scientific 'possibilities'. In theory, at any rate, one can set about solving problems from any of the five main types discussed in the last chapter, in any of three alternative ways: by refining our terminology, by introducing

---

[1] See, e.g., the paper by D. Shapere on 'Plausibility and Justification in the Development of Science' read to the American Philosophical Association (Eastern Division), December 1965, as printed in *Journal of Philosophy*, 63, No. 20 (1966), 611–20.

new techniques of representation, or by altering the criteria for identifying cases to which the current techniques are applicable. Furthermore, the history of scientific thought provides examples to illustrate conceptual innovations belonging to most, if not all, of these fifteen basic types. These fifteen basic forms are, in any case, only a beginning. In actual practice, conceptual problems rarely arise on one level only, and conceptual changes typically affect more than one aspect of our current concepts. More typical real-life examples—like the conceptual innovations leading to the creation of molecular biology—tend to be highly complex. They will frequently explain old phenomena more exactly, and new phenomena for the first time; involve the realignment of conceptual boundaries, both within a single discipline and between neighbouring disciplines; and, in addition, call for modifications in all three aspects of the concepts concerned—i.e. in terminology, in techniques of representation, and in empirical criteria of application equally.

For our purposes, it is enough that a serious conceptual 'possibility' must show promise of yielding a recognizable procedure for attacking some outstanding theoretical problem. This said, we may turn from the intellectual to the professional conditions for the introduction of genuine 'variants'. Here, let us recall the problem that faced Charles Darwin at the same point in his argument about organic evolution. Darwin's difficulty had to do with the inheritability of variations: if every new generation of animals and plants comprised a fresh batch of individuals, with its own entirely independent characters, the process of natural selection would have nothing to work on. The Darwinian mechanisms for the selective establishment of distinct species can be effective only if variations remain 'available' for enough successive generations of organisms, and within suitable forums of competition, for natural selection to have its effect. In short, random variability alone will never lead to organic evolution; what has to be guaranteed is a pool of inheritable variations.

Failing a well-established science of genetics, Darwin himself assumed the existence of such a pool, as a plausible hypothesis which would gain reflected credibility from the overall success of his entire theory; and he candidly conceded the need for this assumption at an early stage in his argument: '*If* variations useful to any organic being *ever do* occur, surely the individual thus characterized will have the best chance for being preserved for the struggle for life; and *from the strong principle of inheritance* these will tend to produce offspring similarly characterized.'[1] Once this presupposition was admitted, on a provisional basis, the argument for his theory could proceed; and its overall cogency would then help us, retrospectively, to confirm the validity of the original assumption. A similar problem arises over conceptual evolution. Conceptual change in a science can proceed effectively, only where transient innovations do not automatically die with their creators. Certainly, individuals must be impelled to reflect about the relevant domain; but their private thoughts will, by themselves, have as little long-term effect as transient organic variations, unless they are taken up by others and so become the raw material of collective evolution. Only then do they become 'available' in the relevant pool of conceptual novelties. While the final, indispensable source of conceptual variations lies in the curiosity and reflectiveness of individual men, these will be without effect unless other circumstances are favourable. While the 'genetics' of intellectual novelty on the individual level may be no great mystery, therefore, the occurrence of full-scale conceptual variations in collective disciplines depends also on those communal factors without which the original ideas of individuals may never get into professional circulation.

A condition for the availability of genuine conceptual variants is, thus, the existence of suitable professional 'forums' of discussion. Serious and methodical development in the ideas of a discipline requires correspondingly well-organized opportunities for the critical debate and amendment of those ideas. If we study the rise of the 'new chemical philosophy' in the fifty years from 1760 to 1810 solely in disciplinary terms, we may concentrate on the development of quantitative methods of analysis, or on the transformation of the idea of 'affinities' between Joseph Black and John Dalton. But from a wider point of view this is only one small part of a larger story, since it takes for granted the existence in England and Sweden, France and Scotland, of those very circles of natural philosophers

[1] See, Darwin, *The Origin of Species* (London, 1859), Chapter IV, 'Natural Selection', Summary.

whose debates were the collective forum for this conceptual competition. If the disciplinary novelties were able to prove their worth, that was because the necessary professional forums were available; and the shared intellectual ambitions of the relevant groups of men shaped the considerations by which those innovations were judged.[1] Conversely, the absence of a suitable forum may, by itself, be a fatal obstacle to the proper consideration of an intellectual variant. Mendel's isolation from his fellow-scientists, for instance, made it difficult for them to recognize the significance of the problems he was tackling and the wider theoretical implications of his ideas[2]. The forum of competition within which effective innovation is possible thus demands that a discipline be professionally organized in ways that permit the novel ideas of individuals to be appraised in relation to a collective set of explanatory ideals; and these ideas can contribute to the collective debate, only if—unlike Gregor Mendel in the 1860s—their authors are in effective touch with the professional groups concerned.

Analogous combinations of factors and considerations—some of them disciplinary or intellectual, others professional or social—are relevant to the second of our questions about conceptual innovation: viz. that of the relative rates at which conceptual variants will enter different sciences at different times. Different sciences do not all change actively, or remain stagnant, in step with one another; rather, the centre of attention shifts from one historical epoch to another. In one decade, the ideas of dynamics may be in rapid development, those of geology comparatively static, while fifty years later the balance may be reversed; astrophysics may lose momentum, just as chemistry begins to pick up speed; physiology may be more active than anatomy, or vice versa. Occasionally, indeed, certain scientifically fruitful techniques may even have flourished in situations which were unfavourable to scientific speculation

---

[1] For a sample of the early history of these professional groups, see Robert E. Schofield, *The Lunar Society of Birmingham* (Oxford, 1963).

[2] On Mendel's isolation, see e.g., H. Iltis, *Life of Mendel* (Berlin, 1924; Eng. tr., 1932), and subsequent commentaries by Elizabeth Gasking, R.A. Fisher and others.

generally. In what terms, then, are all these differences in the comparative rates of conceptual innovation to be explained? We may attack this question, first, in systematic or 'internal' terms—asking how attention comes to be focused on particular fields of science by the character of the outstanding problems—and then go on to consider the sociological or 'external' factors involved—i.e. the economic, political, or institutional aspects of a problem-situation which either encourage or destroy opportunities for research in different areas of science. To begin with the 'internal' considerations: the amount of effort devoted to a particular field of enquiry at any time, and so the rate at which novel ideas are put forward in that area, depend clearly on one intrinsic feature of the field, which scientists themselves describe by using the metaphor of 'ripeness' and 'unripeness'. A problem which scientists judge to be 'ripe' for solution will often be worked on for no more reason than that; while correspondingly little effort will go into 'unripe' subjects unless the questions involved are of overriding interest. Our first step must be to unpack this metaphor in more literal terms.

There are periods in the history of any science when men cannot get a satisfactory grip on its problems. This may happen for several reasons. Its subject-matter may be so varied and complex that it defies analysis: as in meteorology, until quite recently. Alternatively, the more general concepts needed to introduce some intellectual order into the field may still be lacking: thus, the central problems of physiology became 'ripe' only after a general system of chemistry had been developed. Again, the solution of the outstanding problems may demand mathematical methods, instruments or experimental techniques that do not yet exist . . . and so on. For one or more of these reasons, the theoretical problems of a science may hold out no prospect of yielding to available lines of investigation, and so be 'unripe'. Conversely, a moment can arrive—sometimes quite suddenly—when these obstacles can all be circumvented. The precipitating factor may be the development of some new instrument (e.g. the electron microscope) or a novel mathematical technique (e.g. the differential calculus), the importation of general ideas from some other, more basic science (as in the development of biochemistry), the recognition of new principles of classification (in the case of weather-patterns, or plant diseases), or else some combination of these steps. Until this moment, effort put into a science will yield disproportionately little fruit; but, once it arrives, the consequences can be dramatic. A novel alliance of X-ray crystallography with bacterial genetics and the chemistry of the nucleic acids, for instance, made possible the creation of molecular biology;[1] over a few years, or even months, a new and vigorous scientific discipline sprang into existence, with the mission of 'cultivating' the new theoretical possibilities, and 'reaping' the intellectual 'harvest' so created. Instead of confronting problems on which they could get no grip, scientists were now able to state and pursue precise, explicit questions. One after another, new and clear questions presented themselves for investigation, and consistent principles of explanation rapidly established themselves throughout the field.

Notice how, when one describes the efflorescence of a science, agricultural terms like 'fruit' and 'harvest', 'cultivating' and 'reaping', come naturally to one's pen. Scientists are as anxious as farmers not to waste their energies in unprofitable operations, and as careful as farmers to time their activities to the immediate demands of the tasks in hand. So problems that promise to be readily soluble attract their attention more readily than those which hold out no hope of solution. Peter Medawar has put this point in a striking phrase. As politics is 'the art of the possible', he says, science is 'the art of the soluble'.[2] A certain fuzziness of mind tempts some highly intelligent scientists to spend too much time speculating fruitlessly about problems that are one degree too complex for immediate investigation. (Medawar himself has spoken of theoretical embryology as having been in this frustrating state from 1920 right up to the 1950s.)[3] One sure test of a scientist's judgement, by contrast, is his capacity to spot the particular problems that will repay effort now, and to devise the methods of investigation that will 'harvest' the available intellectual 'crop' elegantly and speedily.

[1] See, e.g., *Phage and the Origins of Molecular Biology*, ed. J. Cairns *et al.*, (Cold Spring Harbor, N.Y. 1956).

[2] P. B. Medawar, *The Art of the Soluble* (London, 1967).

[3] op. cit., pp. 106ff.

This is not to say that the easiest scientific tasks are those that attract the most attention. Beyond a certain point further research may lose interest, by becoming too much a matter of routine, for there can be a 'glut' of observations or calculations quite as much as of rhubarb! Next to problems that are completely insoluble, scientists are least interested in problems whose very simplicity means that their solution will shed little new conceptual light. There will normally be little theoretical reason for measuring (say) the specific heat of yet one more rare earth, or the glycogen secretion of yet one more rodent. So the exercise of scientific judgement often consists, in practice, in recognizing just how some quite precise and specific—or 'hard'—investigation can result in new *conceptual* insights.

If all other things were completely equal, then, scientists would distribute their effort between different areas of enquiry in proportion to their estimates of the prospective intellectual returns from each field. Here as elsewhere, however, all other things never are completely equal; and at this point the 'external' or sociological factors must enter our account. At any stage in history, cultural practices and social institutions affect intellectual development in two opposed ways: they provide positive incentives and opportunities for intellectual innovation, and they put obstacles and disincentives in the way of intellectual heterodoxy. Let us look first at the positive factors. The exercise of one's curiosity about the natural world is a 'pure' or self-justifying activity, which pays few immediate dividends other than the intellectual satisfaction of better understanding; by itself, therefore, it has rarely given men a living. In economic terms, as a result, natural science has normally been a 'pensioner', financially dependent on its association with other activities and institutions. Looking back from our twentieth-century standpoint, it may seem that the development of natural science has been one of crucial achievements of the civilization, yet this is a very recent attitude. Sociologically speaking, scientific activities were little more than 'epiphenomena' until the present century, and they could have only a trifling effect on the established pattern of men's other social activities and institutions. When we ask why, within any particular culture, nation, or epoch, different natural sciences develop at different speeds, it will therefore be worth considering on the back of what other activities natural science was riding in that context. What other institutions were, either inadvertently or by design, providing the opportunities then needed by intellectually curious men for devoting themselves to scientific investigations?

In other epochs and cultures, men have been free to criticize accepted ideas, and 'change their minds' about nature, only as a by-product of other professional activities whose outcome was more directly useful, in social, political, and economic terms. So the changing pattern of social institutions has produced, directly, a changing pattern of opportunities for scientific speculation in general and, at a remove, changing rates of development as between the sciences themselves. A very few fundamental problems have (it is true) been so obsessive, throughout the last two thousand years, that scientists have not needed external occasions for speculating about them; elsewhere, however, they have devoted most effort to subjects having some extrinsic interest and support. In this respect, therefore, the rates of intellectual innovation in the different sciences have mirrored roughly, not only their degrees of 'ripeness', but also the patterns of opportunity and social demand. At one extreme, George Gaylord Simpson has described 'exobiology'—the science of whatever life there may be outside the Earth—as a field of enquiry devoid of any known subject-matter, and existing solely in virtue of patronage.[1] While one can trace back a continuing speculative debate about extra-terrestrial life, through Kant and Buffon, Fontenelle and Kepler, at least as far as Nicholas of Cusa, the subject has hitherto been intrinsically 'unripe' for scientific study, and no outside institution has seen any extrinsic advantage in subsidizing it.[2] Yet, in the last ten years, the National Aeronautics and Space Administration of the United States Government has picked on exobiology as having a potential bearing on the problems of space-travel; so institutes and research programmes

---

[1] G. G. Simpson, 'The Non-Prevalence of Humanoids', in *This View of Life* (New York, 1964), pp. 253ff.

[2] See, e.g., Buffon, *Les Epoques de la Nature* (1778); Kant, *Allgemeine Naturgeschichte* (1755); Fontenelle, *Entretiens sur la Pluralité des Mondes* (1686); and so on back to Nicholas of Cusa, *De docta Ignorantia*, II, Chapter 12, in the mid-fifteenth century.

for exobiology have sprung into existence in response to N.A.S.A.'s largesse. Until someone actually finds undoubted extra-terrestrial organisms, the new discipline cannot of course hope to make much progress, since meanwhile there is no way of checking which of its conceptual novelties are well founded. Still, in the meantime, there have been plenty of novel suggestions. Abundant patronage has at least nourished the intellectual imagination, and the would-be science of exobiology has displayed as sharp a burst of conceptual innovation as the depth of our ignorance permits.[1]

At the other extreme, fundamental innovations have sometimes sprung from quite surprising sources. Despite its institutional conservatism, the theocratic society of ancient Babylon provided a niche for major achievements in astronomy; so much so that all subsequent physics and astronomy have been built, in part, on the foundation of Babylonian astronomical computations. In Babylon as in its daughter-societies of Islam, the official calendar was maintained and controlled by State— or Church—astronomers who, over several centuries, devised increasingly accurate procedures of computation for predicting the chief celestial motions. Since the planets were regarded as divine, prudence and piety alike made the keeping of planetary records and the development of astronomical forecasting techniques matters of national concern. In their day-to-day work, Babylonian astronomers such as Kidinnu presumably paid little attention to the supposed divinity of the heavenly bodies, and approached their work in the spirit of all intellectual craftsmen, as being—in Robert Oppenheimer's phrase— 'technically sweet'. But the existence of such professional attitudes was only one of the conditions necessary for the survival of their work. The fact that Kidinnu and his colleagues had a chance to attack their technical problems single-mindedly at all was an indirect product of the public policy of the Babylonian State, which acknowledged a social need for calendrists, or 'monthly prognosticators'. Other contemporaneous cultures possessed no comparable institutions, and the computational

---

[1] This is not, of course, to say that the existence of organic materials outside the Earth's atmosphere is not worth scientific investigation; in fact, there were some intriguing reports to the last meeting of the International Astronomical Union (1970) about spectroscopic observations indicating the occurrence of unexpectedly complex molecules in inter-stellar clouds.

achievements of the Babylonian astronomers were therefore unmatched.[1]

Turning to more typical phases of scientific development, we can again afford to bear in mind the social context of scientific enquiry. In medieval Islam, for example, the natural sciences of Greek Antiquity were kept alive and developed further; but the men responsible earned their living by other means, mostly as Court physicians. In seventeenth-century London, a few of the scholars who founded the Royal Society were men of independent means, but others earned professional incomes elsewhere. The finances for the Society itself were obtained from King Charles II's budget by the Secretary of the Admiralty, viz. Samuel Pepys, the diarist, who was also the first Secretary of the Royal Society. (So scientific curiosity found a paymaster in the Navy, through much the same chain of circumstances that made the United States Office of Naval Research the prime support of academic science in the America of the 1950s.) In eighteenth-century England, educated churchmen —whether Anglicans like Stephen Hales, or dissenters like Joseph Priestley—had enough energy and resources to pursue scientific work on the side. In nineteenth-century France, scientific physiology was a by-product of the hospitals, in Germany it was supported by higher education;[2] . . . and so on. In this respect, N.A.S.A. is only the latest in the long sequence of institutions providing scientifically-minded men with extraneous occasions for exercising their intellectual curiosity.

Still, one must not exaggerate the effect of these social factors. The pattern of intellectual innovations normally mirrors the character of the social context only approximately. While Avicenna earned his living by practising medicine and wrote about the medical sciences, his intellectual interests extended far more more widely, while Stephen Hales' preoccupation with the pneumatic chemistry of transpiration (what he called *Vegetable Staticks*) arose out of his duties as an Anglican clergyman only very marginally. Of course, patronage was rarely without some incidental fruits. The Royal Society

---

[1] O. Neugebauer, *The Exact Sciences in Antiquity* (Princeton, 1959), Chapters 2, 4, and 5.

[2] Joseph Ben-David, 'Scientific Productivity and Academic Organization in Nineteenth-Century Medicine', *American Sociological Review*, 25 (1960), 828–43.

happily repaid its debt to the Royal Navy, by promoting the development of ships' chronometers and the improvement of transoceanic navigation, and the scientific clerics of eighteenth-century England believed—rightly or wrongly—that their scientific work was contributing to religious enlightenment as well.

At this point, we may broaden our questions to include also the absolute rate of innovation in science-as-a-whole. Just as the character of existing social institutions creates or inhibits opportunities for research in different sciences, so it can help to determine whether any opportunities exist for scientific innovation at all. A well-defined scientific discipline requires a common standpoint, associated with shared explanatory ambitions and ideals; yet men will scarcely adopt such a common standpoint in the first place, unless they are already prepared to admit that their current concepts do in fact leave something to be desired. Such an attitude of 'collective modesty' has existed in only a very few civilizations, and for a very small part of human history. The question therefore is, what conditions make such an attitude possible.[1]

To begin once again with an extreme example: Joseph Needham's survey of *Science and Civilization in China* has rightly paid close attention to the question why, despite being for so long pre-eminent in social organization, technology, and the applied arts, Chinese civilization achieved comparatively little for itself in the fields of pure science.[2] In their practical command over natural processes and materials—whether porcelains, glazes and silks, or instruments and machines—the early Chinese excelled, and subsequently taught, the craftsmen and engineers of Europe. Even in practical astronomy, their elaborate water-clocks for controlling the observation of celestial phenomena were unmatched in the West; while, on a purely empirical level, they had an encyclopedic knowledge of the stars and planets, e.g. of the Crab super-nova of A.D. 1154.[3]

[1] Cf. Stephen Toulmin, 'Intellectual Values and the Future', in *Knowledge among Men*, ed. P. H. Oeser (New York, 1966).

[2] Cf. Joseph Needham, *Science and Civilization in China*, esp. Vol. III (Cambridge, England, 1959), sec. 20, and Vol. IV (1965), sec. 27 (a) (3).

[3] Cf. J. Needham *et al.*, *Heavenly Clockwork* (Cambridge, England, 1960), Chapter IX, on the astronomical use of water-clocks in classical China.

Yet, by Western standards, the astronomical concepts of the Chinese remained curiously immune to the influence of their own observations. As Needham puts it, Chinese science never had its Galileo; the astronomers of classical China never saw how to criticize their ideas about the heavenly bodies effectively, in the light of their empirical knowledge. It is not just that they failed to acknowledge the undeniable theoretical bearing of their observations, through muddle-headedness or lack of logic, impatience or over-practicality. It is, rather, that they never formulated an explicit programme for astrophysical theorizing at all. If the Chinese never arrived at Galileo's doctrines, that is to say, this was because they never even recognized his problems: because classical China developed no autonomous tradition of planetary theory, comparable to that which Copernicus and Galileo inherited from Eudoxus, Ptolemy, and Buridan.

The next question is, therefore, why no such autonomous discipline of astrophysics ever grew up in China. Once again, Needham hints at an answer on two different levels, partly intellectual, partly social. On the intellectual level, he recalls that the foundation of Western astronomy was the rational, abstract approach to geometry that originated in classical Greece. From Plato and Heraclides up to Kepler, the ruling theoretical problem of astronomy was to devise a formal, geometrical model representing the familiar planetary motions. In China, by contrast, even the science of geometry never achieved the independent theoretical standing it had in classical Greece; it remained a pragmatic, rule-of-thumb subject—a collection of formulae useful in surveying, rather than a logical network of abstract theorems. So the 'internal' answer to the question, 'Why did China never have its Galileo?', is to reply, 'Because it never had its Euclid.'

Still, we can hardly leave matters at that. If we are puzzled by the fact that the Chinese never had a counterpart of Galileo, is it not even more puzzling that they never had their Euclid? At this point, accordingly, some broader social considerations have to be taken into account. For the conceptual genealogy of an intellectual discipline has to be embodied in the human genealogy of a scholarly or scientific profession, and Needham's question can be generalized; so becoming, 'Why did the Chinese develop neither an autonomous discipline of abstract natural philosophy, nor an independent profession of pure natural philosophers?' Given this reformulation, the sociological elements in his argument come into focus. Chinese technological supremacy was an achievement, not of the ruling élite whose intellectual influence was dominant, but rather of craftsmen and technicians who belonged to secondary, executive grades of society and were frequently illiterate. If the Emperors saw themselves as philosopher-kings, their aims were Confucian, not Platonic. They did not see themselves as intellectual innovators, still less as cosmologists, but as guardians of the moral order in terrestrial affairs; in this mission, heterodox speculations about mathematical astronomy served no useful purpose. So, instead of welcoming new conceptions as a mark of intellectual vigour, they regarded them with suspicion. Social conservatism bred ideological conservatism, and every upsurge of unconventional ideas about Nature led eminent and influential men to deplore the corruption of the public mind. Given such attitudes among the opinion-formers of classical China, the institutions needed for the effective development of science could scarcely be expected to flourish.

In total contrast, the classical Greeks inaugurated not only the conceptual traditions of logic and philosophy, mathematics and natural science, but also—what was even more important —the institutional traditions without which those disciplines could never have developed as they did. To tolerate, for the first time, such independent thinkers as the Milesians and the members of Plato's Academy took great intellectual self-confidence; and it needed greater courage still to see those schools influencing young men from the opinion-forming élite, yet resist the temptation to suppress them. That courage and confidence were not, of course, universal in fourth-century Athens, any more than they have been in seventeenth-century Rome or the super-powers of the twentieth century. The fates of Anaxagoras and Socrates remind us that, even in classical Greece, heterodoxies were easily confused with heresies, and free speculations with dangerous thoughts. Yet the life of the Academy lasted for long enough to prove the virtues of 'academic freedom', and to demonstrate that intellectual innovation could be encouraged—even institutionalized—without becoming a standing danger to either Church or State.[1] Failing this Athenian example, Western civilization would most probably have developed in the same way as Islamic or Chinese. Left to itself, Christian Europe would scarcely have generated the social and institutional framework required if scientific creativity and conceptual innovation were to be woven lastingly into the mental and social fabric. Instead, Europe would have been exposed to the same conservative reaction that subordinated science to theology within the Islamic world, from the time of al-Ghazali on.

To sum up: the broader conditions within which men's reflective curiosity will give rise to authentic science—i.e. to a continuing disciplinary-cum-professional tradition of critically controlled speculation about Nature—have existed only rarely. Taking human history as a whole, heresy-hunting or intellectual conformism has been the rule, tolerance of free conceptual innovation the exception. Political and ecclesiastical authorities have rarely been happy to see men scrutinizing the intellectual foundations of their conceptual inheritance with complete critical freedom, from an ideological fear that the stability of that inheritance might be put at risk. We find authentic scientific traditions and institutions emerging, or re-emerging, only where, for once in a while, this frail confidence is achieved. Only there do we see the shared intellectual ideals and institutions of a collective scientific profession appearing, or reappearing; and only within such contexts do we observe the establishment, or re-establishment, of the crucial public attitudes of conceptual modesty and tolerance of intellectual innovation.

So, at whatever point we study the process of conceptual variation, we find intrinsic (or intellectual) and extrinsic (or social) factors affecting it jointly, like two independently acting filters. Social factors limit the occasions and incentives for intellectual innovation; the scientist's own judgement about the demands of the existing intellectual situation discriminates

---

[1] The Academy at Athens in fact survived until A.D. 529, when it was closed by the Emperor Justinian on the typical politician's excuse that it was 'the last rampart of effete paganism': see A. A. Vasiliev, *History of the Byzantine Empire* (Madison, 1952), p. 150.

between ripe and unripe fields of work. Sometimes extrinsic needs and intrinsic promise coincide, as with much contemporary work in medical science. The effect of the two sieves is then to favour similar lines of research. At other times, the two groups of factors may work against one another. Where all the emphasis is on practical, or 'mission-oriented' research, even the most intellectually promising lines of abstract enquiry may be given a low priority. Elsewhere, the demands of (say) the space programme may divert intellectual and financial resources into intrinsically unripe areas like exobiology. Either way, the types and amounts of intellectual innovation to be found in any culture or epoch reflect the combined action of these two separate filters; and the rate of intellectual innovation to be found in any field and milieu can be explained fully in terms of neither social nor intellectual considerations alone.

Can anything be said about the relative significance of the two filters? One can, perhaps, generalize roughly about the balance between them, by saying, 'The social factors are *necessary*, but the intellectual ones are *crucial*.' Wherever men have the chance to speculate freely and critically in an organized way, they will find some aspects of their experience that are ripe for reflective attention. Intellectual considerations thus focus the theorizing which social incentives make possible. On the other hand, if the institutional, social or ideological conditions are unfavourable, even the most scandalous outstanding problems may remain long unsolved; there was no intrinsic reason, for example, why the transition from Ptolemy's *Almagest* to Copernicus's *De Revolutionibus* should have taken well over 1,000 years. The sheer amount of intellectual innovation going on in any situation thus reflects, first and foremost, its overall social and institutional character, and only secondarily the specific disciplinary considerations of the time. The distribution of innovation between different fields of enquiry depends, by contrast, on a much more even balance of factors. Opportunity and promise are here of equal significance. Given 100 soluble problems, scientists will naturally pay attention to those for which there is broader support and interest; yet beyond a certain point they will weary of fruitless enquiries, however useful, tantalizing, or well-financed.[1] It

[1] For a fuller discussion of this point, see my Sigma Xi lecture on 'The Evolu-

is when we turn to the selection procedures of a scientific discipline—studying the manner in which, and the criteria by appeal to which, the available variants are sifted and judged— that we shall find ourselves at the other extreme, with the balance tilted firmly in the 'intrinsic' direction. There, technical disciplinary factors are primary, even paramount, while 'extrinsic' social and institutional factors play, if not a negligible, at any rate a secondary role.

3.2 : *Intellectual Selection*

As our examination of Collingwood's *Essay on Metaphysics* reminded us, the crucial philosophical difficulty about conceptual change is to explain the relationship between 'reasons' and 'causes' in the process of intellectual selection. When we ask *why* a particular conceptual innovation succeeded in winning a place in science (say), two alternative interpretations of that 'why' straightaway present themselves. It may be asking what 'considerations'—mathematical, empirical, heuristic, or whatever—were, or would have been, advanced by the scientists concerned, to justify accepting the particular conceptual change they did. On this interpretation, the 'why' is a request for reasons, and the reply must be given in intellectual, disciplinary terms, internal to the science itself. (In this case, we explain why any repertory of concepts changes as it does by showing how, as a result of these conceptual changes, the discipline in question is enabled to 'do a better all-round explanatory job'.) But this first interpretation takes it for granted that the change was made 'for reasons' *at all*; whereas, often enough, a concept can drop out of the repertory of a science, or survive only in a modified form, without the scientists concerned taking any deliberate decision to make this change, still less troubling to give an explicit justification.

If, in these other cases, we press our question about the 'considerations' or 'reasons' in the light of which the change was made, we shall get a dusty answer. 'Considerations? The question of considerations does not arise. This particular change was the by-product of other factors, within the science or elsewhere.' If we now continue to ask 'why' the particular conceptual change took the form it did, we must begin interpreting the question in a different sense. This question now becomes one, not about the justificatory reasons available to the scientists in question, but about the scientists themselves. How did they come to stop using such-and-such a concept, in the absence of any specific justification for dropping it from their repertory? What causes led them to do so? Did they just forget about it, as (e.g.) in the case of Newton's speculations about the wave aspect of light? Was all reference to it omitted from fashionable textbooks, as with the concept of quaternions? Were its wider associations found offensive or obnoxious, in extra-scientific respects, as with the Newtonian concept of 'attraction' in eighteenth-century France? Such questions invite us to pay attention to the causes which, in one way or another, may divert or frustrate the 'normal' operation of a scientific enterprise.

Looking at conceptual changes in highly-developed natural sciences dispassionately—as historical phenomena—we thus find them occurring in two contrasted ways. Sometimes they are made knowingly, as deliberate steps in a communal problem-solving activity; at other times, they happen in ways unrelated to this problem-solving function, as the effects of fashion, prejudice, or inadvertance. If we ask about the relations between these two kinds of explanation, and the corresponding implications of the question 'why', it becomes clear that the very existence of an organized rational enterprise, with its own 'normal operation', by itself creates a certain burden of proof. As a result, we can commonly treat the first, or 'rational' interpretation as the primary one; showing that, in the absence of other information, conceptual changes in scientific disciplines are presumed to have been made 'in the light of' relevant intellectual considerations. Explanations of the second type, by contrast, are resorted to only as a second-best: scientists, *qua* scientists, do not have an even-handed choice whether to pay attention to disciplinary considerations or no. (Thus, we would not say, 'Since chemists in the 1850s had no prejudices in favour of the atomic theory, there must have been scientific considerations which led them to embrace it.') The notion of a 'rational enterprise' being what it is, there is a standing presumption that conceptual changes have been made 'for reasons', i.e. in the light of intellectually relevant considerations; and we turn to the alternative, 'causal' possibility—that the changes may have been the effects of (say) prejudice or inadvertence—only as a *pis aller*.

Merely to state this distinction, however, is not by any means to imply that the dividing line between the two kinds of case is always clear. True: there are plenty of examples of concepts being adopted, modified, or dropped in the ordinary course of scientific change, for reasons which were presented explicitly at the time, and which related directly to the explanatory ambitions of the discipline concerned. In a few cases, likewise, we have equally little doubt that certain conceptual changes occurred, without there having been any specific and relevant justification. Yet there are also intermediate or borderline cases, in which we do not at first know whether to regard a conceptual change in rational or causal terms. Taking a broad enough view of science—regarding it not just as a rational enterprise, but as a rational enterprise in *historical development*—we shall indeed argue that such ambiguous, borderline cases not only do arise, but do so unavoidably and in the nature of the case. For, in practice, the very growth of science is continually liable to present us with the question, 'What exactly, in this situation, is to *count* as "rational"—or as a "reason"—in our discipline?' And, when the answer to that question is in genuine doubt, it will be hard for the moment to say whether some conceptual change has been made with adequate rational justification, or whether it is merely the product of (say) modish over-enthusiasm or some other intellectually irrelevant factor.

In considering the varied modes of intellectual selection in science, we must look more closely, first, at some straightforward and unambiguous cases. Many conceptual changes within scientific disciplines take the forms they do, simply because the recognized problems of the discipline concerned impose the intrinsic demands they do. More exactly: certain of the variants available in the current pool of conceptual innovations become established in the discipline as a result of choices made predominantly—or entirely—with an eye to those outstanding problems. In such cases, the selection of one particular conceptual innovation is justified by showing that it best succeeded in resolving the outstanding conceptual problems of the science, and so in increasing its explanatory power; and, if this is done, the conceptual change is sufficiently 'accounted for'.

In order for this sort of account to be appropriate, however, there must be general agreement about the character of the outstanding problems in this discipline, and so about what *counts* as 'increasing the explanatory power' of the science. The selection criteria for judging conceptual novelties thus form an integral part of the 'problematics' of the science, and as such have to be understood in relation to its specific explanatory aims and ideals. From this point of view, the merits or short-comings of a conceptual variant are related, directly, to the general patterns of conceptual problems and modifications discussed in the previous chapter; and so, indirectly, to the current intellectual ideals and ambitions of the discipline concerned. It will count in favour of a conceptual variant, for instance, if its adoption extends the scope of an established explanatory procedure to cover hitherto anomalous phenomena, if it makes possible the unification of explanatory techniques from hitherto separate sciences, or if it resolves inconsistencies between the concepts of a special science and related extra-scientific concepts. Similarly, it will count against a proposed variant if its advantages in these respects are either minimal, or are counterbalanced by other disadvantages from which current concepts are free; if some rival innovation has strikingly greater disciplinary advantages; or if its extra-scientific consequences would saddle us with new and intolerable paradoxes.

Either way, three points need to be made about the intellectual evaluation of conceptual variants within a well-structured scientific discipline. (1) Such evaluations are always a matter of comparison. The operative questions are never of the form, 'Is this concept uniquely "valid" or "invalid"?', nor of the form, 'Is this concept "true" or "false"?' Instead, the operative form is: 'Given the current repertory of concepts and available variants, would this particular conceptual variant *improve* our explanatory power *more than* its rivals?' And, since the business of concepts is to be not true or false, but relevant and applicable, the business of conceptual innovations is—correspondingly—to be 'relevant' more exactly, more precisely, or in more detail, and 'applicable' more generally, more extensively, or more unconditionally. All these general comparative phrases must, of course, be spelled out afresh in detail by relating them to the particular explanatory ambitions of each new discipline in turn. In one field, 'doing a better explanatory job' may mean constructing a taxonomy which is at once more comprehensive, more elegant, and historically more illuminating; in another, it may mean devising graphical or mathematical techniques which can be used to relate quantifiable magnitudes more accurately and in a wider class of cases; in another again, it may mean reordering the general relations between empirical variables into a universal, abstract mathematical system, or else seeing how combinations of mechanisms or processes acceptable in terms of (say) biochemistry could have outcomes displaying the same form as the phenomena of (say) genetics.

(2) In deciding whether a conceptual variant enables us to 'do a better explanatory job', our standards are in all such cases informal ones, expressing the current disciplinary ideals of (e.g.) biochemistry *qua* biochemistry, or atomic physics *qua* atomic physics. Whereas the merits (or 'evidential support') of alternative hypotheses, or propositions, can be judged with the help of formal 'significance tests' and 'probability calculations' so long as we remain within the scope of an established family of concepts, the relative merits of conceptual innovations raise issues of quite another kind. These do not rest on the relations between propositions formulated *within*, or in terms drawn from, any single scientific theory; instead, they must be characterized by statements *about* rival theories—more specifically, by statements about the respective ways in which alternative theoretical changes can help to fulfil the proper intellectual ambitions of the science concerned.

(3) The merits of a conceptual variant can only rarely be stated or judged in simple terms. Conceptual problems and conceptual variants rarely match one another exactly. Even where a conceptual change is proposed with an eye to some one particular shortcoming in the current explanatory repertory, its consequences will more often than not extend beyond that original purpose. In the course of solving the specific problem

for which it was devised, it will have intellectual side-effects; and those side-effects are often stronger witnesses for or against the innovation than its intended consequences. In regularizing the conceptual relations between the theories of electricity and magnetism, Maxwell created—almost inadvertently—a theory of radiation and radio waves embracing the whole of existing physical optics; and this unforeseen by-product of his work testified more convincingly on its behalf than his formal integration of electrical and magnetic theory.

Typically, then, the task of evaluating conceptual changes in science requires us to consider implications of half a dozen kinds; and this calls for the exercise of judgement, in two separate respects. Not only are the relevant considerations frequently incommensurable—not only may we lack any simple index for comparing the respective 'values' of (e.g.) accuracy, scope, and degree of integration—but, in addition, the decisions frequently involve striking a balance between a profit of one kind and a loss of another. The recognized disciplinary criteria of choice are always multiple, and sometimes point in opposite directions; so that a proposed theoretical change may be highly attractive in one respect, retrograde in another. (Just how much coherence or elegance can we sacrifice, in exchange for a small improvement in predictive power? Just how is a gain in simplicity of one kind to be balanced against a loss in simplicity of another kind?) Formally speaking, it would no doubt be tidy if the natural sciences could be expounded, not merely in explicit 'propositional systems', but in formal systems which had all the possible explanatory virtues at the same time: which were at once universally applicable and completely coherent, predictively exact and comprehensive, notationally convenient and intuitively simple, mathematically elegant and straightforwardly computable. . . In actual practice scientists normally face situations in which they must choose between these virtues. They must settle, either, for an unmodified discipline whose current procedures are (say) intuitively simple and moderately predictive though not entirely coherent or consistent, or alternatively for a modified discipline whose procedures are (say) more highly integrated and coherent, but intuitively obscure and predictively exact only in a limited range of cases.

Initially, for example, the Copernican description of the planetary orbits used combinations of circles organized around a static Sun to arrive at a scheme which was physically more coherent and consistent than any description based on Ptolemy's geostatic principles. Yet it achieved that improvement only at a substantial price. Quite apart from the fact that the idea of a moving Earth was counter-intuitive, the original Copernican constructions had technical defects also; they were substantially less simple than the best current Ptolemaic constructions, and at points marginally less exact. The full computational merits of the Copernican approach could in fact be realized only when Kepler replaced the traditional circular representation of the planetary orbits by elliptical constructions.[1] For at least seventy-five years, the arguments over the two rival systems of astronomy were so delicately poised that the hesitation of men like Tycho Brahe and Bacon was not unreasonable. Similarly with the wave theory of light, as advanced by Young and Fresnel in the early 1800s: the corpuscular theory of the orthodox Newtonian physicists had some real advantages, which had to be sacrificed if one were to adopt the new wave theory. (Streams of fast-moving, weightless light-particles would produce sharp-edged shadows quite naturally; and many of the most elementary facts about light and shadows could be reconciled with a wave theory only by quite sophisticated arguments.) The striking advantages of the wave theory had thus to be balanced against genuine losses; and it was some years before a convincing balance of advantage could be struck.

Even in straightforward situations, then, the task of selecting between conceptual variants involves scientists in a complex kind of intellectual accounting. Any single conceptual innovation will normally improve our understanding in certain respects, impair it in others; and it is up to the scientists concerned to decide when that improvement is worthwhile, in terms of some broader set of intellectual priorities. Conceptual problems arise within a discipline in no fixed, universal form,

[1] As a useful introduction to the vast literature on Copernicus and Kepler, one may select T. S. Kuhn's book, *The Copernican Revolution*, referred to above; or, for the general reader, Arthur Koestler's discussion in *The Watershed* (Garden City, 1960).

and their solutions have a multiplicity of dimensions that philosophers cannot afford to ignore. It is therefore vain for us to search for any single index or measure, which will indicate *in all cases* whether a conceptual change is to count as an 'improvement' or not. As philosophers, we can sharpen up our understanding of the various different kinds of merit that are relevant in practice to the evaluation of rival variants—i.e. predictiveness, coherence, scope, precision, intelligibility and the rest—and we can always find examples to illustrate how particular criteria of merit count in favour of some conceptual change. But, whatever criterion we pick for analysis, there will always be other cases to which it does not apply; and, more typically, considerations of several incommensurable kinds will be relevant to any particular choice.

So, an account of the criteria of selection in science given in historical terms—by reference to the substantive intellectual goals of the disciplines concerned—has results quite unlike those which follow from attempting to measure new scientific ideas against timeless logical ideals. For the inductive logician, a well-structured science forms a single system of logically-related propositions, and it therefore seems natural for him to ask what formal features a 'correct' system must display. Our own more historical analysis treats scientific disciplines as comprising informal populations of logically independent concepts, and leads us to ask, rather, in what manner a 'fruitful' discipline must develop. From his Olympian standpoint, the formal logician judges the sciences of every epoch in the same ahistorical scale; whereas our more down-to-earth analysis demands only that the intellectual selections made in a discipline at any time should be adequate to the problems actually outstanding at the time.

Of course, the logician's formal analysis might be defended, as merely presenting an abstract ideal or aspiration—the Utopian vision of an ultimate goal—without offering practical indices for judging the rights and wrongs of actual scientific concepts. If only the entire content of a natural science *could* be set out in a single axiom-system, which had the greatest possible simplicity and coherence yet was also predictively successful, throughout the whole scope of its potential application, to the limit of all future observations, surely the result would be as near to 'scientific perfection' as we could demand, or even imagine! For such an analysis to carry conviction even as a Utopian vision, however, the resulting image of a 'perfect theory' must—like that of a 'perfect society'—have some sort of bearing on the task of constructing better actual theories and societies. It must, at the very least, be based on a realistic understanding of the actual aims and methods of our concepts and institutions, and of the human ambitions and goals they exist to serve. It must help us (that is) to understand better the goals that men actually set themselves in building up their different patterns of life and thought, and the problems that face them in the course of doing so. Otherwise, the timeless Utopian vision—in philosophy of science, as much as in political or social theory—will stand condemned by its own irrelevance.

Working scientists who read philosophical discussions of scientific reasoning often react with impatience and incomprehension: what (they ask) has all this to do with natural science as I know it in real life? The writings of art critics provoked the same reaction in the French painter, Gustave Courbet. 'After all,' he is reported to have said, 'it is hard enough to paint a picture at all, let alone a good picture!' What infuriated Courbet was less the critics' disapproval of his pictures than the ridiculous and irrelevant reasons they gave for their judgements; these betrayed, in his eyes, complete ignorance about the true character of the painter's task. The conscientious artist faces so many obstacles, and has to reconcile so many conflicting demands, that he can scarcely begin to feel anything but dissatisfaction with his own output; and no one who really understood what the art of painting involves could bring himself to impose the critics' exaggerated demands on his products. In dealing with outstanding scientific problems, likewise, it is hard enough to think up possible conceptual variants at all, let alone acceptable ones; and we are exposed, as philosophers, to some of the same dangers as the art critics. Only if we take care to understand fully the true character of the scientist's task, in devising better concepts and explanatory procedures, can we criticize his achievements in an informed way. Otherwise, our legitimate Utopianism is liable to degenerate into mere fantasy.

*   *   *

So much for the straightforward cases, in which conceptual changes are made as the direct outcome of communal problem-solving activities, in a science whose disciplinary goals are sufficiently agreed. In these clear cases, the selective perpetuation of certain conceptual variants can be explained in 'rational' terms, by pointing out how the successful innovation helped the scientists concerned to achieve their collective goals. But these clear cases require that the scientists working in this discipline should share agreed—or sufficiently agreed—conceptions of 'explanation'. (More exactly: their differences about what represents a 'complete' solution to their disciplinary problems must, for the time being, be of no more than marginal significance.) Where this is the case, the men involved can all direct their disciplinary investigations in accordance with a common intellectual strategy. This strategic consensus determines well-defined selection-criteria for deciding between conceptual variants; and the scientists will agree in their judgements on conceptual innovations, simply because this consensus decides for them what kinds of changes will fulfil the agreed intellectual goals of the science for the time being.

Not all conceptual changes are like those, however, and we must now consider the exceptions. These are of two very different kinds. On the one hand, there are cases which involve *failures* of rationality; on the other hand, cases which reflect changes in the very *criteria* of 'rationality'. In the first class of cases, factors such as conservatism or prejudice, lack of professional cohesion or breakdowns in communication, political pressure or sheer inattention may frustrate the normal procedures of intellectual selection. As a result, the disciplinary merits of some new terminology, technique of representation, or method of explanation may for the time being be disregarded, despite the fact that they could make themselves evident—given 'daylight and fair play'[1]—even in terms of currently accepted criteria and standards. In these cases, the scientific issues are concealed, or clouded, by considerations which are, by the present standards of the discipline, entirely irrelevant to the problem under debate.

In the other class of cases, by contrast, the points at issue

---

[1] W. Lawrence, *Lectures on Physiology, Zoology and the Natural History of Man*, Kaygill edition (London, 1822), p. 95.

are—to use a convenient legal term—intrinsically 'cloudy'; and, by coming to see how the historical development of intellectual disciplines gives rise, inevitably, to that intrinsic 'cloudiness', we shall reach the heart of our whole present analysis. For this cloudiness springs directly from strategic disagreements between the scientists concerned, and is—as we shall see—a direct consequence of the fact that our disciplines are in the course of historical change, *even in their deepest rational strategies*. The unity and coherence of a scientific discipline does not require that its intellectual ambitions should be eternal and unchanging; only that they should maintain a sufficient continuity. (Even the half-century's history of atomic physics involved significant changes in the collective ambitions by which the problems of the subject were defined, and to that extent the nature of theoretical judgement within atomic physics was progressively transformed.) When the current intellectual strategies of a science are themselves up for reappraisal, however, the whole process of intellectual judgement quite naturally takes on a fresh form. Given strategic disagreements, there will no longer be well-defined selection criteria on which all professional participants in the science are for the time being sufficiently agreed, since it is precisely those selection criteria which will now be in doubt.

Three historical examples will illustrate the essential differences between 'clear' cases of scientific judgement and these intrinsically 'cloudy' cases. (1) By the end of the nineteenth century, the programme for physics propounded at the end of Newton's *Opticks* had been largely completed, and the period from 1900 to 1914 was a time of uncertainty. The new conceptions of relativity and quantum theory became firmly established only after 1919; for the time being, physicists were casting around for a way ahead, undecided what new intellectual goals they should set themselves. With this background in mind, it is worth re-reading the exchange of papers between Max Planck and Ernst Mach printed in the *Physika-lische Zeitschrift* for 1910-11.[1] In a long essay called 'The Unity of the Physical World-Picture', Planck surveyed the whole historical development of physical thought: he found in it a

[1] This is the exchange recently translated in full in *Physical Reality*, ed. S. Toulmin (New York, 1970), pp. 1–52.

constant direction, involving the progressive elimination of subjective sensations and other 'anthropocentric' elements, and their replacement by quantitative, intersubjective magnitudes and theoretical invariants. This done, he criticized Mach's philosophy of science, arguing that Mach's sensationalism wilfully reintroduced into the heart of physical theory those very subjective elements that the main stream of physical thought had consistently aimed at eliminating. In due course, Mach replied with an equally elaborate paper, which was partly an intellectual autobiography and partly a restatement of his methodological programme. In this, he called for a physics committed to the avoidance of metaphysics, and based solely on 'observables' or—as he regarded them—on 'sensory observations'.

Now, whether Planck's or Mach's conclusions were in the event the sounder, is not the central question here. (In certain respects quantum physicists of the Copenhagen school have, for better or worse, preferred the strategy advocated by Mach. Einstein, on the other hand, though at first attracted by Mach's programme, subsequently changed over to a position closer to that of Planck.)[1] For our present purposes, what is significant is simply the strategic character of the disagreement between the two men. Planck, in particular, analysed in a clear-sighted manner the changing explanatory demands which have historically guided the development of physical theory. As he saw the situation, the new strategies appropriate to the problems of theoretical physics in his own day must make it the 'legitimate heir' of all previous physical investigations; they had, therefore, to be formulated and judged, not in formal or abstract terms, but with an eye to the entire historical evolution of physics, and its ideals of 'physical explanation'.

(2) Our second example is more recent. It comes from the late 1940s, and concerns the emergence of the 'phage' group and the take-over of theoretical biology by men trained originally in physics. (This take-over formed the essential preparation for the development of molecular biology.)[2] In

[1] Einstein's relations to Mach and Planck are illuminatingly discussed in an essay by Gerald Holton due to appear shortly, together with other essays on Einstein, in a supplement to *The Graduate Journal* (Austin, Texas, 1972).

[2] On this episode, see the interesting paper by Donald Fleming, 'Emigré Physicists and the Biological Revolution', in *Perspectives in American History*, II (1968),

1944, Avery and his colleagues had published their classic demonstration that deoxyribonucleic acid (or DNA) was the carrier of a particular hereditary trait in a single bacterium; but they were constrained from claiming too much by their commitment to the currently accepted attitudes of classical genetics. Within classical genetics—which had been one of the great success-stories of early twentieth-century biology— biochemical questions about the material nature of the gene were unimportant, if not entirely irrelevant. As a result, the 1944 paper was, in Donald Fleming's words, 'muffled and circumspect': the authors were 'almost neurotically reluctant' to identify genes with DNA.[1] Eight years later, Watson and Crick were subject to no such constraints; but their success should not be allowed to conceal from us the strategic battle that had been going on within biochemistry in the meanwhile. For the new molecular biologists were the self-confident heirs of a new approach that had been hammered out, in the years between 1944 and 1953, by men like Astbury and Delbrück.

Avery and his colleagues exemplified an attitude which the new physicist/biologists rejected completely. Delbrück has said that biology, as he found it, was a 'depressing' subject: the accepted styles of biochemical interpretation 'stalled around in a semidescriptive manner without noticeably progressing towards a radical physical explanation'.[2] And, once again, what is significant here is the nature of the considerations on which the new approach was based. Fleming quotes Szilard as emphasizing that the new, physically-minded biologists brought to biology 'not any skills acquired in physics, but rather an attitude: the conviction which few biologists had at the time, that mysteries can be solved'.[3] This attitude, characteristic of Delbrück and the entire phage group, enabled them to create a fundamentally new strategy for dealing with the problems of virology and genetics, of which Crick and Watson's molecular biology was the most spectacular fruit.

(3) The third example is taken from contemporary physics. Here again, the subject faces theoretical difficulties that call, not for more elegant mathematics or more ingenious experi-

---

152–89. See also the essays by G. S. Stent, R. Olby and L. Pauling in *Daedalus* (Autumn, 1970), pp. 882–1014.

[1] Fleming, op. cit.   [2] Ibid.   [3] Ibid.

ments alone, but rather for a strategic reappraisal of basic aims and explanatory ideals. Some physicists today—for instance, Geoffrey Chew—believe that 'developments of the past three decades suggest that the capacities of the elementary particle idea may finally have been exhausted, without the identification of an ultimate set of primitive entities'; it is widely conceded, for example, that none of the known set of hadrons could be elementary. This being so, Chew argues, it may be that 'an end to the elementary particle road really has been reached', and in that case physics 'must find an alternative'.[1] In this situation, we may proceed in either of two directions. On the one hand, we can refuse to accept that 'the end of the road' has been reached, and continue a 'vigorous search for new entities which might conceivably be identified as fundamental constituents of matter'.[2] The first step in such an approach, as Julian Schwinger has recently suggested, would be a phenomenological analysis of the existing body of particle theory, analogous to Mendeleev's table of the chemical elements.[3] Unfortunately, as Chew points out, 'the elementary particle idea never gets to an end. . . . If there are elementary particles, why those particular particles? . . . It always leaves you with the problem of understanding the last particle you have identified.'

Chew's own alternative is to look for a theory of an entirely different kind: one in which existing sets of particles are explained, not by appealing to other, yet smaller particles, but by showing that 'the particles are as they are because this is the only possible way they could be'. Unfortunately, this so-called 'bootstrap' idea has its own difficulties. 'By its very nature', Chew concedes, 'it cannot be formulated through equations of motion in the time-honored tradition as all previous physical theories, because in principle there are no entities which could conceivably appear in the equations of motion.' So acceptance of this alternative strategy will compel us to modify our ideas about the ultimate form of any satisfactory 'physical explanation'. Whether such an alternative

[1] G. Chew, 'Crisis in the Elementary-Particle Concept', *Publications of the University of California Radiation Laboratory*, no. 17137 (Berkeley, Calif., 1967).
[2] Ibid.
[3] See, e.g., the letter by J. Schwinger in *Science,* 166 (1969), 690.

approach will prove sounder than continuing the familiar search for ever smaller particles, subsequent history will show. If it does, Chew concludes, 'the dilemma with the hadrons' will turn out to have been 'the precursor of a wholly new kind of science'.[1]

Notice that these three examples differ substantially in several respects. Only one of the three, for instance, can be said to have involved a 'crisis' in the development of the science. If men like Delbrück and Luria argued for a new strategy in theoretical biology in the late 1940s, for instance, they did so not in order to remove intolerable 'strains' or to exploit a 'revolutionary situation', but rather from a desire to bring physical and biological understanding into closer relation. Still, the examples do have one feature in common. In each case, there had ceased to be a collective, agreed conception of 'explanation' in the science concerned, and the normal unambiguous criteria for selecting between new ideas were suspended. In each case, as a result, the fundamental theoretical question at issue had ceased to be, 'What conceptual innovations will, in fact, best solve our outstanding problems, and so help us towards our agreed intellectual goals?', and had become, 'What general goals should we be aiming at here? What new types of explanatory task are there to be tackled in this field?'

A dispute over intellectual strategies is thus a dispute for which no established decision procedure exists. In 'clear' cases, the agreed aims of a science determine also agreed procedures of judgement; but, in cases that are intrinsically 'cloudy', scientists are obliged to reappraise the goals of the whole theoretical game and, along with them, their standards of judgement also. So the clear cases are also 'routine' cases, in which one can play to an old familiar set of rules; while the cloudy cases are also 'moot' cases, which confront scientists with questions of a kind that they can normally beg.

Faced with such cloudy issues, what procedures can the parties to the dispute invoke in the hope of reaching agreement? Consider again the prolonged debate in recent physics over the status of quantum-mechanical explanations: this debate has repeatedly brought to the surface strategic differences that

[1] Chew, loc. cit.

had earlier been passed over tacitly, and the course of the resulting discussions is revealing. The fundamental question at issue in the debate has been: 'Can we continue to make the existing structure of quantum mechanics, as conceived by Heisenberg and Bohr, the final court of appeal in physical theory? Can we continue, that is, to give it sovereign intellectual authority in our physical explanation?' Or, to invert the question: 'Has the time come to appeal behind the established framework of quantum mechanics, and to build up a more refined form of theory on a finer level of analysis, as different from Bohr's own theory as the quantum-mechanical picture itself was from that of nineteenth-century physics?' Clearly, this problem arises on a plane quite unlike that on which questions in theoretical physics are commonly resolved. Typically, physicists who disagree about one or another phenomenon will nevertheless agree, at least tacitly, about the sorts of evidence, argument, or discovery that would resolve their disagreements. They will agree, for instance, that the issues between them would be settled if it could be shown that only one of their interpretations conformed to the accepted principles of quantum mechanics; this interpretation would then be the victor. In the current debate, no such straightforward procedure can settle the issue. To put the point in political terms: the question under discussion is now a question of 'sovereign authority'. No formal demonstration conforming to the patterns of argument accepted as finally authoritative hitherto can yield an agreed solution, since what is in question is the authority of those very patterns.

Yet it would be an exaggeration to conclude from this that no rational procedures at all exist for resolving such disagreements. These cases call for appeal, not to the codified rubrics of an established theory, but to broader arguments involving the comparison of alternative intellectual strategies, in the light of historical experience and precedents. That is how the actual discussion has in fact turned out. During the 1950s there was a series of head-on confrontations between critics and defenders of the sovereign authority of quantum mechanics, which frequently degenerated into vituperation and cross-purposes. The critics advanced their objections; the defenders of the faith demonstrated, in reply, that these objections flouted the orthodox interpretation of quantum-mechanical principles; the critics answered 'Just so, since it is your claims for those very principles that we are rejecting'; the orthodox retorted in turn, 'In that case you heretics are plainly misguided' . . . and so the dispute went on, with growing bitterness. After a while, however, the full profundity of the issues became evident to both sides, and the style of argumentation changed. From then on, each side attempted to explain, in less doctrinaire terms, what general kinds of mistake their opponents were making, and as a result a new sort of consideration entered the debate. Released from commitment to any codified procedure, the disputants found themselves (so to say) in a 'common-law' world, which compelled them to discuss their disagreements in terms of 'precedents', 'consequences', and 'public policy'. Orthodox quantum-theorists—the critics proceeded to argue—were making the kind of mistake Pierre Duhem had made in the 1890s, when he dismissed J. J. Thomson's hypotheses about 'sub-atomic' electrons as incompatible with sound (i.e. classical) physics. Not so—the defenders replied—the heretics, rather, were at fault in the same kind of way as William Prout, who challenged Dalton's brand-new atomic theory as early as 1815, with his premature speculations about hydrogen, as the universal material from which other chemical elements were composed . . . .[1] And so the search went on, for the most exact and revealing historical precedents.

Our own task, here, is to make the implications of this switch from 'codified' to 'common-law' argument absolutely explicit. Before it, the argument over quantum mechanics had run into barren cross-purposes, just because both parties still assumed that some formal procedure or pattern of argument could be found, on whose authority they could jointly agree. (Given this, they could simply have agreed to 'calculate' and accept the result.) After the switch, they no longer appealed to formal arguments since, as they now saw, there were no longer any formal arguments carrying conviction to both parties. Now, all the solid arguments were informal, consequential ones: designed, not to invoke or apply particular calculative procedures, but rather to come to terms with their

---

[1] See the exchanges between Rosenfeld, Bohm, Vigier, and others in the symposium, *Observation and Interpretation*, ed. S. Körner (New York, 1957).

strengths and weaknesses, their range and limitations. This meant appealing to considerations of an essentially historical kind: using the theoretical experience of earlier physicists as a precedent, in estimating the most promising lines for future theoretical development. By this switch from formal (or codified) to historical (or common-law) arguments, the true character of the issues in debate was brought out into the open. The question now became at just what point it was legitimate to challenge claims to sovereign intellectual authority on behalf of existing standards of scientific judgement, and to begin looking for a Young Pretender to take over the throne. Such a question must certainly be resolved in a rational manner; but that could be done only by setting aside the formal demands of all theoretical principles, and treating the matter in broader, disciplinary terms.

At such points as these (one might say) a theoretical dispute in science ceases to be a matter for routine judgement, and resembles rather a Supreme Court or House of Lords case. The question at issue ceases, that is, to be a question of applying established procedures correctly, and takes on the character of a 'constitutional' question. The judicial analogy is, indeed, worth pursuing further. If we have repeatedly slipped into the use of legal terms, in exploring the distinction between 'clear' and 'cloudy', or 'routine' and 'moot' questions, that is no accident. The special problems which arise, and the special decisions which are called for when scientific strategies are in doubt, show quite genuine parallels to the judicial problems and decisions arising in (say) constitutional law, when a 'court of last resort' like the United States Supreme Court reinterprets the provisions of a sovereign constitution and, in doing so, is compelled to reanalyse the social functions of the law in its application to some novel historical situation.

We referred, in an earlier chapter, to Oliver Wendell Holmes's analysis of this question. A court of last resort, he said, is sometimes faced by cases for which no unambiguous decision procedure is available, or in which the accepted principles—as interpreted hitherto—lead to manifest anomalies and inequities.[1] When this happens, the judicial task is no

---

[1] On Holmes and Pound, see once again H. S. Commager, *The American Mind* (New Haven, 1950), Chapter 18.

longer to reapply pre-existing procedures to fresh cases. Rather, the judges now have to take one step back, and reconsider the overall justice of the accepted legal principles and constitutional provisions, as viewed against a larger sociohistorical background. In the final juridical context, logic thus becomes the servant of those fundamental human purposes that are constitutive of the law itself. In this way, theoretical jurisprudence and judicial practice alike are based on our developing understanding of the historical sociology of the law.

The sociological jurisprudence of men like Mr. Justice Holmes and Professor Roscoe Pound is relevant to our present argument, simply because Pound and Holmes put their fingers, very precisely, on the point at which legal reasoning ceases to be a formal or tactical matter—of applying established rules and principles to new situations—and becomes concerned with strategic issues: the point (that is) at which the 'just' way ahead can be determined, only by reappraising the fundamental purposes of the law in a new historical context. Over such strategic issues, they insisted, judicial decisions can no longer be treated clearly and definitively, as 'right' or 'correct'; yet such decisions are, in their own way, none the less rational for all that. At such points, a Supreme Court Justice can no longer speak for the law as it is; instead, he must pass judgement in the light of a longer-term vision, both of what the law has been and is, and of what it should become. To that extent, the resulting judgement will inevitably be, not 'the judgement of the law', so much as (say) Justice Holmes's or Justice Frankfurter's best individual judgement of the way in which the law should now develop, at this point in historical time, in order to fulfil most completely its general ideals of equity, humanity, and certainty. Cases of this kind take us to the 'rational frontier' at which fallible individuals, acting in the name of the human enterprise that they represent, have to deal with novel and unforeseen problems by opening up new possibilities; and, at this frontier, we can no longer separate the rational procedures of law from the judicial purposes of the men who are reshaping them, or from the historical situations in which these men find themselves.

Similar problems arise at moments of strategic uncertainty in science, also. When the basic intellectual goals of a science are up for reappraisal, its rational procedures and decision-criteria can no longer be defined in unambiguous terms. So the intellectual decisions of a Mach, a Delbrück, or a Chew can no longer be judged, clearly and definitively, as 'right' or 'correct', by any standards available in their times; yet, in their own way, such decisions are—once again—none the less rational for all that. At such points, for instance, a Delbrück can no longer speak for biology as it is; instead, he must do his best to pass judgement in the light of a longer-term vision, both of what biology has been and now is, and of what it should become. To that extent, the resulting decision will inevitably be, not 'the judgement of biology', so much as Max Delbrück's best individual judgement of the way in which biology should now develop, at this point in historical time, in order to fulfil most completely its general ideals of scientific understanding. Here again, strategic uncertainties take us to a 'rational frontier', where men must deal with novel types of problem by developing whole new methods of thought; and, at these frontiers, we can no longer separate entirely the rational procedures of a science from the intellectual purposes of the men who are reshaping them, or from the historical situation in which they find themselves.

The cloudiness of the issues arising at such 'rational frontiers' is therefore inescapable. There can be general agreement about selection-criteria, only for so long as the current aims and strategies of a discipline are sufficiently agreed upon; and, in a discipline which is also in historical development, this state of affairs cannot be maintained for ever. Any profound redirection in the strategy of a discipline thus has to be justified by appeal, not to previously established patterns of argument, but to the overall experience of men in the entire history of the rational enterprise concerned. So long as this redirection is in progress, there will be corresponding doubts about the validity of considerations which were previously authoritative and agreed; and these doubts often provoke bitter polemics, couched in terms that reflect uncertainty about the proper boundaries of the discipline. ('That's just not physics!') Yet such strategic re-directions can be reasoned about none the less, provided it is clearly understood what kinds of reasoning are required. As Wittgenstein used to say in his later years, 'My arguments may not be "philosophical", on any previous definition of the word, but they are the "legitimate heirs" to what was formerly known as philosophy.' In the same way, at moments of strategic re-direction in other disciplines and enterprises, the basic question will no longer be, 'Is this law/physics/music . . .?' Rather, it will be, 'Is this the legitimate heir to what has been known hitherto as "law"/"physics"/"music". . .?' And the crucial question of intellectual method is then, how the overall experience of man's history can be mobilised, in deciding what counts as a 'legitimate redirection' of a discipline or enterprise.

3.3: *The Objective Constraints on Scientific Change*

The proceedings of any scientific enterprise, therefore, provide a range of different occasions for rational choice and judgement. At one extreme, the current strategy of a discipline establishes clear criteria and unambiguous procedures for choosing between conceptual innovations; and in these cases there is some plausibility in the traditional empiricist picture of science, as a search for 'objectively true' propositions. At the other, we have to go beyond the scope of all existing established rules and pro-cedures, and take strategic decisions which can redirect the whole discipline. In the latter kinds of situation, we saw, the task of appraising alternative future strategies in the light of past disciplinary experience leaves considerable scope for individual judgement; and this conclusion may easily be misunderstood. For it might be taken as implying that, on this deepest level, our criteria of conceptual choice become 'sub-jective'. In direct opposition to this, of course, most scientists and philosophers alike would claim that the development of the natural sciences, unlike that of law or politics or the fine arts, is governed by *objective external constraints*. And what objective external constraints can there be, they might ask, in a discipline whose basic strategies are decided by the 'individual choices and judgements' of particular scientists? If this is the conse-quence of turning the analysis of science away from questions about the truth of theoretical propositions to questions about the adequacy of theoretical concepts, we may seem to have opened a Pandora's Box.

This alarm would be premature. In arguing that, at moments of strategic reappraisal, the choice between disciplinary goals or strategies is a matter for the judgement of authoritative and experienced individuals, we are not suggesting that these decisions are matters of personal taste; nor are we implying that the resulting scientific developments are arbitrary products of human idiosyncrasies, uncontrolled by external requirements or constraints. On the contrary: in one way or another, all the judgements on which the historical development of an intellectual discipline depend represent the outcome of mankind's accumulated experience in dealing with problems raised by the corresponding aspect of the 'external' world. Enunciating true propositions is an 'objective' task of one kind; but formulating well-founded concepts, in terms of which further such true propositions can be stated, is an 'objective' task of another kind; while devising fruitful new strategies capable of yielding further sets of well-founded concepts (and so of true propositions) is also, in a third—but none the less demanding—sense, an 'objective' task, subject to the 'external' constraints of scientific experience. The empiricist mistake is to suppose that there is one and only one way of mobilizing our accumulated experience of objective problems, as a constraint on the intellectual development of an empirically based science: namely, by matching 'particular propositions' against 'individual facts'. Statements about the adequacy of our concepts are, by contrast, 'meta-empirical', so that no question arises of matching them directly against 'individual facts'; while scientific strategies, being general policies for the judgement of conceptual changes, are even further removed from such elementary fact-matching. Yet this in no way entails that conceptual and strategic problems are any the less objective. It means only that our conceptual and strategic judgements are exposed to criticism in the light of experience in rather different ways from judgements about particular empirical propositions.

To some extent, but only to some extent, scientific propositions can be verified one at a time. To some extent, but only to some extent, accumulated human experience can be condensed into well-defined rules and procedures, so that we can develop methods of representation for handling certain types of phenomena and systems whose scope and reliability have been thoroughly checked. To some extent, and only to some extent, the resulting rules, procedures, and methods of representation are grouped into compact disciplines, whose conceptual development is itself governed by sufficiently agreed strategies. But even where none of these conditions holds, so that the best way of advancing scientific understanding is at present unclear, the issues confronting scientists are none the less 'objective'. Thus two authoritative scientists may propose different strategic directions for the future development of their science, basing themselves on their individual readings of historical experience, and of the current problem-situation. These proposals may not, as they stand, entail any true empirical propositions, or establish any well-founded concepts. Yet the intellectual policies which they respectively propose may, none the less, prove objectively sound or unsound, fruitful or unfruitful, by making it possible, in due course, to recognize and establish new and more powerful sets of concepts and explanatory procedures. Initially, the two proposals may be the products of individual judgement; but we shall decide which of them was the 'better judged', in retrospect, not by personal considerations but in the light of their actual practical sequels. For though neither proposal can, in the nature of the case, have been based on a formal, rule-governed argument, neither of them is directed, either, merely at satisfying the tastes or prejudices of the scientist concerned. Each of them, rather, is directed to the same general and objective task, of suggesting how our intellectual understanding of Nature can best be improved.

In this respect, the old legal controversies about judge-made law have parallels in the philosophy and sociology of science. In the field of jurisprudence, strict constructionists attack those who too readily go beyond the direct implications of current statutes and precedents, on the ground that to do so is to usurp the proper authority of the legislature: for their part, liberals retort that, failing explicit statutes or precedents, basic constitutional principles must be reconstrued to meet the needs of new situations, with an eye to their deeper social function. So, in exercising its authority as interpreter of the Constitution, the United States Supreme Court cannot but take into account larger considerations of public policy and social ideals, in addition to binding rules and formal precedents. The resulting arguments may look as much like 'lawmaking' as like 'law-finding', but the issues they raise are still—basically —questions of fact. Suppose that a basic constitutional provision, such as that of freedom of speech and opinion, is invoked to forbid the government from some course of action involving a twentieth-century invention, such as secret telephone wire-tapping: will this extension, or will it not, promote in our present situation those general social interests for which the eighteenth-century Fathers of the Constitution designed the provision in question? This may be a profoundly difficult question to answer, since the actual consequences of deciding it in one way or the other may be very hard to estimate; so there may well be individual differences about it between different Justices. But it is in no sense a 'personal' or 'subjective' question—still less, one which the Justices are free to decide as they please, regardless of the facts.

In the natural sciences, we risk similar misunderstandings: e.g. the criticism that, on the strategic level, we are making scientists free to build their sciences as they please. (On this view, 'scientist-made truth' becomes a bogey corresponding to 'judge-made law'.) Once again, this accusation rests on a misinterpretation. The basic strategic decisions of science, which redirect both the conceptual development of scientific disciplines and the institutional development of scientific professions, may take the forms they do because a Delbrück or a Heisenberg persuades his colleagues to turn their collective attention in a new direction. But, though the temperaments and personalities of the scientists concerned may play a legitimate part in arriving at these decisions, the ultimate verdict on them remains an objective, and even a factual matter. For the ways in which Nature will actually respond to our attempts at understanding her is something that goes beyond all human tastes, and all human power to alter.

For lack of any formal decision procedure, the ultimate strategic choices may thus be a matter for individual judgement by the authoritative scientific 'judges'; but these judgements are none the less judgements about thoroughly 'external and objective' issues. They are arrived at, not by the accumulation of bare 'facts of Nature'—whether facts about sense-impressions, or facts about material objects—but rather in the light of all our experience in the enterprise of explaining such facts. The 'objective' constraints which govern the fruitful development of scientific concepts and theories thus have to do not with logically unsophisticated questions (whether every last raven is black, or every last robin's egg greenish-blue) but with issues of a far more complex, yet none the less factual, variety: e.g. questions about which new strategy for conceptual change will in fact make it possible to develop the more powerful new explanatory procedures, and so deepen our scientific understanding in this particular field.

Judgements of this kind involve prospective estimates of the consequences to be expected from alternative intellectual policies, and so amount to 'rational bets'. As such, of course, they are concerned less with uninterpreted Nature—regarded as a world of neutral objects, coexisting with the human race—than with the possibility of making the natural world itself a more intelligible object of human understanding. So, in the nature of the case, these strategic issues have an epistemic aspect; and it is no more surprising that twentieth-century writers like Planck and Mach, Einstein and Heisenberg, have found the fundamental questions of physical theory merging into those of epistemology than that their seventeenth-century forerunners found 'natural' philosophy inseparable from 'metaphysical' philosophy. After all, the ultimate question for science has always been in what terms we can make the world of Nature most fully intelligible to ourselves. With all respect to Galileo and Descartes, Nature has no language in which she can speak to us on her own behalf, and it is up to us, as scientists, to frame concepts in which we can 'make something' out of our experience of Nature. Whether this can be done at all and what kind of intellectual construction will prove most effective, in (say) zoology or electromagnetism, may not be matters of straightforward prediction and verification. But they are in no sense 'subjective', and remain in their own ways authentically factual questions about our objective experience in dealing with the world of Nature.

The doubts about the 'objectivity' of science raised by an account such as this do, nevertheless, have a genuine and understandable foundation. It is not that our account makes the content of scientific judgements any the more 'subjective' or 'personal'; only that it recognizes once again the full 'relativity' of the concepts and standards accepted as authoritative for the time being in different milieus. However much the actual issues raised in a science, at every level, may be authentically factual, scientists in different periods and with different backgrounds may well end by tackling them in very different ways. After all, there is no more reason to assume that the intellectual demands governing conceptual development in a science will *in fact* be everywhere absolutely identical, than there is to assume that the environmental demands governing organic development in a species are absolutely identical throughout its living-range. All that matters is (i) that differences in their intellectual demands should not lead different groups of scientists to judge conceptual variants in directly contradictory ways, and (ii) that, for all the differences in their longer-term expectations, scientists working in different countries or centres should continue to find each other's theoretical arguments intelligible.

In actual practice, there may well be substantial differences between the explanatory goals of different men working in the same discipline, even at one particular time. Pierre Duhem, for instance, wrote a striking essay about national styles in scientific theory.[1] Contrasting the manner in which the problems of electrical theory were handled by theoretical physicists in nineteenth-century France and Britain, he demonstrated that at certain crucial points there were systematic differences in the ruling strategies current among physicists in the two countries. In France, the accepted ideal was to cast all physical theories into axiomatic mathematical form. In Britain, it was quite as much an ambition to develop models—even working models— by which physical phenomena could be made intelligible in a visible or tangible, rather than a mathematical way. These differences (Duhem argued) had parallels in other fields of thought and life. The same contrast—between the French *ésprit géometrique* and the British *ésprit d'ampleur*—shows itself, also, in literature, in the individual characterization of Shakespeare's heroes and heroines as contrasted with the personified virtues of Racine; in the law, in the concrete particularity of common-law precedents as contrasted with the abstract generality of the *Code Napoléon;* in the philosophy of science, in the

[1] P. Duhem, *La Théorie Physique: son But, sa Structure* (Paris, 1903), Eng. trans. by P. P. Wiener (Princeton, 1954); see Chapter 4, 'Abstract Theories and Mechanical Models'.

contrast between Francis Bacon's method of empirical generalization and Descartes' mathematical rationalism; or, we might
add, in the differences between the deceptively natural sweep of
the traditional English garden and the geometrical precision
of a French *parterre*. The contrast between the two national
styles in nineteenth-century electrical theory were, thus,
further expressions of a broader difference between two long-
standing national temperaments, as applied to the problems of
a common intellectual discipline.

What attitudes should we ourselves adopt to such national
differences? Duhem himself had no doubt: he wrote without
apology or reservation as a French physicist. Faced with this
contrast, he did his best to understand the approach of his
British colleagues, but he could not bring himself to condone it.
The continual quest of men like Kelvin after mechanical
interpretations struck him as a kind of addiction, or intellectual
frailty, irrelevant to the 'proper business' of theoretical physics.
At its most extreme, it roused him to heavy irony:

Here is a book [Oliver Lodge's popular book, *Modern Views of
Electricity*, 1889] intended to expound the modern theories of
electricity and to explain a new theory. In it there are nothing but
strings which move around pulleys which roll around drums, which
go through pearl beads, which carry weights; and tubes which
pump water while others swell and contract; toothed wheels which
are geared to one another and engage hooks. We thought we were
entering the tranquil and neatly ordered abode of reason, but we
find ourselves in a factory.[1]

The sheer limitations of the human mind, reinforced by the
British preoccupation with the tangible and the individual,
might excuse a resort to such imagery; provided only that it was
accepted merely as an 'aid' or 'prop', for those who could not
digest the essential content of a physical theory in its pure form.
And, in his own mind, Duhem was quite clear what that 'pure
form' was. It was the axiomatic, mathematical, Cartesian form
in which his nineteenth-century French predecessors had cast
their own science of electricity. If one had argued, in reply,
that physical models might provide an alternative technique
of representation having merits of its own, capable of counter-
balancing the timeless systematicity of a quasi-Euclidean

[1] Op. cit., Eng. tr., pp. 70–1.

axiom-system, that would have struck Duhem as an outright betrayal of 'rationality' itself: 'We thought we were entering the tranquil abode of reason but we find ourselves in a factory.' Even in physics, the term 'rationality' could evidently mean, to a nineteenth-century French academic, nothing except Cartesian 'rationalism'.

Yet here, too, there are—surely—genuine options. Granted that Duhem has brought to light a real contrast between the theoretical strategies, and selection-criteria, of physicists in different countries, how can he then hope to produce definitive arguments to prove that one strategy is uniquely in the right, and the other inescapably misguided? Polemics apart, one must doubt whether any such arguments could have any validity. For they would have to rest on one or other of two grounds, neither of which is available: either, they must appeal to universal epistemic principles, of the sort we have here renounced, or else they must rely on an assessment of comparative consequences, i.e. on comparing the intellectual harvests to be expected from the two strategies. As to this latter calculation, there was no *a priori* guarantee that the Cartesian strategy would be the more fruitful. In all their substantive conclusions about electrical theory, Duhem and Lodge were in little real disagreement. Whatever electrical phenomena a Kelvin explained using mechanical models and analogies, his French colleague could explain equally by appeal to mathematical theorems, and vice versa; whatever theoretical connections there were, linking electricity to magnetism and optics within French physics, they had exact counterparts within British physics; so the mechanical models used by Maxwell, Kelvin, and Lodge to explain electrical phenomena simply served as alternative representations, or projections, of those same intellectual relationships that figured as formal entailments in French treatises of mathematical physics. The disagreement between Duhem and Lodge arose (that is) only after the entire current explanatory harvest had been gathered in, at a point where future theoretical policy had to be decided, on the deeper strategic level.[1]

---

[1] On the conception of rival physical theories as alternative representations, see the discussion by Hertz in the Introduction to his book, *The Principles of Mechanics* (English tr., New York, 1900), reprinted with a new preface and bibliography by

Over such deeper strategic judgements, men working out of quite different cultural and philosophical traditions might, quite legitimately, adopt conflicting intellectual priorities and see the 'essential' rationality of science as embodied in quite different policies. Certainly, the rival French and British strategies in electrical theory focused attention on different types of problem, and altered the whole emphasis of subsequent investigation. From nearly a century later, we can see retrospectively that both strategies had particular strengths and their weaknesses. Duhem's commitment to the Cartesian model of physical theory, for instance, made him unduly distrustful of J. J. Thomson's ideas about electrons and sub-atomic structure, as relying too much on despised mechanical analogies and not enough on careful mathematical analysis.[1] Brought up to a rougher, more mechanical way of thinking, Ernest Rutherford was better placed to exploit the new theoretical possibilities created by J. J. Thomson's ideas; though Rutherford himself was to see the resources of his own strategy exhausted twenty-five years later, in turn, when quantum mechanics reintroduced a Cartesian order into the new physics that he and Thomson had first hewn out in more mechanical terms.

The historical unity and continuity of a discipline is thus compatible with a substantial range of methods and strategies. Nor is this diversity confined to historical epochs or national styles; we may find similar differences of emphasis between the characteristic judgements of different research centres or schools, even in the same country at the same time. There are Cambridge geneticists and Edinburgh geneticists, Columbia operant psychologists and Harvard operant psychologists; and, while one may sign on under either flag, it would be partisan to claim a monopoly of insight for either school. For the maintenance of a coherent discipline at any time, no more than a 'sufficient' degree of collective agreement is required about intellectual goals and disciplinary ambitions. And by this word, 'sufficient', we must here mean, 'sufficient for the actual demands

---

R. S. Cohen (New York, 1956): especially Hertz's remarks about 'logical permissibility', 'correctness' and 'appropriateness', pp. 39ff.

[1] P. Duhem, op. cit., pp. 98–9.

of the present situation'; this is consistent with substantial disagreements about matters that go beyond the problems at present up for determination. In the 1870s and 1880s, for instance, all theoretical chemists were prepared to use the terms, 'atom' and 'molecule', when keeping track of chemical reactions and transformations; and the unity of nineteenth-century chemical theory was thereby preserved. But, as to the status which these atoms and molecules could be allowed, in the longer run—about that different chemists had quite contradictory ideas. Were they intellectual fictions; were they ultimate material units; or were they provisional entities, which might eventually prove to have a sub-structure? Faced with these further questions, Ostwald, Maxwell, and Kekulé held quite different views.[1] For the moment, however, these differences were not of active importance; the current tactics and conceptual problems in chemistry were neutral as between these longer-term strategic points of view, and which answer you gave made no immediate difference to your chemical analysis.

Considering different historical periods, instead of different countries or research centres, we can again find similar variations in the standards of scientific judgement. Viewed in historical terms, indeed, the selection-criteria employed in scientific disciplines, at different times, have had quite as much of a historical development as the particular theories, concepts, and variants on which they have been exercised. This conclusion might seem the final death-blow to any Cartesian ambition, of putting arguments in natural science on the same intellectual basis as those in mathematics; yet, paradoxically, this turns out on closer examination not to be the case. Certainly, it undercuts any hope of putting natural science on the same footing as Descartes's own Euclidean ideal for mathematics. But it establishes no essential difference in intellectual status between natural science and actual mathematics; on the contrary, mathematics itself is by now having to travel the same historical road. Descartes's commitment to Euclidean geometry involved in fact an unshakeable belief, not just in the superior certainty of mathematical methods in general, but also in the finality of

[1] See, e.g., the paper by I. B. Cohen referred to earlier (p. 153 n. 1).

Euclid's particular concepts and axioms. Yet, as Imre Lakatos has shown, the development of mathematical disciplines exposes their concepts and methods to transformations as profound in their own way as any in the natural sciences. Such fundamental mathematical concepts as 'validity' and 'rigour', 'elegance', 'proof', and 'mathematical necessity', undergo the same sea-changes as their scientific counterparts, 'soundness', 'cogency' and 'simplicity', 'relevance', and 'physical necessity'. Even the basic standards of 'mathematical proof' have themselves been reappraised more than once since Euclid's time. The result is that the concepts, methods, and intellectual ideals of mathematics are no more exempt from 'the ravages of history' —as Descartes and Frege hoped and supposed—than those of any other intellectual discipline.[1]

So the intellectual strategies employed by the scientists working in any field, at one time or another and in one research centre or another, themselves form an overlapping population, with its own internal variety. The extent of current agreement is a measure of strategic consensus, only towards the immediate problems of their common discipline. Beyond that point, the variety of longer-term views is exposed to no active ecological demands, or selection processes. In the historical transition from epoch to epoch, accordingly, the development of our intellectual disciplines creates a diversity of approaches limited only by the selective influence of active conceptual problems. Within the resulting historical genealogies, major scientific innovators commonly leave their marks, as much by their idiosyncratic conceptions about the overall demands to be made on any satisfactory theory whatever, as by their specific and detailed contributions to particular theories. Hagiography aside, Galileo's contribution to physics can be understood much better now that we are in a position to relate it back, by way of Benedetti and the Padua mathematicians, to the unsolved problems of Buridan, Bradwardine, and Oresme. It was Galileo's example and exhortations that convinced seventeenth-century physicists how rich a theoretical harvest could be reaped, if only one combined geometrical analysis with corpuscular ideas. In large measure, therefore, Galileo's originality con-

---

[1] Imre Lakatos, 'Proofs and Refutations', *British Journal for Philosophy of Science*, 14 (1963–4), 1–25, 120–39, 221–43, 296–342.

sisted less in his separate discoveries than in his vision of the intellectual fruit to be won for physics from a single-minded pursuit of the new 'geometrico-corpuscular' strategy.

At this strategic level, indeed, one could write the history of entire sciences as a story of scientists' changing conceptions about what 'physics' (say) has been, is, and could become. At first sight, the physics of Buridan and Oresme may, apparently, have nothing to do with the physics of Einstein and Heisenberg; yet the disciplinary genealogy linking them is in fact quite direct. The 'physical' problems of the 1330s—the scholastic analysis of the terms available for characterizing different varieties of 'change' or 'motion'—have been transformed bit by bit, through six centuries of intellectual development, into the contemporary problems of relativity theory and quantum mechanics. At no point, however, has this process been rationally discontinuous. What makes twentieth-century 'physics' appear so different from its fourteenth-century progenitor is simply this: that the principles of intellectual selection for sifting the conceptual variants available at every stage have themselves been 'the daughters of time'. They themselves have, quite legitimately, developed and changed in the course of historical development, along with the rest of physics. And any historical reconstruction of the process by which conceptual variation and intellectual selection have jointly shaped our intellectual disciplines must include, at its heart, an account of the manner in which the ruling disciplinary and intellectual strategies have themselves been transformed.

One final question about the objectivity of science still remains to be faced at this point. If the process of conceptual change in scientific disciplines cuts as deeply as we have suggested, it is not clear what possibility remains of drawing an acceptable distinction between the 'intrinsic' considerations that are genuinely relevant to the current problems of a science, and the 'extrinsic' considerations of (say) political prejudice or ideology that can distort or frustrate the normal procedures of disciplinary judgement. Certainly it is not clear how such a distinction can be drawn in terms that will have any lasting validity. Some such distinction is surely necessary and legitimate; but difficult problems arise the moment we try to contrast the two sorts of consideration at all sharply and permanently. How, then, can we lay down objective and permanent tests for deciding what considerations are—and what are not—'intrinsic' or 'relevant' to intellectual judgements in a scientific discipline? And will it always be clear in practice how this distinction should be applied? It would be convenient for philosophy if we could demarcate sharply the various factors operative in theoretical speculation, and distinguish the strict requirements of disciplinary development clearly from extraneous considerations like fashion and national style, metaphysical taste and political ideology. Our analysis leaves it unclear how far this can in fact be done; and this feature, too, may be seen as impugning the objectivity of science.

For the last thirty or forty years, this question has frequently been read as calling for a 'demarcation criterion', which will finally separate science from metaphysics, theology, and ideology. The hope has been that some test could be found for distinguishing genuinely 'scientific' propositions from others, either on the basis of their content, or on the basis of their methods of checking. Given such a test, we could then mark off those propositions that were authentic candidates for 'scientific' status from those others which were insufficiently supported by evidence, or mere expressions of a religious or ideological attitude, or even literally empty. So understood, however, the search for a demarcation criterion has been exposed to the same criticisms as the search for formal significance-tests and verification procedures. If we confine our attention to propositions stated in terms of previously agreed concepts and theories, we may be able to frame a test for telling genuinely 'scientific' propositions from others. (The proposition, 'This electron has an energy of 3.10^6 electron-volts', is then a demonstrably scientific one: the propositions, 'This electron is a material object' and 'This electron is sad', are not.) But a test that is applicable, only if we accept pre-existing concepts and theories, may cease to be relevant in those situations where our concepts are subject to radical change. For our own purposes, accordingly, the crucial question arises about concepts rather than propositions. When scientists modify their repertories of explanatory concepts, and select novel conceptual variants for incorporation into the content of their discipline, are there any permanent or hard and fast tests for deciding whether their reasons for this selection are truly 'scientific'?

Let us ask, for instance, what considerations have led physicists to opt, at one stage, for 'corpuscular' theories of heat, magnetism, and light, at another for 'subtle fluid' theories, and for 'field' theories at a third; certainly, it would be hard to discover compelling mathematical or experimental reasons to justify those preferences. Looking back at these changes of view, indeed, we may feel that the choice between these different theories finally turned on local and temporary tastes for certain fashions or styles of scientific thought. Yet such fashions as these have rarely been matters of 'mere' fashion. The decision to treat magnetic phenomena as a manifestation of fields or of fluids, or corpuscles or of emanations, may not have been forced on science either by experimental tests or by mathematical necessity; to that extent, it represented the arbitrary theoretical judgement of the particular physicists in question. Yet this decision rested on something more than a transient fad. Rather, it demonstrated once again how shared conceptions of 'mechanism', 'explanation' and 'intelligibility' underlie the current explanatory ideals of a science. Instead of relaxing comfortably on the assured testimony of the past, these novel choices once again laid bets against the future—voting for patterns of future theory in whose intelligibility and fruitfulness their advocates could feel a rational confidence.

Forecasts of this kind can have origins and grounds of many different kinds. These may lie in analogies with the results of other sciences; they may involve longer-term considerations of disciplinary strategy of a kind that is as yet only partly articulated; or they may even come from outside the whole area of what we now call 'science'. Either way, the intellectual fashions and theoretical styles of one generation have often acquired a more substantial significance for later generations, when the disciplinary goals of the science have expanded to deal with a new range of experience. On the level of concepts, in short, the search for a permanent demarcation criterion is incompatible with the fact that the intellectual goals of our disciplines are subject to historical development, along with all their specific theories and concepts.

Once we admit that the scientists of other times and other

cultures not only saw, but were entitled to see, the missions of their disciplines in quite different terms, we must modify our demands. From now on, it may not always be clear whether conceptual decisions taken in other milieus were indeed the product of 'extrinsic'—and so irrelevant—factors and considerations; or whether, alternatively, they reflected different ideas about the current explanatory missions of the sciences. The motives that led naturalists like John Ray to develop the first satisfactory classifications of organic species were theological as much as scientific; these taxonomies were seen as an expression of the 'rational order' imposed on Nature by the Divine Wisdom at the original Creation of the World. In nineteenth-century Germany, again, early cell-biologists developed their theories of fermentation and tissue-growth on analogy with current ideas about inorganic catalysis and crystal-formation. Are we to regard all such arguments as marks of sheer irrelevance, or rather as evidence of intellectual strategies which—though different from ours—were reasonable enough at the time? In asking these questions, we must be careful neither to be misled into anachronism, as a result of learning from subsequent experience, nor to suppose that such speculations were necessarily unsound from the start, as having been influenced by 'unscientific' prejudices or 'metaphysical' considerations. Unless we must take the trouble to reconstruct the actual scientific choices confronting the men of other times, we shall be in no position to recognize what considerations they could quite properly regard as 'relevant' to their decisions.

To say this is not to imply that anything whatever can be relevant to a scientific decision, if only we choose to regard it as such: far from it. It is merely to insist that questions of disciplinary relevance—like questions of 'adaptation' generally —have to be decided in the context of some particular problem-situation, and at some particular time. The variety and variability of our conceptual problems being what they are, it is scarcely possible to generalize, in any hard and fast way, about the kinds of considerations that are necessarily 'relevant' or 'irrelevant' to our conceptual choices. To go even further, one might actually say: No consideration whatever is wholly and completely irrelevant, on intrinsic grounds, to all scientific judgements whatever.

What does mark a man's beliefs as prejudices or superstitions, on the other hand, is not their content but his manner of holding them. In this respect, prejudice and superstition are the converse of 'reasonableness'; both have to do less with what our opinions are than with how we seek to enforce them. And in almost all the *causes célèbres* of science—where we are tempted to denounce the clouding of scientific discussion by the intrusion of 'extrinsic' theological and ideological irrelevances—indignation is better directed at the tyranny and dogmatism by which unconventional ideas were suppressed, and conservative views enforced. Whether in the Galileo affair or in the hostility provoked by La Mettrie and Priestley, in the theological criticism of James Hutton, the opposition to Lawrence and Darwin or the Lysenko episode: in each case, there were genuine intellectual points at issue, and the 'conservatives' had a possible position to defend.[1] The manner in which Huttonian geology clashed with the accepted ideas about the time-scale of the Universe, for instance, posed quite authentic conceptual problems for a scientist like Lyell; nor was there anything necessarily 'anti-scientific' in the fact that educated men in the early 1800s should have regarded the suggestion that the Universe had existed for an unimaginably long time with scepticism.

We can legitimately speak of 'prejudice', or 'dogmatism', only at the point where this scepticism expressed itself in sweeping denunciations and appeals to fear and suspicion, rather than in the presentation of genuine difficulties to be dealt with on their merits; and particularly at the point where novel scientific speculations began to be countered, not by arguments, but by political, religious, or legal sanctions. If Galileo had not been put on trial; if malicious politicians had not incited the Birmingham mob to burn down Priestley's house; if Lawrence had not been denied copyright protection for his lectures, on

[1] On Galileo, see G. de Santillana, *The Crime of Galileo* (Chicago, 1955); on Joseph Priestley, see J. T. Rutt, *Life and Correspondence of Joseph Priestley*, Vol. II (London, 1832), pp. 116ff.; on Lawrence, see June Goodfield-Toulmin, 'Some Aspects of English Physiology: 1780–1840', Part II, *Journal of the History of Biology*, 2 (1969), 307–20; for a light-hearted account of Darwin, Huxley, and Wilberforce, see e.g., W. Irvine, *Apes, Angels and Victorians* (London, 1955); and, for a more serious first-hand report on the Lysenko affair, Z. Medvedev, *The Rise and Fall of T. D. Lysenko* (New York, 1969).

grounds of blasphemy; if Wilberforce's attack on Darwin had not descended to abusive personalities; if Lysenko had allowed his theories to take their chance against other genetical views, without political interference . . . if, in each case, the disciplinary debate had proceeded without resort to rhetoric, prejudice, or forcible sanctions, the question of 'irrelevance' would never have arisen in just the way it did. One way or another, some accommodation was required between the new astronomy of Copernicus and the religious cosmology of the mediaeval Church. One way or another, the new natural history of the nineteenth-century geologists and palaeontologists had to be weighed against the implications of the Biblical tradition. But, in each case, an understanding could have been reached without taking the issue to the Courts or to the streets. In each case, therefore, the 'extraneous' aspect of the episode was a failure, not of logic, but of 'due process'.

Conversely: if prejudice, conservatism and the other forms of unreasonableness have to do with the manner of an intellectual debate rather than with its content, there is nothing to stop them from being manifested *within* the professional conduct of a discipline as much as on its intellectual frontiers. Anyone who has studied scientific institutions with an impartial eye knows very well that the 'normal procedures' of disciplinary development are frustrated more frequently than is publicly admitted, not merely by sanctions external to the science in question, but as a result of prejudice and dogmatism internal to it. The conflict between new scientific ideas and established religious orthodoxies provokes the peculiar rancour we know as *odium theologicum*; but theoretical innovators whose ideas conflict with firmly established scientific orthodoxies are just as frequently engulfed by *odium professionale*. Concepts and methods which have long been established—in the disciplinary sense of having earned their place in the science long ago— frequently become 'established' also in the sociological sense of having acquired an entrenched position in the institutions of the science. When this happens, the sanctions against change need not invoke considerations which are 'unscientific' or 'irrelevant', because they are extrinsic to science; yet they may fail just as completely to deal with the real intellectual demands of the issue in debate.

All in all, then, the search for a permanent and universal demarcation criterion, between 'scientific' and 'non-scientific' considerations, appears a vain one. Little useful can be said, in completely general terms, about the directions of conceptual change which are intrinsically 'permissible' in scientific disciplines. On as completely general a level as this, there is room for little more than quasi-political slogans. There may be libertarian slogans: like Paul Feyerabend's 'first principle of intellectual morality' (or *Wissenschaftstheoretische Anarchismus*) according to which all new positions are equally entitled to a fair chance of proving their merits in actual practice.[1] Or there can be authoritarian slogans, like those implicit in Kuhn's claim that the 'normal', or cumulative, progress of science depends on channelling innovation within the limits defined by some authoritative master-theory. And such slogans add little to our general understanding of scientific development, since in each case a question-begging word destroys its effective substance: 'Just how fair does a "fair" chance have to be?' 'Just how normal is the "normal" progress of a science?' Clearly, it is not 'fair' for a Velikovsky to be silenced by the professional astronomers; but it is neither tyranny, nor an offence against intellectual morality, if they fail to give him 'equal time' with a Fred Hoyle.[2] Clearly, there are situations in which existing theories must be rethought, and in which it would be improper to restrict the conceptual imagination of scientists, in the name of a current 'paradigm'; but we must have some kind of intellectual priorities, and cannot consider all conceptual variants in an entirely random order. The attempt to arrive at a completely general demarcation criterion thus ends up with an empty formula.

The root of the trouble is this. What is 'sound' in science is what has proved sound, what is 'justifiable' is what is found justifiable, what is 'internally relevant' is what turns out to be internally relevant. Since any strategic redirection of a science may lead to a redrawing of its boundaries, none of these

---

[1] See especially P. K. Feyerabend, 'Against Method', *Minnesota Studies for the Philosophy of Science*, 4 (1970).

[2] On the Velikovsky affair, see particularly the article by Alfred de Grazia, 'The Scientific Reception-System and Dr. Velikovsky', in *American Behavioral Scientist* (September 1963), pp. 45–68.

discoveries can ever be absolute or final. The actual application of such distinctions, and the resulting decisions about what can be allowed to be 'relevant' in any situation, are matters for scientific judgement, to be determined afresh in the light of the details of each particular case. As with all matters of judgement, these decisions are based on the experienced interpretation of historical precedents, which may nevertheless prove, in the longer run, to have been in certain respects misguided or misleading. Yet here, as elsewhere, the art of scientific investigation and theory-building must learn from its own history as best it can; and there is no question of arriving at some infallible 'method' by philosophical legerdemain.

At the very least, then—it may finally be asked—can we not distinguish the objective concerns of natural science from the arbitrary social interests of (say) the law? Even here, we must be careful not to exaggerate the difference between these various kinds of rational enterprise. Certainly (we may allow) the basic ideals of the law are 'regulative', whereas the basic ideals of science are 'representational'; and the question what is to count as 'just regulation' may well leave more scope for disagreement than the question what is to count as 'faithful representation'. Yet this contrast rests not so much on an absolute opposition as a matter of degree. Before A.D. 1600, ideas about justice and equity were much more uniform and generally agreed on than ideas about physics and biology; and even now the question, how a change of juridical strategy would work out in practice, is a question of fact—calling for a prospective estimate of actual consequences—as much as the corresponding question about a new scientific strategy. In each case, this prospective estimate must take into account the past history and continuing goals of the relevant enterprises; and in each case this estimate will be used as the basis for a decision, in what strategic direction the further development of our scientific or legal concepts is to be pursued.

CHAPTER 4   Intellectual
Professions: Their
Organization and
Evolution

Historians sometimes chronicle the changing ideas of science as though scientific thought developed quite autonomously, and succeed in disregarding entirely the human embodiment of those ideas: the names, identities, and characters of the men who collectively created and criticized them. And it is true that, in some cases, we can legitimately expound the historical growth of scientific ideas as a continuing rational argument, at least up to a point, without troubling to specify who it was that did the actual arguing; relying on the impersonality of the passive voice—'It was known . . .', or 'It has been demonstrated . . .'— as a grammatical veil hiding the idiosyncrasies of the scientist involved. If this kind of account carries us as far as it does, that is just further evidence showing the 'disciplined' character of scientific thought. Eventually, however, we have to acknowledge that these passive forms are no more than a veil. Beyond this point, we are forced to invert all such bland, passive expressions as 'chosen for perpetuation', and press the more candid, active questions: 'Who does the choosing in science?', 'How do they come to employ the criteria of choice they do?', and 'By what means do they get their selections accepted by their most influential professional colleagues?'

If the development of the natural sciences had—after all— been an application of fixed and universal principles of human understanding, we could have postponed those further human questions indefinitely. For it would in that case have been possible to state permanent and unambiguous criteria of scientific choice, which would have been the same for all historical and cultural situations; and all questions of 'who', 'where', and 'when' would—in principle—have been needless. The only human questions would then have been questions about human frailty: whether a particular group of scientists in fact applied those universal criteria correctly and, if not, what led them astray. But, since there is no longer any general basis for believing in such universal criteria, and a great deal of evidence that our criteria of intellectual choice are significantly variable, that option is not open to us here. Rather, we must be prepared to go behind all abstract accounts of scientific change to another level, where the questions have to do with the people whose concepts, theories, and explanatory ideals are under debate, from whose point of view the problems in question are 'problematic', and in whose eyes novel variants must find favour before being accepted as 'established'. In this present chapter, accordingly, we shall be representing the rational enterprise of a natural science not as a changing population of *concepts*, associated together in more or less formally structured theories, but as a changing population of *scientists*, linked together in more or less formally organized institutions. For the life of science is embodied in the lives of these men: exchanging information, arguing, and presenting their results through a variety of publications and meetings, competing for professorships and presidencies of academies, seeking to excel while still vying for each other's esteem. And the structure of these activities gives rise to the recognizable and distinct professions and sub-professions of science, which serve as the institutional embodiment of its disciplines and sub-disciplines.

4.1: *The Professional Embodiment of Science*

Scientific professions, like scientific disciplines, have a recognizable unity and continuity despite historical change, which falls short of an unchanging identity through time. So a scientific profession needs to be regarded as an 'historical entity', or 'population', whose institutional development parallels the intellectual development of the discipline for which it accepts responsibility. Furthermore, the same set of collective concerns and ambitions confers unity and continuity on both aspects of a rational enterprise—both on the discipline which represents the intellectually organized product of that enterprise, and on the profession which is the socially structured human agency by whose activities it is carried forward. The central question for our analysis is how the institutions and roles, publications and incentive-systems characteristic of an organized scientific profession reflect, in their structure and historical development, the collective intellectual concerns and ambitions of the men working in the discipline concerned.

The very closeness of the relationship between the professional and disciplinary aspects of a science creates certain initial difficulties for us. For there are standing ambiguities in many of the terms we use when we discuss the content and historical development of a natural science: they are used in two distinct senses, one disciplinary, the other professional. We speak, for instance, of the intrinsic 'authority' (or validity) of ideas, and also of the professional 'authority' (or power) of institutions and the magisterial 'authority' (or dominance) of individuals. Novel concepts, similarly, win an 'established' (or validated) place in scientific disciplines, while institutions and publications win an 'established' (or influential) place in the corresponding professions; and both ideas and institutions achieve this position by the favour of those influential individuals who form the 'authoritative' professional reference-group, or 'establishment'. How is it, then, that institutional authority or power comes to be distributed as it does between the individual members of a scientific profession? And how does this professional distribution of power take into account, protect, and preserve the public ideals of a discipline—its collective aspirations and goals, procedures and criteria?

At this point, the relationship between the disciplinary and professional aspects of science raises again, by implication, one of the central problems of political theory: viz., how certain individuals can become 'authoritative spokesmen' for whole professions, social groups, or even nations. Suppose that we are puzzled by some phenomenon in genetics (say) and ask, 'What does biochemistry tell us about this?' Two things are then clear. Firstly, biochemistry—or any other discipline— can tell us nothing, except through the mouths of human spokesmen; and, secondly, not every biochemist speaks *qua* biochemist with equal authority. In such a context the term 'biochemistry' becomes, in effect, a grand political abstraction —like 'France', the 'American people', or the 'international working class'. The question, 'What has biochemistry to tell us?', has accordingly to be understood obliquely, like the questions, 'Where does France stand?', and 'What is the attitude of the trade unions?'

In scientific as in political affairs, the easiest course is to ignore this question, and to assume that all 'authoritative' or 'established' judgements express an informed consensus on which the entire profession is agreed; whether for brevity or for more tendentious reasons, we are tempted to talk as though the rubric, '*It* is known that so-and-so', meant simply, '*Everyone* in the profession knows that so-and-so.' In science as in politics, however, this view rests on a convenient fiction. Pure syndico-anarchism and participatory democracy are no more of a practical reality in the actual life of a scientific profession than they are anywhere else. In this respect, intellectual professions are no different, in practice, from other theoretically demo-cratic institutions. All accredited members of a scientific profession may, in theory, be equal; but some turn out to be 'more equal' than others. On the one hand, there are the men whose word carries weight in the profession—the men whose judgements are accepted as authoritative by other workers in the field, and who come to speak 'for and in the name of' the science concerned. On the other hand, there are the men who have no such influence, either because their opinions and attitudes are regarded as of no consequence, or because they are dismissed as heretical—and these men are in no position to act as spokesmen for the science that they serve.

When we say, 'It is known that so-and-so', or 'Biochemistry tells us that so-and-so', therefore, we do not mean that everyone knows, or that every biochemist will tell us that so-and-so. We normally imply, rather, that this is the 'authoritative' view: both in the disciplinary sense, i.e. the view supported by the best accredited body of experience, and also in the pro-fessional sense, i.e. the view supported by the influential authorities in the subject. And the central problem about the relationship between these two aspects of a science can be restated, in terms of that ambiguity, as follows:

How is it that, within a well-organized intellectual enterprise, those ideas on which collective experience confers intellectual authority also acquire institutional authority? And what ensures that institutional authority shall be exercised predominantly on behalf of views that are also entitled to disciplinary authority?

Evidently, it is the proper business of influential scientists to lend their professional weight only to concepts and doctrines which are, from the disciplinary point of view, sufficiently well-supported. The problem is, to explain how it is ensured that (i) the intellectual authority of scientific ideas, (ii) the magisterial authority of individual scientists and (iii) the professional authority of scientific organizations shall remain as closely related to one another as the needs of the science demand. This question requires us to extend to the institutionalized activities of natural scientists the same type of theoretical analysis that Max Weber introduced into general sociology: it is, in effect, one further application of the general theory of dominance (*Herrschaft*) within human institutions.

Suppose, then, that we accept the parallels between disciplinary and professional development in science. If the intellectual authority of concepts and theories rests not on universal, but on variable criteria and standards of judgement, the same is presumably true, also, of the magisterial authority of individual scientists and the institutional authority of scientific organizations. Novel theories and concepts win their intellectual place within a discipline by contributing to the solution of its current outstanding problems, and we can specify the exact demands of those problems only in terms of the current intellectual ambitions and explanatory capacities of that discipline. Similarly, individual scientists win their right to speak on behalf of their subject, only on account of the judgement that they have previously demonstrated in tackling the current conceptual problems facing their profession—and this judgement, too, must be assessed in terms of the current explanatory powers and ideals. Scientific professions, in short, are like all other social organizations. They have their 'reference groups', comprising the men whose individual choices become—in effect—the choices of the whole profession. And any serious sociology of science must, in due course, explain by what processes the reference-groups of influential judges that exercise authority within a scientific profession constitute themselves, achieve dominance, exercise their authority, and are eventually displaced.

A new concept, theory, or strategy, for example, becomes an effective 'possibility' in a scientific discipline, only when it is taken seriously by influential members of the relevant profession, and it becomes fully 'established', only when it wins their positive endorsement. Conversely, an innovation which the current reference-group declares 'totally unsound' is, for the time being, as good as dead. There are, of course, two alternative ways of escaping from this condemnation: either to abandon the innovation or to change the reference-group. So the idea in question may be successfully revived under the auspices of a later generation of scientists, whose other theories can accommodate it more easily. Meanwhile, however, an 'authoritative' judgement of unsoundness can be the death-sentence on a new hypothesis, on a would-be scientific institution, or even on an individual's scientific career.

By what career-sequence of fellowships and publications, editorships, university chairs, and committee memberships, then, does an individual scientist reach a position of influence, and enhance his ability to make his voice authoritative? Through what errors of judgement, failures, or defeats may he prejudice his earlier power to speak for his discipline? And what learned societies and 'invisible colleges', journals and congresses, define the forum of competition within which the effective disciplinary contest between intellectual innovations is conducted? If all these questions were answered in sufficient detail, this would shew us just how far the scope and opportunity exist for the operation of familiar political mechanisms, even within the most high-minded of intellectual professions: how (that is) the institutions of a science, like those of any collective human activity, develop through the actions of parties and pressure-groups, through *coups d'état* and unilateral declarations of independence, and are the scene of a continuing tug-of-war between Old Guards and Young Turks, autocrats and democrats, oligarchies and gerontocracies—so displaying the characteristic forms, varieties, and instruments of political life and activity generally.

To put the problem in these terms is not to speak disrespect-
fully, or cynically, of the learned professions. It is merely to
recognize that power remains power, and institutions remain
institutions, whether serving economic, political or intellectual
functions. Until recently, of course, scientists have cultivated a
public image of disinterestedness; and this has carried with it
a pretence that the institutional activities of scientists—forming,
as they do, the professional face of a 'rational' enterprise—are
somehow exempt from the general principles of political and
social action. Happily and more realistically, we are no longer
obliged today to suppose that the conduct of scholars and
scientists, when assembled into professional bodies, is emanci-
pated from the general laws governing the collective actions of
other institutions. Individuals and organizations in fact exercise
as real a power and influence over the development of science
as they do in any other sphere of human life. Correspondingly,
the roles, offices, and positions of influence within a scientific
profession are worth fighting for—and are, in practice, fought
for—as singlemindedly, methodically, and even deviously,
as in any other sphere.

However much a scientific profession turns a stainless and
unanimous face to the outside world, therefore, its internal
organization is as much a field for political action as that of any
other institution. Here as elsewhere, the tranquil, orderly
conduct of ballots and elections is only the culmination of a
more complex and fluid process, involving lobbying and
caucusing, procedural manoeuvring and occasional skulldug-
gery. And since, in academic politics, power and influence form
the only freely negotiable currency, the competition for them
is—arguably—liable to be even more intense there than outside.[1]

The central problem facing us in the political theory of the
scientific professions accordingly resembles that of traditional
laisser-faire economics. The development of a rational enterprise, like that of an economic community, can be analysed on

---

[1] I. A. Richards likes to point out that spleen and intrigue are traditional
features of the scholastic world: recall Robert Browning's *Soliloquy in a Spanish
Cloister*. Harold Wilson is also reported to have remarked that, after the rough-
and-tumble of an Oxford Senior Common Room, it is a relief to return to the
gentlemanly atmosphere of the House of Commons!

three different levels. We have the 'common good'—represented by the historical development of the collective discipline, with all its accredited concepts, explanatory procedures, and strategies. We have the professional institutions of the enterprise—which exist, normally, with the sole function of serving the discipline, but which soon develop other independent interests, some of them unrelated to those intellectual concerns. And, finally, we have all the individual scientists—who have lives to lead and careers to pursue, as best they can, with an eye both on the ideal demands of their chosen discipline and on the practical realities of their professional situation. The problem then is to give a historically convincing account of the science, as seen from both professional and personal points of view, which makes it clear how intellectual, institutional, and individual factors interact; and how, in the course of pursuing their own legitimate interests, individual scientists and scientific institutions can at the same time promote the 'common good' of their collective disciplines.

By analysing the scientific professions as 'populations', we can avoid a sociological trap parallel to that with which the cult of systematicity faces the philosophy of science. Just as it is a mistake to treat the intellectual content of a complete science as a single coherent formal system, whose rationality resides in its internal articulation—and therefore to overlook the rational aspects of its historical development—so, too, it is a mistake to treat the social organization of an entire profession as a single tightly connected 'social system', whose functionality lies in the static equilibria between its component roles and institutions. This is a mistake which some sociologists of science have in fact, been ready to make. Having fallen into the theoretical habit of analysing 'society-as-a-whole' as comprising an integrated network of roles and offices, institutions and reward-systems, they have naturally assumed that men's various collective activities, on a smaller scale, are all of them taken care of by lesser 'social systems', characterized by similar equilibria. On this view, the 'social system of science' is merely one of several coexisting 'systems', which together make up the integrated totality we call 'society-as-a-whole'.[1] To explain the workings of some particular professional organization, the

[1] M. W. Storer, *The Social System of Science* (New York, 1966), esp. pp. 29–30.

influence of some particular journal, or the significance of some particular reward (e.g. the Nobel Prizes), it is then sufficient to show, first, how it enters into the 'social equilibrium' of science itself, and, secondly, how it plays an indirect part in the overall equilibrium of the entire society.[1]

From our point of view, this analysis has two serious defects. An equilibrium model of society once again confuses the historically variable interactions between different institutions with the systematic connections holding between different roles within any one institution; and, having 'frozen' those historically variable relations, this confusion prevents us in turn from understanding how inter-institutional relations can change. The component positions in the organizational chart of an institution may have fixed status-relations: so that any one position (e.g. a vice-presidency, or senior research fellowship) has a significance which is intelligible only in virtue of its relations to the structure of the whole institution. But the same is not true of the coexisting institutions which go to make up an entire functioning profession. Such coexisting institutions originate separately, develop in parallel, and are not linked by systematic status-relations. This, in fact, is the thing that makes it possible for the internal organization of a scientific profession to adapt to historically changing situations.

Sociologically and philosophically alike, the systematic or structural relations within a particular theory or institution are one thing; the historically changing and functional inter-actions between different institutions and theories are something quite else. The formal institutions of a profession represent, at most, pockets of 'organizational systematicity' within the larger institutional populations of the entire profession. And we can give an historically convincing account of the professional development of the sciences, only if we respect this fundamental distinction between organizational structure and functional interactions. This done, we shall be able to explain why the notion of 'authority' is—at the same time—so central to our analysis, and also so ambiguous. For, instead of supposing that a scientific profession has a static 'organizational chart' with a fixed network of institutional status-links, we shall then have to

[1] On the role of Nobel Prizes, see the article by H. Zuckerman in *Scientific American*, 217, No. 5 (Nov. 1967), 25–33.

recognize that the coexisting institutions of any profession are in continual competition for 'establishment' or 'authority'. Instead of working together harmoniously at all times within a single 'system', the rival institutions of a profession can easily act in ways that frustrate one another's goals. And the resulting conflicts or rivalries provide much of the impetus for historical change in the professional sphere, as theoretical inconsistencies do in the disciplinary sphere.

Such a populational analysis at once explains several kinds of sociological phenomena that a systematic account does not so easily accommodate. Instead of competing for influential positions within the established institutions, for instance, individual scientists can alternatively increase their influence on their fellow-scientists—and that of their novel ideas—by setting up rival centres of power or establishing professional splinter-groups, which manage their own affairs independently of older-established institutions. Similarly, instead of fighting to control the most respected existing journals, it may be more effective for them to inaugurate a new journal, relying on the merits of their novel approach to win an influence comparable to that of older periodicals. The functions of the periodical literature epitomize, in fact, the functions of scientific institutions more generally. Within an enterprise that is so dependent on the exchange of results between different scientists, 'authoritative' journals have a particularly important part to play. So the very *raison d'être* of many scientific societies lies primarily in the journals they sponsor, only secondarily in their formal meetings. In practice, indeed, the editor of an influential periodical acts in his own person as a disciplinary 'filter', sifting out those papers which deserve publication in his journal, and so embodying the disciplinary selection of accredited 'possibilities'. And, in view of the part played by this 'filtering' in the selective perpetuation of new ideas, one can therefore speak of scientific periodicals as among the most powerful 'institutions' of a science.

The current institutional population of a science comprises elements of several different types: learned societies, career positions, journals, reward-systems, conferences etc. Within each type, 'older-established' and 'upstart' members normally coexist; and the newer, less authoritative elements will normally have a more informal mode of operation than their older-established counterparts. But, in due course, history brings its changes in these relations. To-day's 'invisible college' or duplicated newsletter contains the germ of a society or journal which will be dominant in the next generation—not to mention the superannuated institutions of a century hence. Whether we consider learned societies, books or journals, annual meetings, university chairs, or research directorships: in each case, it has to be shown not only how particular organizations function within the current operations of the profession, but also how competing organizations come into existence and supersede one another in its historical development.

In short, the crucial question is once again the question of historical change: not how institutional authority is currently exercised within a scientific profession as it is, but how that authority is redistributed between successive stages in the development of the profession. Alongside the disciplinary question, 'How are conceptual variants selected for incorporation into the established content of any science?', we can thus state its professional counterpart: namely, 'How do new individuals, associations, journals, and/or centres of research displace one another in the competition for authority within any scientific profession?' And, just as the disciplinary population at any time comprises not only well-established concepts but also other conceptual variants which have, as yet, a lesser intellectual authority, so too the professional population of institutions at any time comprises both the existing possessors of prestige and influence and also those other newer organizations which have, as yet, only a lesser institutional authority. The sciences of the future will no doubt have new spokesmen, new journals, new societies; and these will only rarely take over from their forerunners by an abrupt *coup*, or revolution. So our problem is to see (i) how it is that professional authority is transferred gradually, from one decade to another, between the constituent institutions of a profession; and (ii) what guarantees there are that the redistribution of institutional authority from older-established to upstart institutions will protect, in the process, the legitimate intellectual claims of disciplinary authority.

We are now in a position to cash in those veiled passive-voice statements from which we began: 'It was known . . .', 'It has been demonstrated . . .', and so on. Restated in professional terms, such disciplinary statements implicitly involve a great variety of individuals and institutions. Suppose we say, for example, 'By the year 1860, *it was known that* Heat is a form of Motion'; this means, in disciplinary terms, 'By the year 1860, *it had been demonstrated that* the temperature of a material body is a direct measure of the mean kinetic energy of its constituent molecules, so confirming the old maxim that Heat is a form of Motion.' Translating into professional terms, we can now fill out the first implications of this statement, and make it explicit just who had to do the necessary 'demonstrating', and to whose satisfaction: 'By the year 1860, Clausius and Maxwell had demonstrated, to the satisfaction of Tyndall, Helmholtz, and Thomson, that the temperature of a material body etc. . . .' Yet even this transcription leaves the situation still obscure in some essential respects. In order to explain why—after so many years of debate—the kinetic theory won a conclusive place in physics during the later 1850s, we must make it clear, also, (i) how Clausius and Maxwell obtained attention for their ideas so quickly, (ii) why scientists like Helmholtz, Tyndall, and Thomson were so readily convinced, and (iii) in what respects the blessing of these men accelerated acceptance of the new theory.

Clearly, it is no coincidence that this particular argument was presented by two men, both of whom had already earned attention as authoritative physicists in the forum of their peers. Coming from a lesser hand, or presented to a less influential forum, the same argument would have won neither attention nor acceptance so quickly and easily.[1] While, as theoretical arguments, the papers of Clausius and Maxwell may have possessed all the necessary cogency and power by the disciplinary standards of their time, the best argument in the world could 'carry professional weight' at all quickly only if its origin and presentation brought it to the attention of the influential 'reference-group' at the right time, and in the right manner. Or, to put the point more exactly: even the best

---

[1] The kinetic theory of heat had in fact been expounded at length, independently, by Waterston in 1845, without attracting anything like the same attention —still less winning general approval. Even then, the 'weight' actually carried by an argument depended on whether it was advanced by an 'insider' or an 'outsider'.

argument in the world could win the institutional authority merited by its intrinsic intellectual authority, only if the professional circumstances were favourable in other respects. So, transcribing our disciplinary statement into professional terms more fully, therefore, we must now say:

By the year 1860, the repertory of concepts and explanatory procedures current in European theoretical physics had reached a point at which Clausius and Maxwell (being influential physicists) could give a comprehensive reanalysis of the thermal properties of gases, according to which the temperature of a material body etc. . . .; and at which this reanalysis could quickly win endorsement from the authoritative theoretical physicists of the time (e.g. Helmholtz and Tyndall), become incorporated into the standard literature (e.g. the influential textbook of Thomson and Tait), etc. . . .

Unpacking the ambiguities concealed in the passive-voice statement, 'By the year 1860, it has been demonstrated that . . .', we can thus replace it by the trite but more explicit active-voice statement, 'By the year 1860, well-established scientists (exercising institutional authority) had satisfied themselves that there were well-established arguments (entitled to intrinsic intellectual authority) for the view that the temperature of a material body etc. . . .' And the question now is, how the institutional machinery of the scientific profession keeps the actual distribution of institutional authority reasonably in line with the intrinsic claims of intellectual authority.

Let me sketch briefly how a populational analysis applies to scientific institutions of different kinds. At a given time, any scientific profession includes a number of longer-standing learned societies, with comparatively well-established structures and traditions. These old-established societies tend to embody a somewhat conservative view of the scientific discipline concerned; and this tendency is reinforced by the very elaborateness of their internal organizations, since upstart young scientists with heterodox ideas will take correspondingly longer to win influential positions within their hierarchies. Alongside these established and conservative institutions, however, there exist other more flexible and informal groupings, which lack something in the way of social stature and degree of establishment, but serve as more effective channels of collaboration between the men engaged in front-line research. (Such informal groupings of correspondents and collaborators are referred to nowadays as 'invisible colleges'—a name originally applied in the seventeenth century to the informal precursor of the Royal Society of London.)[1] In due course, the shared problems and intellectual approach, which served as an occasion for the original crystallization of an 'invisible college', will normally lead to the development of a more formal organization, having the power to sponsor journals, congresses, and all the other trappings of a learned profession. And, in time, this new learned society will itself turn into an 'established' scientific institution, with its own traditional interests to conserve . . .

There is a corresponding population of career positions in any scientific profession. From graduate fellowships on, by way of junior lectureships or assistant professorships, to senior university chairs and research directorships, the working scientist makes his way as much by the posts he occupies, and the institutions in which he occupies them, as by the papers he writes and the periodicals in which they are published. Yet here again, in any field of science, coexisting institutions and departments differ greatly in professional standing, and this 'standing' is essentially a historical variable. The relative standing of different research teams or laboratories, within a given science, is at the same time one of the more significant variables underlying the redistribution of professional authority, and also one of the subtler and more confidential variables, about which every working scientist knows much more than he will ever commit to writing. In this sense, professional standing makes its effect without being registered in any formal titles. As with a 'pecking order', it displays itself through the practical relations between the men and institutions involved; and those who feel a need to lay claim to such standing explicitly are most often reacting to a situation in which they are in danger of losing it. So, in any thorough study of the manner in which professional authority passes between scientific institutions, the question of 'standing' will be especially worth investigating.

Similar historical changes affect the relations between the different communication channels of any science, between the

1 See, e.g., D. J. de S. Price, *Science since Babylon* (New Haven, 1961); and also J. Ziman, *Public Knowledge* (Cambridge, Eng. 1968).

various types of meeting in which the profession conducts its proceedings, and within the reward system of the profession.
Different media of publication—textbooks, monographs, quarterlies, abstracts, and 'review letters'—have been introduced,
one after another, to meet new professional needs; and the
historically changing operations of a scientific profession are
reflected once more in the transfer of influence from one medium
to another. The 'invisible colleges' of seventeenth-century
Europe were initially linked by the circulated correspondence
of men like Henry Oldenburg. With the foundation of national
academies, emphasis shifted to their *Transactions* and to treatises
such as Newton's *Principia*, which were published under their
auspices. In subsequent centuries, the balance has again
shifted several times: to quarterlies, to twice-monthly periodicals,
weeklies, and even shorter-term publications. The proliferation
of journals and the acceleration of publication are effects, in
part of the fragmentation of sub-disciplines, in part of the
sharpened competition for priority; but they are associated also
with a great decentralization of scientific authority. Where no-
one can hope to master all the available concepts and theories,
scientific professions were bound to move towards a pluralistic
pattern of authority. On the very frontiers of research, indeed,
we are now back not only with 'invisible colleges' but with a
multiplicity of Oldenburgs, who circulate duplicated 'pre-
publication' material in highly specialized subjects to an
international circle of equally specialized devotees. In the
more self-consciously original branches of science—it has even
been suggested—only out-of-date ideas ever actually get into
print!

Scientific meetings, likewise, have taken different forms at
different times, and have provided correspondingly different
occasions for presenting original results. So we can plot a
similar redistribution of emphasis, as between international
congresses and annual professional meetings; between open
sessions of Associations for the Advancement of Science and
'invitation only' seminars; and those historical changes again
reflect changes in the degree of specialization. Over the last
few years, indeed, there has even been a slight reversal in this
trend, with a shift of emphasis in the pattern of scientific meetings from ever more specialized sub-sub-disciplines to fields of

interdisciplinary concern, e.g. environmental studies and the neurosciences.

As to the reward systems of science: at one extreme, the marks of professional success and 'authority' take obvious forms, in the Nobel Prizes, fellowships and presidencies of the national academies, and the like. At appropriate stages in their careers, the prospects of these rewards gives working scientists a genuine —sometimes a very powerful—incentive. They can even end, as the Nobel Prizes seem to have done, by distorting the balance of scientific work: distracting attention away from 'non-prize-worthy' subjects (e.g. ecology and systematics) or rewarding restricted aspects of scientific work (e.g. ingenious experimental techniques) at the expenses of others (e.g. profound theoretical reappraisals). At less exalted levels, the reward system of science is less tangible, but not less powerful. Long before he is in the running for his National Academy, a young man must establish his professional credentials with his peers, as a 'sound' worker. For this purpose, he must demonstrate publicly, not only that he has mastered the critical standards of his chosen discipline, but also that his commitment to the discipline is single-minded and absolute, to a degree of exclusiveness demanded at other times only in the monastic orders.

Any lapse from this commitment—frequent appearances on television, lucrative government consultancies, incautious excursions into popularization—will put his professional reputation at risk. So, until his personal standing in the profession is secure, the apprentice scientist will view such outside activities as professional poison, if not as outright sin. Once safely established in a position of authority, he may perhaps allow himself a few such distractions; until then, his 'rule of life' requires exclusive devotion to the aims of his discipline. (It is rumoured that J. D. Bernal once invited his scientist son to collaborate on a paper about some slightly unorthodox ideas, only to get the response, 'It's all very well for you, Dad; but I've got my career to consider.') The analogies between the scientific professions and the monastic orders, pointed out by John Ziman, are in fact very close to the mark.[1] In particular, the value system underlying both modes of life has been essentially 'unworldly' and 'single-valued'.

[1] See Ziman op. cit., pp. 138ff.

Meanwhile, alongside the obvious, public rewards of science there is a much more influential set of informal rewards, based on the esteem in which the members of each profession hold one another's work. Like the 'pecking-order' of rival scientific institutions, this personal order of merit among individual scientists is rarely committed to paper; but is none the less powerful for being tacit. It is a glorious thing to be President of the Royal Society, just as it is a glorious thing to be an Archbishop: but what novice will not prefer to win a reputation for brilliance (or sanctity) among his own peers, even if this later proves to be a disqualification for worldly office? And what middle-aged scientist is wholly free from the fear that, by accepting an institutional promotion, he will be admitting to his younger colleagues that he is past his intellectual best? Standing within this implicit hierarchy is not marked by formal titles or offices, but it is well understood by all those directly involved. Within the Order of molecular biochemists (or neo-Darwinian systematists, or particle physicists) everyone from the novices up soon develops a precise—if unwritten— sense of *Who's Who*. And any serious attempt to map the historically changing patterns of authority within a scientific profession must face the delicate problem of assessing, not only the external signs of professional rank and the evident allocation of public offices and positions, but also the more significant internal pattern of esteem, which largely determines just whose novel ideas are taken seriously by other members of the profession.

One final locus of authority within a science consists in the 'standard texts' of the subject; and, by looking at these, we can see more clearly how the intellectual transmit of a discipline becomes the collective property of a profession. Once accepted by Helmholtz and Tyndall, the kinetic theory of matter may have won an established place in front-line physical science; but what crowned its new authority was incorporation into a new generation of standard textbooks of physics—both elementary, e.g. Maxwell's own *Matter and Motion*, and more advanced, e.g. Thomson and Taits's *Treatise on Natural Philosophy*. Whereas the 'micro-evolution' of scientific ideas is manifested in the most up-to-date research discussions (whether the letters of Darwin and Wallace, or the *Physical Review Letters*), its 'macro-evolution' is embodied in the standard texts accepted as authoritative in each successive generation. From Rohault's exposition of Cartesian physics and Isaac Newton's *Opticks*, up to Ernst Mayr's *Animal Species and Evolution* and Feynman's *Lectures on Physics*, these standard works define the successive bodies of doctrine that form the accepted starting-points for the next generation. By digesting the specialized literature of the preceding generation, indeed, these comprehensive expositions create a 'conceptual platform' on which the next generation of budding scientists can stand firm, in defining and attacking their own disciplinary problems.

In brief: the career-structure and administrative organization of a scientific profession, its channels of publication, meetings and reward systems all display the same pluralism and the same tendency for authority to pass from one group or individual—editor or president, professor or iconoclast—to another. This pluralism makes historical change within a scientific profession possible; yet, at the same time, it exposes the current professional authorities to criticism. And, if we consider the arguments used to justify such transfers of authority, we shall begin to see in outline how our own crucial question is to be answered: that is, how the organization of a scientific profession comes to serve the proper interests of the discipline it represents.

The workings of scientific institutions exemplify, in fact, not only the same mechanisms and processes that govern political relations in all human institutions; they illustrate also the deeper principles and ideals on which all political power rests, by appeal to which it is justified, and in terms of which it is subject to legitimate criticism. The ultimate source of the power that office-holders in a scientific profession wield is the implied consent of their professional colleagues in the same discipline. But this consensus is also the ultimate sanction, by which their power is controlled and their conduct kept within reasonable limits. For the professional authorities in a science are open to a standing challenge: this consists, precisely, in questioning their claim to act as the current spokesmen of their discipline. However drastically the socio-political situation of a profession may change, however completely a single magisterial individual (a Newton, a Cuvier, or a Helmholtz) may dominate the science of his age, one crucial fact remains. Such positions of power are occupied, and utilized, 'in the name of' particular communal disciplines, and so on pain of public forfeiture. In this respect, the plurality of coexisting societies, publications, and meetings is just one of many institutional devices for ensuring that power is used, patronage dispensed, and authority exercised with a sufficient eye for the collective goals of the discipline concerned.

Acting 'in the name of' a discipline means acting as the guardians, for the time being, of its collective intellectual ideals; and the exercise of professional power is circumscribed within limits set by the collective commitment of these ideals. The men who exercise professional power *ex officio*—whether as editors or referees, presidents or professors—must be ready to meet challenges based on appeals to the current principles, procedures, and aims of their discipline. Indeed, if the institutions of a profession are to continue serving effectively the discipline in the name of which they act, their structure must provide means for the conduct of its officers to be challenged in this way. Such institutional opportunities will, of course, never be perfect. As in any other organization, Young Turks will often accuse the Old Guard of claiming more power than they deserve, or of hanging on to office after their legitimate authority is exhausted. Still, any professional institution must provide, in some way or other, for all the sociological functions implicit in its collective ambitions; and must accordingly reflect, within its own structure, the full variety of activities necessary for the development of the corresponding discipline.

So the collective ambitions which define the current intellectual preoccupations of a scientific discipline also have a crucial—indeed, a critical—institutional part to play in the working of the corresponding profession. The word 'critical' is here used advisedly; at one and the same time, these collective ambitions define the current intellectual selection-criteria of the discipline, and also serve as the basis for judging the current performance of professional office-holders and institutional arrangements. If the research reports of one generation become the archives of the next, this change is to be justified by the current disciplinary purposes of the science in question. If the current preoccupations of the Nobel Committees are criticized as distorting the balance of current research, this objection rests equally on disciplinary judgements about the comparative urgency and 'ripeness' of different lines of investigation. If the current organization of National Academies is denounced as a 'gerontocracy', this again implies that a profession dominated by older men inevitably lags behind the true needs of a discipline. (Did the Royal Society really need until 1961 to discover that molecular biology was 'ripe' for encouragement?)[1] And if, in retrospect, historians deplore the overwhelming power exercised by (say) Georges Cuvier in the French scientific establishment of the post-Napoleonic era, their criticisms too imply that Cuvier had a retrograde influence on the development of French biology: e.g. that he used his authority to pursue his earlier rivalry with Lamarck, and so imposed on French biology a hostility to evolutionary ideas which long out-lasted his own death in 1832.[2]

No century in the history of science has, of course, been exempt from the tyrannous misuse of professional power, even at the highest level. So the dark brooding autocracy of Isaac Newton himself, during his years as permanent President of the Royal Society, launched the 'Newtonian' world-picture, not just as a creative framework for physical speculation, but also as a dogma which later bore down oppressively when such men as Thomas Young set out to revive unfashionable aspects of Newton's own ideas. And, on a smaller scale, a multiplicity of similar tyrannies is continually being exercised 'in the name' of science. Papers are refused publication, academic posts are denied, professional honours are withheld—even from an Ohm, a Mayer, or a Helmholtz—not for lack of worthwhile disciplinary arguments, but through professional disagreement with the editor, the research director, or the influential professor.[3] The only thing—but the essential thing—that redeems the scientific profession is the fact that all its claims to act 'in the name of'

---

[1] See, e.g., Stephen Toulmin, 'The Complexity of Scientific Choice: a Stocktaking', reprinted in *Criteria for Scientific Development*, ed. E. Shils (Cambridge, Mass., 1968), p. 71.

[2] On Cuvier's political dominance, both as a zoologist and as a national statesman, see W. Coleman, *Georges Cuvier: Zoologist* (Cambridge, Mass., 1964), pp. 126–39; and also F. D. Adams, *The Birth and Development of the Geological Sciences* (New York, 1938), pp. 263–7.

[3] Even Helmholtz had his original paper on the Conservation of Energy rejected by Poggendorf's *Annalen*: on this episode, see Y. Elkana (forthcoming).

a discipline are subject to appeal: if not within the framework of any one particular institution, then before the profession at large, and if not before a tribunal of the current Old Guard, then in the eyes of the up and coming Young Turks.

In all these cases, then, the operative question is: 'How far do the structure, performance, and distribution of power within the professional institutions concerned enable them to meet the proper needs of the discipline for which they are acting?' This question can be widened still further, to become a question about the proper scopes of those disciplines themselves. For scientific professions 'embody' their disciplines, not just by providing institutions and channels of communication for all the necessary activities of those disciplines; at a deeper level, also, their organization shows how the boundaries between different disciplines are currently conceived. Where, for instance, two prospective sub-disciplines have not yet developed separate channels of professional expression—whether separate journals, separate university posts, or separate research laboratories—they will not yet wholly have established their claim to be distinct, even on the disciplinary level. The intellectual question whether the demands of biological understanding require us to recognize 'molecular biology' (say) as a separate discipline, alongside general biochemistry and cell-biology, entails also such institutional questions as whether molecular biology is entitled to its own separate societies, journals, and university posts; these institutional questions, likewise, can be answered satisfactorily, only in the light of intellectual judgements about the needs of the relevant discipline.

4.2: *The Generations of Judges*
Treating a scientific profession as a historically developing entity, we can ask both how novel elements enter the relevant population and also how certain of these novelties subsequently achieve an established position within it. These questions can be raised about societies, or journals, or meetings; but they can be raised most usefully about the scientists themselves. The succession of human generations continually draws younger recruits into the scientific profession, and eliminates their elders. From generation to generation, as a result, men with new and different attitudes of mind are continually taking over

positions of institutional power and individual dominance within their profession. So the changing character of a science is embodied, first and foremost, in the changing attitudes of its scientists.

It is first necessary to decide what exactly, in this context, is to be meant by the term, 'generation'. If we are to study the manner in which scientific disciplines or professions develop at all illuminatingly, just how far apart in time should our successive cross-sections be taken? Just how fine-grained must our analysis be, if it is to reveal the significant aspects of historical change in a science? For most purposes, a separation of between (say) six months and ten years will presumably give us the kind of picture we require, though there is evidently nothing magical about the exact figure. And the first thing to get clear about is why this time-unit has the right order of magnitude: why, in the case of a science, a mean period of some five years is a realistic interval for measuring significant historical changes, whether professional or disciplinary.

The content of a science is (we said) a 'transmit'. It comprises a repertory of currently established explanatory procedures, together with a pool of more tentative conceptual variants; and its development is governed by a general consensus about the selection-criteria for judging variants from that pool, and about the explanatory ideals in the light of which this is done. From the disciplinary standpoint, all that matters is that this transmit should in fact be transmitted. From the professional standpoint, by contrast, we need to consider also, in human terms, the processes by which that transmission takes place. How, then, does the historical succession of representative scientists embody the historical succession of their explanatory procedures? And what changes in the professional sphere are the normal human counterparts to the conceptual and strategic changes in a developing scientific discipline?

About these questions there are some long-standing myths. For instance, it is often supposed that all truly Great Scientists have the special capacity, and courage, to change their minds; and that the progress of science has been largely owing to this intellectual courage. But this view (as we shall see) puts too much emphasis on the individual, not enough on his professional colleagues. Here it is essential to recall, rather, that the new 'theoretical physics' of every ten or twenty years comprises not merely a fresh set of disciplinary concepts and procedures, but also a group of scientists within which the balance of professional authority has shifted. After twenty years, the reference-group of authoritative spokesmen and judges for theoretical physics will have changed; so if we raise the question 'What has theoretical physics to tell us?' again after twenty years, we shall have to consult different physicists.

Looking at scientific change from the professional standpoint, accordingly, we have an immediate clue to the natural 'time-unit' of change. For one central aspect of this professional evolution is the process by which the ruling reference-groups of the different sciences change. To the extent that some group of men can be identified, whose judgement carries dominant weight with professional colleagues in the science concerned, the approval of these men does more than anything else to ensure the success or failure, not only of new societies, journals, and meetings, but also of new ideas. In the professional development of a science, therefore, the significant time-interval or, 'generation', is the time it takes for one reference-group to be displaced by a successor-group.

How is this interval actually determined? We must first sharpen this question up a little. How exactly we set about measuring our time-unit will depend, in part, on the purposes of our analysis, and so on the context in which it is defined. For different scientific activities operate on different time-scales: those which involve internal, intellectual developments, for example, differ in significant respects from those which involve also interactions with the larger culture or society. When concentrating on the formal, public institutions of science, accordingly, we may have to pick on one natural unit, but on quite another when considering the developing internal character of the science itself. The sequence of Presidents of the U.S. National Academy of Sciences, for example, can tell us as much —or as little—about changes of intellectual attitude and emphasis within the natural sciences, as one could learn about religious attitudes in England from the sequence of Archbishops of Canterbury. Inevitably, presidents of national academies, like primates of Churches, are men who deal on behalf of the entire enterprise with the heads of other major social institutions in the society; and their role is not merely a disciplinary one. As a result, the dynastic sequence of presidents is only the roughest of clues to the changing character of the intellectual reference-groups within science itself.[1]

To come a little closer to the internal dynamics of a science: each scientific profession (as we remarked earlier) has an informal, unwritten 'order of merit', alongside the formal career structures of professional positions. This 'order of merit' is itself a significant historical variable. For while there will normally be general agreement in a given profession about a few supremely authoritative names, there will also be some systematic differences of opinion, and these will tend to reflect 'generational' differences. A Rutherford or a Bohr, for instance, can retain the respect of his colleagues throughout his career; but more often an older scientist will tend to lose authority with his younger colleagues, and will continue to be 'taken seriously' only by his own contemporaries. Who is accepted as speaking 'in the name of' any science thus depends, in part, on the seniority of the scientists whose opinions you obtain.

So let us look again at the transmission process by which responsibility for the intellectual content and development of a science passes from one generation of scientists to another. During their intellectual apprenticeship, younger men gain a grasp of the explanatory procedures, disciplinary ambitions, and overall picture of nature accepted by their immediate predecessors, and use these as one source of material in developing their own ideas about the subject. But the replication involved in the transmission of ideas is never absolutely exact. Within a scientific discipline, or any other genuinely 'rational' enterprise, the teaching process is necessarily imperfect. Or, rather, the quality of the teaching process in such cases is measured not by the exactitude with which specific concepts are handed on, but by the critical attitudes with which students learn to judge even the concepts expounded by their own

---

[1] It has been remarked, with some justice, that the political slogans of the 'England-returned' men, who took over power in the Anglophone countries of Africa and Asia during the 1950s and 1960s, bore less relation to their actual policies—which were decided, very largely, on pragmatic grounds—than they did to the views circulating among Harold Laski's research students during the particular years when they were themselves studying at the London School of Economics.

masters. From the point of view of the apprentices themselves, it is as though the intellectual content and programme of their science had to be re-created afresh in every generation. Certainly, the views of the senior professors at whose feet they sit, or at whose benches they work, carry substantial weight with them; but their respect, even for the most eminent of their elders, will be highly selective. In due course, they will re-appraise all questionable issues for themselves, criticizing their teachers' views in the light of the rest of their reading, of the reports they hear from their fellow-students elsewhere, and of the general intellectual climate of their time . . . So each generation of apprentices pieces together the established—and variant—concepts and procedures of its discipline into a pattern of its own.

We can now reanalyse, in more concrete terms, the reasons why the 'natural time-interval' of scientific change has the magnitude it does. Within the graduate schools and research laboratories in which the intellectual content of the natural sciences is transmitted, one soon learns to recognize those shifts of approach, and even of strategy, that mark off successive generations of bright younger research scientists. One particular technique of research or style of theoretical interpretation may, very occasionally, hold its own among the most promising and talented novices for as long as seven years; at other times, there will be striking changes of approach or interpretation as little as two years apart. Either way, most significant changes will be allowed for, provided that we sample successive 'generations' of research workers at intervals of three to five years. Correspondingly, we shall find significant variations in the reference-groups accepted as authoritative among professional scientists, if we choose our respondents in groups aged some five or more years apart. Is molecular biology a wild and exaggerated new fad; is it a fruitful extension of classical biochemistry; is it the beginning and end of all biological wisdom; is it a first step towards a larger theory of cellular function and morphogenesis; or is it already old hat? Which answer you get to that question will depend, very largely, on whether you talk to scientists who are themselves seventy, sixty, fifty, forty or thirty years old.

From the outside, it sometimes appears as though the transfer of intellectual authority from an older to a younger scientific group was made necessary solely by increases in the sheer amount of factual knowledge in a science. Our present analysis leads to a different conclusion. If after a time working scientists often find it hard to keep up with current developments in their subject, that is because succeeding generations are liable to approach the problems of their common discipline from a quite novel point of view, putting the old intellectual pieces together in a quite novel pattern and redistributing their emphasis in the light of fresh explanatory ideals; with the result that their interpretations may prove uncongenial, or even abhorrent, to scientists of an earlier generation. At their blandest, such changes may take place without rancour. Ernest Rutherford, for instance, came so close to creating atomic physics single-handed that his successors (e.g. Bohr and Heisenberg) retained their intellectual respect for his views, despite his confessed inability to grasp the mathematical methods of quantum mechanics. For his own part, Rutherford bowed out from his authoritative position with good grace—admitting candidly that his own training left him unequipped to master the new abstractions, and hopelessly prejudiced in favour of a material model of atoms and fundamental particles, as 'little hard billiard-balls—preferably red or black'.[1] At the other extreme, these intellectual takeovers involve a kind of hostility and spleen, on the side of the victors and of the defeated, that on-lookers can only find distasteful and offensive. In science as in politics, the authors of a successful *coup* can afford to be magnanimous, but too often fall victims to the corruptions of power: falling for the temptation to dismiss their predecessors, however distinguished, as obtuse and unteachable old fogies.

For instance, James D. Watson's superficially light-hearted account of the decipherment of the DNA molecule, in *The Double Helix*, often comes close to crowing over older biologists and crystallographers, who had at first failed to see just where he and Crick were going.[2] But, now that Watson is secure in the knowledge of what molecular biology has become, this

---

[1] See A. S. Eve, *Rutherford* (Cambridge, Eng., 1939), p. 384, reporting an after-dinner speech made by Ernest Rutherford in 1934.

[2] Cf. J. D. Watson's passing remarks about Lawrence Bragg and other senior scientists in *The Double Helix* (New York, 1967; London, 1968).

triumphant manner comes easily to him, and it is salutary to compare it with the cautious tone of a much earlier letter to Max Delbrück reprinted at the end of his book, written at a time when the fate of his ideas still depended upon the good opinion of his seniors. Still, history brings its own revenges. Those who compete in the Sacred Groves of science should do so knowing that the Golden Bough will be in their keeping only for a limited time. Here as elsewhere, every transfer of authority helps to prepare the ground for the next. Having captured the key professional positions for themselves, the radicals of one generation are outflanked before long by the *fronde* of still younger men, for whom their radical novelty is already half-way to being stale news. In the quasi-political power plays which professional scientists quite rightly undertake 'in the name of' their respective disciplines, institutional victories can never be more than temporary. Each new generation of apprentices, while developing its own intellectual perspectives, is also sharpening up the weapons for an eventual professional takeover. Five, ten, or twenty years hence, their word will carry weight in the profession, their authority will guide and reshape the discipline; and meanwhile, at their heels, still other younger men are coming along, who will in due course form the generation of their own successors . . .

What light does this process of professional apprenticeship and generational competition throw on the institutional conditions required for the effective intellectual development of a science? We may touch on two aspects of this general question: (a) the role of the individual scientist—particularly, the relationship between the personal ideas and activities of scientists taken singly, and those which have an established place in the collective transmit of the science—and (b) the broader sociocultural context within which a science can develop most rapidly and effectively.

As to (a), here there are two separate questions to be considered: firstly, under what circumstances the collective content of a science is transformed and, secondly, how far this transformation involves changes in the ideas of individual scientists. If we take care to distinguish these two questions, the result is to cast grave doubt on the familiar Great Man view of science. For it is one thing to enquire whether Great Scientists ever really Change their Minds, and quite another to ask whether it is essential for science that they should. In intellectual as in political history, the view that all major human achievements depend on the supreme genius of a few Great Men has a certain romantic attraction. For historians of this temperament, the salvation of nations is due, above all, to the ambition of a Napoleon, the flamboyance of a Garibaldi, the steadfastness of a Churchill; while intellectual success in the sciences is to be credited, likewise, to the obstinate self-critical honesty of some individual scientific genius—whether Galileo or Lavoisier or Darwin. Such men were prepared (we are told) to approach Nature with eyes unclouded by inherited prejudices, had the integrity to abandon their received opinions in the light of contradictory first-hand observations, and spoke out the truth as they saw it with the unmuffled candour of a young George Washington. And these Great Scientists have been able to advance men's collective ideas another irreversible stage only because, when others were still fudging along with old ideas, they had the courage to think afresh, look again, and speak honestly.

This romantic saga of the great scientific innovator unfortunately leaves some crucial parts of the story untold. For it implies that there is the same sharp contrast in science that we find in the Great Man theory of political history: between the active, dominant Great Man and his passive, imitative followers. Just as Napoleon 're-shaped' and 'gave new pride to' the French Nation, Newton 'dominated' eighteenth-century physics and Lavoisier 'created' modern chemistry. For reasons left unexplained, other scientists working in the same tradition were content simply to follow his lead, and sign on his dotted line. And such an account clearly oversimplifies to the point of travesty the actual relations between the work of any individual scientist—however 'great'—and that of his colleagues and successors. It would, in fact, be much nearer the truth if we asserted the contrary exaggeration: viz., that Great Scientists almost never change their minds, and certainly never need to do so. For it is of little consequence to the collective human enterprise of a science whether or no any single individual involved in it—however eminent—alters his personal ideas in the light of experience. Such changes of mind may, occasionally, help in modifying the collective tradition, especially where some influential scientist's hesitations have been discouraging others from taking a radical innovation seriously enough. Thus Charles Lyell's belated acceptance of organic evolution was greeted by Darwinists with undisguised relief, since his continued opposition had prejudiced evolutionary ideas in Britain, just as the shade of Cuvier had been doing in France.[1] Yet it was not Lyell's change of mind that either originated, or established, the intellectual claims of the Darwinian theory. The essential loci of conceptual change remained—and still remain—not the opinions of individuals, but the collectively attested repertories of concepts that form the intellectual transmit of disciplines. Taken on the personal or individual level alone, the question who exactly accepted which concept, or when, or for what reason, is relevant only incidentally to collective questions about the established content of the science concerned.

It is by now a historical commonplace that, even in his early twenties, Isaac Newton had already conceived in outline possible solutions to the outstanding problems of dynamics and planetary theory, about which he wrote a full-scale, detailed treatise only in his forties, in the *Principia* (1687).[2] Yet there is something to be said about this fact in the light of our analysis here. If we compare the general ideas about the physical world with which Newton was working in his early years with the full-scale corpuscular 'philosophy of matter' which he finally put into print after the death of his great adversary, Leibniz, as an appendix to later editions of his *Opticks* (from 1717 on), it is clear that even his most mature theories were simply elaborating and reapplying, with fresh ramifications, the same central notions that he had already accepted in his own mind some fifty years before.[3] True, in the middle of Newton's career,

---

[1] Lyell's support for evolution came at last—and somewhat grudgingly—in 1864, with the publication of *The Antiquity of Man*, in which he distressed Darwin's associates by giving most of the credit for evolution-theory to Lamarck.

[2] On the motives for Newton's delay in publishing his theory of motion and gravitation, see, e.g., F. E. Manuel, *A Portrait of Isaac Newton* (Cambridge, Mass., 1968), Chapter 7.

[3] Newton's famous Query 31, in which he finally brought the fundamental tenets of his 'natural philosophy' into the open, was added only to the new edition

we find him soft-pedalling his commitment to atomism, and concentrating in public on purely mathematical procedures; but there are reasons for thinking that he did so out of diplomacy, rather than through any loss of basic conviction. (Atomism had a bad name, on account of its association with Epicureanism, and Newton did not willingly land himself in bitter public controversies that could otherwise be avoided.)[1] For the time being, he laid low and withheld his comprehensive world-system from publication, until his most dangerous opponent was safely out of the way. So, if Newton succeeded in transforming the scientific ideas of mankind, what mattered was not that, in the course of his career, his own personal ideas changed in any fundamental respects; in fact, they did not. Rather, the changes produced by his work in the collective minds of his contemporaries and successors created a new starting-point from which all later theoretical construction could begin.

Max Planck can serve as an even more striking example.[2] The story runs, in outline, as follows. When a young man, Planck was trained in electromagnetism, and became caught up in the debate about the spectrum of radiation from heated bodies. In December 1899 he introduced into this debate his novel idea of a 'quantum' of action, according to which a material body was assumed to emit radiation only (so to say) in 'gulps' of energy, whose size was directly proportional to its frequency; at one stroke, this idea resolved all the major difficulties of the existing theory. Five years later, Albert Einstein added a rider to Planck's theory, based on his reinterpretation of the so-called 'photo-electric effect', in which light of a high enough frequency sets an electric current flowing in a metallic wire on which it falls. This phenomenon (he claimed) could be explained as an effect of the impact on the metal of particle-

---

of his *Opticks* published in the year 1717: cf. the latest edition of the *Opticks*, ed. I. B. Cohen and D. H. D. Roller (repr. New York, 1952), pp. 375–406.

[1] The evidence that Newton was afraid of being accused of supporting Epicureanism, which was in bad odour in the late seventeenth century, has been analysed by Henry Guerlac. See, e.g., his lecture, 'Newton et Epicure', *Éditions du Palais de la Découverte*, D-91 (Paris, 1963).

[2] Certain aspects of this episode are discussed by Gerald Holton in his essays on Einstein (forthcoming): see also the correspondence in *American Scientist*, 56 (1968), 123Aff.

like 'photons' of light, each of which carried energy equal to the corresponding Planck quantum. To younger physicists it seemed clear that this convergence between the results of the two men's work was no mere coincidence. The same 'photons' that released electric currents in metallic wires presumably served also to carry 'quanta' of energy away from the bodies by which they were originally radiated; in short, during the transmission of radiation across the space between material bodies, electromagnetic energy presumably existed always in the form of particle-like photons. This interpretation became a central element in the new quantum-theoretical picture of the physical world, which took definitive shape during the 1920s and 1930s.

It is a surprise to discover, in retrospect, that Max Planck began by rejecting Einstein's idea of 'photons' entirely. He had his reasons for doing so. Having grown up as a fervent supporter of Maxwell's electromagnetic theory, he had introduced 'quanta' as an ingenious device for rescuing Maxwell's analysis from the discredit which was threatening it, because of its failure to explain the familiar spectra of heated bodies; and one integral part of Maxwell's account was the conclusion that radiation travels in the form of continuous wave-fronts. Einstein's radical new photon hypothesis, on the other hand, was a step away from sound Maxwellian ideas, and betrayed the very purposes for which Planck had introduced his own quantum concepts in the first place. For years, he fought against Einstein's interpretation, hoping to find some way of restoring the Maxwellian picture of continuous wave-fronts; only much later did he regretfully admit that matters had gone too far along the Einsteinian road for any such rescue operation to be a serious possibility. In this sense, Max Planck did eventually 'change his mind' about photons. Yet, by doing so, he did nothing at all for the progress of physics; and, by the time he reached this point, it was too late for him to make any major contributions in the new direction. This 'mind-changing' was a purely personal phenomenon with no collective consequences. In this respect, he was not a leader, but a follower; and he ended by merely bringing his own views into line with the collective position of his successors.

In the continual process of renewal and self-transformation

by which the content of a science develops, what is important, therefore, is not any individual man's readiness to change his mind. Rather, the significant thing is his ability to keep his mind open about questions which have not yet been adequately answered, and to find ways of adapting the general framework of ideas to accommodate new results and discoveries. A lifelike picture of the relations between the individual scientist and his profession is accordingly quite at variance with the romantic Great Man view. Very often, a scientist spends his whole career elaborating on conceptual foundations laid down in his youth; if he is compelled, by any chance, to dismantle and lay afresh any substantial part of those foundations, his final construction commonly lacks the coherence and stability it would otherwise have had. And this is as true of Galileo as it is of anyone else. Maybe Galileo did 'change his mind' about the exact analysis of 'acceleration'. But it would be more accurate to say that, in A.D. 1605, he was still very confused about the difference between defining this quantity in terms of time-intervals (δt), and defining it rather in terms of increments of distance (δs);[1] it was this distinction which he had come to see clearly by 1630, when he finally arrived at his classic statement of kinematic theory.

Whether in the seventeenth century or the twentieth, physics has evolved as the product of a *collective* debate. In the course of this debate, the contributions of individuals may have been striking and original, but any individual can play an effective part in the development of a discipline only by submitting his ideas to the collective judgement of the current reference-group. If anything, the collective professional concerns of a science exert a more powerful influence on those of individual scientists than vice versa. The intellectual ambitions and conceptual difficulties of a discipline are the focus around which professional scientists organize their joint intellectual efforts; and these ambitions filter out from their individual ideas and interests those conceptual issues that are relevant to the collective development of the science. How necessary this filtering

---

[1] There is, for instance, a letter from Galileo to Sarpi, written in 1604, in which he still assumes that 'uniform acceleration' involves equal increases in speed per equal *distance* travelled.

action is, we shall discover if we investigate all the personal
reasons which lead scientists to choose particular areas of
specialization. These personal preoccupations are often related
only very remotely to the actual disciplinary business of the
science, and they achieve an effect they would otherwise lack,
as a result of being channelled into a collective enterprise.

To illustrate this point, we may leave aside such marginal
individuals as Teilhard de Chardin, whose work on human
palaeontology was so clearly a by-product of a more central
interest in the theological nature and destiny of Man.[1] Rather,
it is interesting to look at a case like that of Jacques Loeb, who
played a major part in the development of physiology during
the early twentieth century. Loeb explained quite candidly
that his personal motive for taking up physiology had been a
quite general concern with the traditional idea of 'free will'. He
had, in fact, begun by studying philosophy, and turned to
physiology because he decided that the problem could be
better resolved by looking to see whether, within organisms,
the laws of physics and chemistry held with the same universality as they did outside.[2] And Loeb was exceptional only in
making his external preoccupations so clearly apparent. In
fact, many of the most distinguished scientists probably sustain
themselves through the single-minded labours of their professional careers by similar hopes and expectations.

These personal preoccupations do not normally become
evident, however, until the men in question retire from active
research, and turn from writing research papers to composing
essays on the wider 'implications' of science. Then, at last, we
find them publicly taking up with theology or ethics, politics
or metaphysics, and invoking the specialized results of their
particular sciences as the key to larger-scale intellectual or
practical problems.[3] As we read on, we suddenly realize that

[1] See, e.g., Stephen Toulmin, 'On Teilhard de Chardin', *Commentary*, 39
(March 1965), 50–5.

[2] The immediate philosophical influence on Loeb was that of Schopenhauer:
see the introduction by Donald Fleming to the latest edition of Loeb's book,
The Mechanistic Conception of Life, originally published in 1912 (Cambridge, Mass.,
1964), pp. xii–xv.

[3] These cross-connections cover the board, from G. H. Hardy's Platonist philosophy of mathematics, by way of the 'evolutionary ethics' of C. H. Waddington
and Julian Huxley, to the near-mysticism of E. Schrödinger's book, *My View of the
World* (Cambridge, Eng., 1964).

their professional commitment to (say) biochemistry has been nourished, all these years, by a silent conviction that the science will enable us—after all—to prove that God does (or does not) exist, and that we do (or do not) have free will; or by the hope of finding some fundamental sanction for ethics in the world of Nature itself; or even by a private obsession with problems of sexuality . . . Meanwhile, their professional work has to be shaped by the collective demands of the discipline, and by the literary conventions of scientific publication—passive voice and all. For it is these that make it possible—as we now see—for theists and atheists, believers and disbelievers in free will, puritans and libertarians alike to collaborate scientifically on a well-defined group of shared problems.

(This divergence between individual and collective preoccupations helps, incidentally, to explain the attitude of scorn frequently shown by scientists towards 'mere philosophy'.[1] For such scientists speak of philosophy, not as one more legitimate field of intellectual activity capable of its own styles of rigour and self-criticism, but rather as a field for unbuttoned intellectual self-indulgence, of a sort that is automatically exempt from the critical discipline and standards of natural science. The trouble is, of course, that within the actual careers of most working scientists—as opposed to professional philosophers—this is all that a concern with 'philosophy' ever really amounts to!)

As to (b): the effective development of a science, both professional and disciplinary, depends also on certain other broader socio-cultural considerations.

(i) For instance, intellectual adaptation, as much as evolutionary adaptation of other kinds, can take place effectively only in somewhat special circumstances. A protected 'forum of competition' is needed, if 'better adapted' variants are to demonstrate their merits without being swamped in the larger population from which they were originally derived. The 'ecological barriers' tending to isolate this forum must, at the same time,

[1] Max Born's text book on *Atomic Physics* (Eng. trans., London, 1935), for instance, closed with a sweeping dismissal of the 'irridescent fancies' of metaphysics; Born evidently regarded these as a delusory attempt to guess at truths which science cannot yet disclose by its own more solid experimental methods.

be low enough to permit the specialized 'niche' to be colonized in the first place, yet high enough to prevent the resulting variants from fading back into the former population. In the most successful scientific cultures, therefore, the effective development of more adequate scientific concepts has gone along both with the creation of largely autonomous disciplines, and also with the emergence of institutionally distinct professions. This is no accident. Only in a situation with which the specialized 'intellectual demands' of a science are clearly recognized and agreed can the adequacy of conceptual innovations be assessed with any hope of agreement. The professional 'forums' of science thus play a significant part in creating the local 'niches', surrounded by institutional barriers, within which conceptual variants can be publicly and critically tested against the theoretical requirements of the discipline concerned.

Failing this professional isolation, and the critical control exercised by professional reference-groups, it will be much harder for novel ideas to stake out precise claims, and so establish their place in a well-founded body of knowledge. Instead, they will be lost in a welter of speculative debates and polemical objections, in which their characteristic virtues and implications can no longer be identified and explored. And, once the essential balance on which conceptual evolution depends is destroyed, there will be no stable equilibrium short of intellectual conformism or conceptual anarchy. Either—as at some times in Imperial China—an unspecialized opinion-forming group will end by exercising thought-control over general ideas and vetoing intellectual novelties simply for their novelty.[1] Or else there will be no effective control over intellectual innovation at all, and the speculative debate will proliferate in a way that provides no scope for critical judgement. In the former case, new conjectures will not survive for long enough to show what their real potentialities are; in the latter, intellectual life will fall into an incoherent toleration for all conjectures, without any authoritative judges responsible for deciding that some are more meritorious than others. (I recall hearing a well-educated Moroccan justify some utterly fanciful speculation by

---

[1] Cf. J. S. Needham, *Science and Civilization in China*, Vol. 3 (Cambridge, Eng., 1959), see especially pp. 192–4 on the secrecy about astronomy in Chinese society.

quoting the maxim, 'Where nothing can be proved, every man is entitled to his own view.') Specialized scientific professions are, therefore, the institutional price we pay, in order to keep the twin activities of speculation and criticism—which are the human embodiments of conceptual variation and selection—working together in harmony. Just because the establishment of inter-disciplinary boundaries and the delegation of authority to distinct reference-groups results in the isolation of specialized professional niches, it is possible for conjectures to be put forward, tested, and judged in a selective, discriminating way, with an eye to the well-defined requirements of a correspondingly specialized problem-situation.

(ii) Conversely, the ecological success of novel forms requires that the 'forum of competition' should not be too isolated. If the professional isolation of scientists were ever to become total, that itself would hamper intellectual evolution, in two respects. To begin with, the specialized problems and concepts of the sciences sprang originally from the broader intellectual concerns of people at large, and they are continually capable of reacting back on them. The task of deciphering the biophysical processes by which norepinephrine (say) can cross nerve-membranes originally specialized out of—and can always reflect back onto—the more general problem, how the central nervous system can function as the organ of thought and experience.[1] So, on the boundary between scientific disciplines and extra-scientific concerns, questions are always arising about the broader relevance of specialized disciplinary novelties; and the full potentialities of new ideas can be kept clearly in view, only if this interaction between specialized and extra-scientific concepts is allowed to continue.

If the barriers around the scientific professions are too high, however, novel ideas which have proved their specialized scientific merits cannot spread outwards into the general culture, or win acceptance in the 'public mind'; nor—in reverse —may the specialized disciplines of science attract enough outstanding recruits from each new generation to maintain their own professional vitality. A science which cuts itself off entirely from the broader intellectual debate will thus retain

[1] On this subject, see my essay, 'Neuroscience and Human Understanding', in *The Neurosciences*, ed. G. C. Quarton *et al.* (New York, 1967).

only a localized significance; its professional technicalities will
have no power to influence 'common sense' or 'common knowledge', and the science itself will be in danger either of expiring,
or falling into the hands of second-rate men, for lack of good
new recruits to cultivate it. Between A.D. 1290 and 1340, for
instance, scholars at Paris and Oxford were forging an important link in the conceptual genealogy connecting the ideas of
Aristotle and Archimedes to those of Galileo, with their scholastic analysis of motion.[1] At certain points, indeed, the fourteenth-
century mathematicians came tantalizingly close to Galileo's
starting-point: so much so, that we may wonder why—even
after allowing for the effects of the Black Death—theoretical
mechanics did not develop faster during the next 250 years. If
physics developed so slowly during the period after 1350, this
may well have been because the analytical methods in question
were the possession of a narrowly-defined group of professional
scholars alone, so that nobody else was in a position to appreciate their potential wider significance. Certainly, the scientific
implications of the fourteenth-century theory of impetus, with
its graphical analysis of quantitative change, were fully worked
out only from the mid-sixteenth century on, after the Renaissance had led to a general secularization of intellectual life.
Meanwhile, the contribution of these fourteenth-century
scholars had been largely forgotten, and remained hidden until
twentieth-century historians of science demonstrated the
significance of their work.

The story of Babylonian astronomy is even more extreme.
From 750 B.C. on, the 'prognosticators' of Babylon formed an
entirely isolated and specialized professional guild; indeed, their
professional methods and ideas were official secrets, which they
were forbidden to reveal to outsiders. Though some of their
results eventually reached the Greek world, after the capture of
Babylon by Alexander the Great, the full sophistication of their
methods was unsuspected until modern archaeologists re-
covered their records, and the detective-work of twentieth-
century scholars deciphered them.[2] Without this rediscovery,

---

[1] This phase in the development of ideas about motion has been fully documented by M. Clagett in his book, *The Science of Mechanics in the Middle Ages*
(Madison, 1961).

[2] The relevant material has been discussed at length in O. Neugebauer's classic

Babylonian astronomy would have been destroyed as totally and finally as the institutions of the Babylonian State. Here, a locally successful discipline suffered the fate reserved for isolated but over-specialized 'populations' whose original niche has disappeared. And, for all that we know, the lost cultures of Mexico or Cambodia may have had comparable intellectual achievements, which were equally the property of isolated professional guilds, but which were later forgotten as completely as the techniques for producing the stained glass of Chartres Cathedral.

In such cases as these, the 'resonance' between the specialized discipline and the larger public mind is so weak that innovations within the professional guild—however striking in their local effect—woke no echoes in the broader ideas of the culture. By contrast, the fact that science has developed with such vigour and fertility in Western Europe since A.D. 1600 is a consequence, not least, of an active resonance between scientific specialists and the general public, and of the interaction of ideas between the newly emerging special sciences and the wider culture of the time. On the one hand, this resonance has helped to confer dominance on the new scientific ways of thought, by incorporating their outlines into the general world-picture of 'common sense'; on the other hand, it has helped to fuel the development of the sciences themselves, by focusing attention on conceptual problems of wider importance, so maintaining a flow both of intellectual innovations and of talented apprentices into the sciences concerned.

(iii) Finally, the broader social conditions for maintaining the unity and continuity of a scientific discipline are relevant also to the profession that embodies it. A term like 'scholastic' can be applied, not only to a body of concepts and doctrines which is transmitted uncritically on the basis of mere authority, but also—in a more literal sense—to the tradition-bound institutions of the 'schools' in which this authoritarian teaching takes place; while the term 'anarchic' applies equally to the process of intellectual variation, when uncontrolled by any effective selection procedure, and also to an institutional situation which permits such a proliferation of uncriticized

volumes on *Babylonian Mathematical Ephemerides*, and summarized conveniently in his more popular account, *The Exact Sciences in Antiquity* (Princeton, 1952).

ideas. In each case, the consequences are the same for both discipline and profession; stasis in the former case, loss of definition in the latter.

Corresponding parallels can be found, for instance, where one earlier science gives rise, by fission and specialization, to two or more successor-sciences; or, alternatively, hybridizes with another science. There, disciplinary changes in the patterns of problems and methods are once again associated with professional changes in (e.g.) the institutions and journals of the science. Ecologically, however, it is clear that this relationship between disciplinary and professional changes, however normal and even desirable, is by no means necessary. So, the actual conduct of scientific professions creates still further occasions for self-criticism. At one time, the existing media of publication can be criticized, as failing to meet the actual disciplinary needs of a science; and, at another, the institutional question can now be raised, how far the current professional sub-division of the sciences accurately reflects the real pattern of intellectual problems at the time concerned.

There may (e.g.) be excellent intellectual reasons for making some group of interdisciplinary problems the concern of an independent new sub-discipline; yet influential members of the existing professions may see this as an implied threat to their own interests, and block the establishment of the professional organizations which would give institutional effect to the new sub-discipline. In that case, it can be argued that the institutional arrangements of the existing professions are not 'adequate to' the requirements of the new sub-discipline, so that anyone who refuses institutional autonomy to the new professional organizations will be hampering the intellectual development of the science also. Conversely, there have been times when the number of scientific periodicals was deliberately multiplied for commercial reasons, which had no real relevance to the intellectual needs of the disciplines concerned, but simply conferred the trappings of independence on artificial sub-divisions of larger but perfectly coherent disciplines. Here again, the very needs of the science itself provide the basis for evaluation and self-criticism. The question, 'Is your journal really necessary?', then means, 'Does your journal fulfil an authentic disciplinary need?'; and this question must be answered in the light of the strategic and conceptual problems characteristic of the actual scientific situation.

In this way, the proliferation of periodicals can run ahead of genuine disciplinary needs, just as the conservatism of professional organizations can hold back proper disciplinary development. Either way, the parallels between intellectual and institutional change, while close, are not inevitable. And, either way, the sanctions by which the convergence of disciplinary and professional considerations—of ideas and institutions—is ultimately determined are not those of logical necessity, but those of ecological need.

4.3: *Intellectual Ecology and Historical Understanding*

The phrase 'historical understanding' requires careful elucidation. At this point, we are up against one of the central issues in our entire present analysis: namely, the general nature of the relationship between the disciplinary (or intellectual) aspect of a scientific enterprise and its professional (or human) aspect. We can best clarify that issue in three steps, by breaking it down into a number of more specific topics and dealing with each of these in turn. So let us, first, take up two unresolved problems of methodology which still afflict our historical understanding of scientific change. One of these is the long-standing divergence between two styles of historical approach to science—the 'internalist', which concentrates on the changing content of a scientific discipline, and the 'externalist', which focuses rather on the relations of the science to its broader social context. The other is a more strictly philosophical problem—that of sorting out the respective roles of 'reasons' and 'causes' in the historical development of a science. Having reconsidered these two problems in the light of the argument we have gone through up to this point, we shall be in a position to explicate the crucial notion of 'intellectual ecology': i.e. to show how, by comparing (i) the intellectual demands of the problem-situations which are the occasions for conceptual change with (ii) the ecological demands of the niches which are the loci of adaptation in the organic sphere, we can throw light in turn on the whole process of conceptual development in a collective rational enterprise.

To begin with the historiographical problem; over the last

half-century, academic historians have been analysing the development of the natural sciences from several directions independently—statistical and biographical, intellectual and sociological. Up to now, however, they have been unable to agree at what point the results of these separate analyses converge, or how they can be fitted together to provide a single coherent picture of scientific change.[1] On the one hand, internalist historians study the intellectual changes by which scientific ideas displace one another from a strictly genetic or 'genealogical' point of view. For them, scientific development is to be seen as a dialectical sequence: Problems lead to solutions, which in turn lead to new problems, whose solution pose further problems . . . Within such a 'problematic' approach to the development of science, two further points of view can be distinguished, one socio-historical, the other logico-philosophical. Some scholars study the problematic changes within science directly, with the aim of showing by what successive steps they have actually taken place.[2] Others aim rather at a logical analysis (or 'rational reconstruction') of the arguments by which scientific progress has properly been made and justified.[3] Either way, this genetic approach to science is subject to a certain self-limitation. Internalist accounts of scientific development refer to factors outside the immediate concerns of the discipline in question only marginally, if at all. As a result, they encourage us (so to say) to study the morphogenesis of any science in isolation from its ecological environment.

Meanwhile, other historians have been considering scientific change as a social phenomenon, and have concentrated their attention on the interactions between natural science and the larger context within which the scientist operates—its institution, social structures, politics, and economics. The results of these investigations have been interesting and in some ways

[1] See, e.g., J. Agassi, 'Towards an Historiography of Science', *History and Theory*, Beiheft 2 (1963). A useful collection of historical texts illustrating this contrast is included in the book, *The Rise of Modern Science*, ed. G. Basalla (Lexington, Mass., 1968).

[2] This approach is illustrated, in more or less extreme forms, by the works of M. Clagett, A. R. Hall, H. Butterfield, and others.

[3] For a discussion of this approach, see the useful paper by I. Lakatos, 'History and its Rational Reconstructions', due to appear in *Boston Studies in Philosophy of Science*, ed. R. S. Cohen and R. Buck, Vol. 8 (1971).

unexpected. As a result of their work, we understand better how the scientific professions are organized into institutional 'guilds', and how the character of scientific work during the last 100 years has been affected by novel economic, political, and social factors.[1] Yet the methods of enquiry required in studying the sociology and social administration of science have once again restricted the range of questions we can profitably ask. Specifically: they have left it unclear whether, and in what respects, there is any feedback from the social context of the scientist's work to the intellectual content of scientific thought itself. It may seem that the content of nineteenth-century thermodynamics must have been influenced by contemporary developments in the technology of steam locomotives; yet just what form did this influence take? And what are we to make, in turn, of (say) Bernal's more extreme claim, that the content of Darwin's theory of natural selection somehow reflected contemporary beliefs about *laissez-faire* economics?[2]

The problem is to state, in an acceptable form, how it is that these complementary approaches to the history of science bear on one another. For this purpose, we cannot stay securely inside either one of the approaches; instead, we need to fit their results into a broader account, embracing the development of scientific concepts and activities alike. So let us use one classic example to show how our own distinction between the 'disciplinary' and 'professional' aspects of scientific change permits us to reconcile the two styles of historical approach.

Recent historians of science have explained in quite different kinds of terms how Newton's *Principia* ended up with the intellectual content it did. (1) The late Alexandre Koyré embarked on his Newtonian research with limited, and largely genealogical aims. As a result, he was able to demonstrate how much more profoundly Newton's methods were influenced by the ideas of Descartes than English scholars had supposed; made out a strong case for examining more closely Newton's relations with the Cambridge Platonists; and, by putting Newton's motto *hypotheses non fingo* back into its original context, demolished all positivistic interpretations of his procedures . . . The one

---

[1] See, e.g., Don K. Price, *Government and Science* (New York, 1954), and *The Scientific Estate* (Cambridge, Mass., 1965).

[2] J. D. Bernal, *Science in History*, 3rd edn. (London, 1965), p. 483.

thing that Koyré's account of Newton's ideas scarcely mentioned was seventeenth-century technology and industrial economics.[1] (2) By contrast, the late Boris Hessen's essay on *The Social and Economic Roots of Newton's Principia* mentions scarcely anything else. All the basic themes of the *Principia* (Hessen argues) had roots in contemporary technological demands: artillery or canal-building, transoceanic navigation or iron-mining. In England, the seventeenth century was 'the period in which merchant capital was becoming the predominant economic force and manufactures began to develop'. The problems of statics and mechanics were of practical significance, as 'relating to raising equipment and transmission equipment important to the mining industry and building art'; the mathematical theory of trajectories and free fall constituted the 'basic physical task of ballistics'; while the principles of hydro-statics were 'of fundamental importance to the problems of pumping water from mines and of their ventilation, the smelting of ores, the building of canals and locks'. Even Newton's central topic, viz. planetary mechanics and the theory of the tides, was central to the practical techniques of navigation. At the same time, the institutional framework of Newton's research—at Cambridge University, in the Royal Society, and at the Mint—illustrates well the liberation of intellectual life from the Church resulting from the rise of the bourgeoisie. True, Newton chose to dress up his novel theory in abstract geometrical language; but the 'earthy core' of his work still lay close to the heart of seventeenth-century technological needs.[2]

The divergence between Koyré's 'internalist' genealogy and Hessen's 'externalist' analysis is striking enough by itself. As Hessen's latest American editor puts it, 'The exact relationship between science and the socio-economic structure has not yet been revealed'.[3] For good measure, however, we now have a third account of Newton's work, too. More recently (3) Frank Manuel has painted a psychoanalytical portrait of Isaac Newton the man, and used this to throw light, not merely on

[1] Alexandre Koyré, *From the Closed World to the Infinite Universe* (Baltimore, 1957), and *Newtonian Studies* (Cambridge, Mass., 1965).

[2] Boris M. Hessen, 'The Social and Economic Roots of Newton's *Principia*', originally published in *Science at the Cross-Roads* (London, n.d.), pp. 151–76, and reprinted in Basalla op. cit., pp. 31–8.   [3] See Basalla, op. cit., p. 31.

Newton's deeper personal motives, but on the actual structure of his ideas. Newton's life—on Manuel's account—displays a pattern of fantasy, introversion, and sublimated homosexuality to be expected after the psychic disasters of his early years. (Newton's father died before he was born, and his mother abandoned him to foster-parents in infancy.) As a result, he imposed a self-contained authoritarian structure on his own intellectual system, and this was reflected in his sense of being a divinely appointed spokesman called on to reveal the overall character of the entire cosmic system. Like Koyré, Manuel spares no time for technology, apart from Newton's tyrannical attitude towards forgers and money-clippers, as Master of the Mint; nor does he discuss the social structure of capitalist society, except for the outlets it gave to Newton's quirks of temperament. So his *Portrait of Newton*, however plausible in itself, does even less than Hessen's socio-economic analysis to clarify the 'rational' development of natural science; on the contrary, it gives—as it is intended to do—a new twist to Collingwood's claim that fundamental changes in our presuppositions are the product of 'unconscious thought-processes'.[1]

How, then, is it possible for a single piece of work in theoretical dynamics and astronomy to lend itself to such varied interpretations? Far from being inconsistent or in conflict, these different strands can all be woven into a single consistent fabric. A purely 'internalist' history of scientific thought can, in fact, be written quite consistently, so long as we confine our attention to conceptual genealogies alone—with appropriate provisos, e.g., that we ignore all questions about differences in the rates of conceptual change in different epochs. Like the evolution of organic species, conceptual change displays striking changes of pace: in some epochs, major changes follow one another rapidly, at other times centuries may go by without significant developments. Despite the fourteen centuries separating them, the ideas and methods of Copernicus are far nearer to those of Ptolemy than they are to those of Newton a mere 150 years later. Yet an 'internalist' analysis of the astronomical argument leading up to Newton's work must ignore these changes of pace, and consider only the steps by which the argument itself was advanced.

[1] F. E. Manuel, *A Portrait of Isaac Newton* (Cambridge, Mass., 1968), pp. 117ff.

Conversely, a purely 'externalist' history of scientific change can be written quite consistently, if we concentrate our attention on the historical relations between scientists and their socio-economic situations—on appropriate conditions. We may, perhaps, ask why the physicists of a particular period illustrated their theoretical arguments with examples from (say) hydraulics or navigation; but this is quite different from asking what those arguments succeeded in proving. Similarly, we may ask why, at some stage, so many scientists were working on one particular topic; but this is quite different from asking what their research actually achieved. We may ask what external patronage was supporting research in this or that science; but this is quite different from asking what made certain problems ripe for solution . . . If we look again at Boris Hessen's paper, indeed, we find that he chose his words with great care. Nothing he said threw light on the explanatory achievements of Newton's *Principia*, but he did not claim to have done so. On the other hand, the things he said did throw some real light on the reasons why seventeenth-century England provided an effective 'forum' for Newton's research, as well as on his choices of particular examples to illustrate his general, abstract principles. Terrestrial and celestial mechanics, hydrostatics and tidal theory (as Hessen's arguments remind us) then had an interest and appeal going far beyond their intrinsic mathematico-intellectual content; while the reasons why the activities of the Royal Society won support from King Charles II's Admiralty were clearly political rather than intellectual.[1]

So it need not be hard to work out terms for a compromise between these different approaches to the historiography of science. A comprehensive account of scientific change must raise questions of many different kinds; some of these will respond—within limits—to internalist analysis, while others can be attacked—within limits—by externalist methods. Neither method is self-sufficient or exhaustive, but their results do all have a quite genuine—if restricted—validity, when seen as dealing with complementary aspects of the entire scientific enterprise. Nor does this compromise rest on an arbitrary subdivision, by which the different schools of historians carve up the available territory into 'fair

[1] Hessen, op. cit.

306 Rational Enterprises and Their Evolution

shares'. Rather, it has a deeper rationale in the complex relationships actually linking different aspects of that enterprise—intellectual, individual, and institutional. One virtue of our analysis, indeed, is the capacity it gives us to demonstrate on what conditions, and in what cases, the abstractions involved in each of the current historiographies are legitimate and valid.

We can here spell out two of these conditions in outline. In the first place, although both current approaches to the history of science involve initial abstractions, the accepted dichotomy between 'internalist' and 'externalist' accounts of scientific change is misleadingly sharp. In actual practice, the questions we can raise about conceptual change in science lie along a continuous spectrum, with those which turn predominantly on 'internal' (i.e. rational or dialectical) considerations at one extreme, and those which predominantly reflect 'external' (i.e. political or socio-economic) factors at the other. Questions about the selection-criteria for judging conceptual variants fall nearer to the 'internal' extreme, while questions about the occasions for scientific innovation fall nearer to the 'external' end. But there is no class of questions which lend themselves to analysis in one set of terms to the absolute exclusion of the other. Even the most purely intellectual argument can explain a conceptual change in science, only on the assumption that some social occasion existed with which this argument could have an effect. Likewise, even the most generous external patronage will produce intellectual results, only given a topic of research which is ripe for investigation. From our point of view, the question is no longer whether or no we can legitimately study (say) the conceptual genealogies or the external relations of a science in isolation from one another; rather, it is to what extent, and at what price, we can do so. And this is something about which no universal generalization is really possible—something which must be considered afresh, and in detail, for every culture, period, and discipline.

In the second place, we can put this compromise into effect cleanly and straightforwardly, only in cases where the scientific disciplines concerned are sufficiently well-defined, or 'compact'. In a mature, fully-fledged scientific discipline like atomic physics, there is little difficulty in distinguishing factors and considerations 'internal to' the discipline from those which are 'external to' its intellectual concerns. Elsewhere, the distinction between internal and external factors is not so easily drawn. Outside the 'compact' disciplines, all major arguments will involve disputes about intellectual policy and priorities, as well as arguments about concepts, tactics, and techniques; and all shifts in the 'rational boundaries' of argument will immediately change the direction and ground-rules of the whole enterprise. In such cases as these, it is no longer possible to isolate the 'internal' aspects of a discipline from its 'external' context, to anything like the same extent. Failing the clear, self-defined boundaries of a compact discipline like atomic physics, there is nothing for internal and external factors to be *internal or external to*.[1] The abstractions on which the division between the internalist and externalist historiographies depends are, accordingly, legitimate only in a restricted class of cases. Outside the 'compact' sciences, one can no longer disentangle the dialectical sequence of the disciplinary argument from considerations about the social context within which it is professionally pursued.

Let us now turn to the outstanding philosophical problem about scientific change: viz., the problem of 'reasons' and 'causes', which gave R. G. Collingwood such trouble.[2] As we have seen, conceptual changes always find their rational justification within a historical context, and the disciplinary standards of justification are themselves the outcome of a historical process. Conversely, the very fact that scientists have changed their intellectual strategies, standards, and demands can itself be a historical factor of great influence; so the professional activity of justification becomes an inescapable element in the causal process of history. This being so, are we obliged to say that the very 'reasons' to which scientists appeal in justifying their conceptual changes are—at the same time— the 'causes' in terms of which we ourselves must explain those changes? And are we, correspondingly, obliged to say that the 'causal factors' by which a historical change in science is explained must—at the same time—be regarded also as 'rational considerations', when it comes to justifying this change? Or, if we are not to equate 'reasons' with 'causes' in this way, how else

[1] Lakatos, op. cit.   [2] See essay 1.2 above, esp.pp. 75-8.

do the rational (or disciplinary) and causal (or professional) stories of scientific development overlap and connect up?

This is a point of some delicacy. As we saw in our discussion of Collingwood, the line separating reasons from causes becomes very thin at certain points. It may seem innocent enough— when speaking colloquially—to describe the reasons justifying a conceptual change as 'explaining' why the scientists concerned changed their minds, or alternatively to refer to the activities of a research team in collecting new evidence as 'justifying' the explanatory procedures in support of which their research was undertaken. Yet such colloquial expressions are potentially misleading and, for philosophical purposes, we shall have to choose our words more carefully.

To begin with, then: considered as an entire human enterprise, a science is neither a compendium of ideas and arguments alone, nor a population of individual scientists alone, nor a system of institutions and proceedings alone. At one point or another, the intellectual history of a scientific discipline, the institutional history of a scientific profession, and the individual biographies of the scientists involved evidently touch, interact, and merge. Scientists grasp, develop, apply, and modify their intellectual methods 'in the service of' the intellectual demands of their science, and their institutionalized activities take the forms they do so as to operate effectively 'in the name of' that science. The disciplinary (or intellectual) and the professional (or human) aspects of a science must therefore be linked by close ties of mutual relevance, yet neither of them can be totally prior, or subordinate, to the other. It is a significant fact about well-developed scientific disciplines, for instance, that internalist historians have been able to say so much about their theories, concepts, and explanatory procedures, and about the ways in which they change, without referring to the actual men who devised, tested and/or accepted those procedures, or to where and when they worked. And it is an equally remarkable fact about the sciences that externalist historians can say so much about the professional activities of individual scientists, and of scientific organizations, without needing to refer to the specific theories, concepts, or explanatory procedures employed in the course of those activities. Yet both abstractions hold good only up to a certain point. If we study the professional organization of a science in enough detail, for instance, we shall find ourselves having to bring in also considerations of an internal, disciplinary kind. The dominant concerns of one science may call for field-work in many different countries, those of another may call for elaborate computer calculations; and these special disciplinary interests will necessitate correspondingly different patterns of professional activities.

At the last, then, we are driven back onto the idea of a science as being, first and foremost, an integrated 'rational enterprise', and of the intellectual and institutional features of science as complementary aspects of that single enterprise. This rational enterprise provides the framework or 'form of life' within which, in their different ways, both disciplinary and professional accounts become relevant and operative; only against this background, indeed, can we give an unambiguous sense to all the familiar terminology of 'disciplines' and 'professions', or pose relevant questions about their historical development. So we may restate our present problem by asking: How do the disciplinary and professional approaches to a science differ in emphasis? How—that is—do we recognize one set of factors or considerations as relevant to the intellectual content of a science, another to its human activities and institutional organization?

These differences in emphasis and relevance, as we can now demonstrate, spring directly from the contrasted preoccupations with which we embark on each kind of story: the disciplinary and human approaches to scientific change have quite different points and purposes. When we ask about the historically developing content of a discipline, our interest is in the change from one representative set of concepts to another, regarded as an *intellectual achievement*. In this case, we are concerned with the intellectual outcome of that change: what consequences it had, what it did for the science, how it helped to fulfil its disciplinary ambitions—in short, what contribution it made to the explanatory powers of that science. These questions about the outcome of an intellectual change are normally quite independent of the question, how the change occurred in the first place. Disciplinary accounts of scientific development are thus concerned primarily with judgement and rational justification, not with diagnosis or causal explanation. Under-

HU—L

stood in a strictly disciplinary sense, for instance, the question, 'How did Copernican astronomy develop out of Ptolemaic?', should be re-stated so as to read, 'In what respects, and by appeal to what arguments, did the procedures of Copernican astronomy *build upon and go beyond* the intellectual achievements of Ptolemaic astronomy?' Disciplinary accounts aim, in other words, at the 'rational appraisal' of conceptual changes, regarded as contributions to the success of a rational enterprise. So it is no wonder that such accounts make as much use as they do of the vocabulary of rational appraisal and justification—viz., 'reasons', 'considerations', and 'arguments'.

When we approach the development of a science from a professional or human direction, by contrast, our concern is now with the change from one representative set of individuals, institutions, and explanatory activities to another, regarded as a *historical process*. In that case, we have to consider how the later set came to supersede the earlier, who was responsible for initiating the change, when and where it took place, and how the various men involved made their influence felt. All those questions—we must note—will arise equally, whether the change in question made any substantial contribution to the content of the science or not. Whereas disciplinary accounts of scientific change are concerned primarily with the outcome, rather than with the process itself, the reverse is now true: professional accounts are preoccupied with the process rather than its outcome. Analyses of intellectual change in human or professional terms are thus aimed primarily at *diagnosis and causal explanation*, rather than at *judgement or rational justification*. Understood in strictly human terms, the question, 'How did Copernican astronomy develop out of Ptolemaic?', must be re-stated to read, 'At whose hands, within what institutions, and under what kind of patronage, were those successive intellectual steps taken which eventually *led to the replacement* of Ptolemaic astronomy by Copernican?' Whether a change improved the explanatory powers of the science concerned is no longer the primary issue. The question now is, how it came to occur at all. So it need be no surprise that professional accounts of scientific development make as much as they do of the vocabulary of explanatory or diagnostic analysis—viz., 'causes', 'factors', and 'forces'.

In the former case, we approach every change in a science *prospectively*, with an eye to the differences it made for the future: characterizing the change in terms of the novel discoveries which were its end-product, and so showing how far it was 'justifiable'. In the latter case, we consider it *retrospectively*, with an eye to the factors in the past that led up to it: focusing on the causal sequences of which those discoveries were the outcome, and so showing how far they were 'intelligible'. The disciplinary aspect of intellectual history is rational, justificatory and prospective, the professional aspect causal, explanatory and retrospective; and, in the nature of the case, these two aspects are complementary rather than equivalent. In the course of any rational enterprise, experience of previous explanatory achievements is continually being mobilized to influence current intellectual decisions, while the results of those decisions are, in turn, modifying the rational verdict on our accumulated experience. Either way, however, the relationship between the retrospective diagnoses and their prospective consequences involves an element of interpretation. In the course of our historical explanations of the processes leading up to some scientific change, we may well have occasion to refer to the 'reasons' justifying that change. But those reasons cannot operate directly as 'causal factors' in the historical explanation; they are causally operative at all, only to the extent that the scientists responsible for bringing about the change were aware of, and paid attention to, those particular reasons. What is causally relevant to such an explanation, therefore, is not the reasons themselves, but the scientists' *awareness* of those reasons.

Conversely, 'causal factors' do not, by themselves, serve directly as the 'reasons' justifying scientific changes. When it comes to defending some theoretical innovation, our task is to demonstrate in disciplinary terms what is achieved by that change; the human activities of particular scientists are relevant, only when interpreted as evidence of this disciplinary achievement. What bears directly on the question of rational justification, therefore, is not who did those experiments or calculations, or what exactly they did, or how they came to do it, but the prospective improvements in explanatory power resulting from their work. Research activities become rationally relevant, in short, not on account of their causal efficacy alone, but only to the extent that they are seen as leading to new explanatory achievements as their outcome.

We may accordingly view a collective intellectual enterprise, either as a network of causal processes or, alternatively, as a field for rational achievements. These two viewpoints are alternatives, rather than competitors, and we may move from the one to the other by an appropriate switch of interpretation. If the two resulting types of history overlap in practice, that is because both interpretations are built into the life of the rational enterprise itself. As scientists, for instance, we may observe the intellectual activities of our colleagues *qua* historical events and processes, and then go on to appraise their outcome *qua* rational achievements. Subsequently, we may both redirect our own activities, and modify our view of our colleagues' work in the light of that appraisal. So the practical conduct of scientific enquiry obliges us to switch frequently and easily, and it may seem pedantic as a result to insist on the differences between them. Yet the very fact that this distinction does not have to be insisted on in the course of normal scientific work can make it that much the more dangerous to disregard it for philosophical purposes. And, in fact, the consequences of this distinction—between 'explanatory' causal factors and 'justificatory' rational considerations—will, at a later stage, have far-reaching consequences for our whole analysis of human understanding.

To hint at the philosophical confusions that lie in store for us, if we ignore this particular distinction, we may just remark on the fact that terms like 'force', 'weight', and 'power' are idiomatically used both in the rational assessment of arguments, and also in causal or diagnostic contexts. These ambiguities have been current for as far back in history as we might care to go— certainly since well before the development of modern scientific notions about causality, force, and mechanical determinism. Yet their effect is to blur over the very differences that are important for us here: e.g. between rational considerations which are intellectually 'weighty' on their own account, and causal factors which—regardless of their intrinsic merits— 'carry weight with' the actual participants in a scientific debate. No doubt a 'weighty' argument deserves to 'carry weight with' all informed reasoners, but it may not succeed in doing so. Likewise, the arguments which 'carry weight with' men in any milieu may fail when rationally challenged. The (causal) 'force' of 'powerful' rhetoric may sweep us away; but the (rational) 'force' of 'powerful' arguments requires us to assent of our own choice . . . And so on, and so on . . .

We can now begin to see why it is so hard, in practice, to distinguish the rational cogency of scientific arguments from their causal efficacy, and to discuss these two topics independently. Philosophers have traditionally handled questions about rational judgement and justification as though these were abstract, detached activities, taking place in some external (or eternal) realm quite separate from the actual investigations whose outcome is up for judgement; but the results of the present analysis make this separation impossible. For our own part, we have been increasingly compelled to regard 'judging' and 'justifying' as internal, constituent activities, within the working life of rational enterprises themselves. Standards of judgement, criteria of relevance, rational ideals, and intellectual ambitions emerge from, develop along with, and are refined in the light of the explanatory, practical, and/or judicial activities in which scientists, lawyers, and other 'rational craftsmen' are engaged. The making of discoveries may be one facet of the scientist's professional work, but the justifying of his discoveries—by the presentation of 'acceptable' supporting arguments—is another, complementary facet of this same work. And we can understand either of these two facets adequately, only by considering how they are both related to one another, and to the entire 'rational enterprise' of science. Indeed, a comprehensive account of conceptual development in any rational enterprise must talk, at every stage, both about the framing and about the justifying of the conclusions which were the products of that enterprise.

We are now, at last, in a position to explain what substance lies behind the allusions to 'intellectual ecology', 'niches', 'demands', and the rest, which—up to this point—we have thrown out in passing, without elaboration or elucidation. To come to the heart of the matter: the duality we have recognized here, between the disciplinary and human aspects of intellectual change, is a direct counterpart to the duality which can be found in accounts of organic evolution, as between 'ecological' and 'genealogical' idioms. These two accounts of change in organic populations are, again, complementary and overlapping; yet neither can be wholly translated into the terms appropriate to the other. The genealogical story explains speciation as a causal succession of interlocking processes, while the ecological one interprets it rather as a biological sequence of functional successes; and the resulting histories of organic development, though interdependent, are by no means equivalent.

Suppose that we study organic speciation with an eye, at each stage, to the circumstances in which a later species developed out of earlier ones. This process can be characterized in terms of the sequence of changing 'niches', which were available to be filled by fresh populations of living things; in this way, we show how, in that environment, the 'selective perpetuation' of particular novel characteristics contributed to the biological 'success' of the new species. Ever since the original exchanges between Louis Agassiz and Asa Gray, Darwinists have emphasized that their account of organic speciation—far from being 'brute, blind and mechanistic'—displays at every stage an adaptive or 'teleonomic aspect.'[1] From the ecological point of view, organic change is always to be explained by recognizing the fresh territories, environments and 'ecological demands' to which the selective perpetuation of 'better-adapted' variants was the response.[2]

Alternatively, however, the same process of organic speciation can be characterized in quite different terms, by focusing on the genealogical links between earlier and later populations. Such an account will attempt to reconstruct the succession of episodes in evolutionary history as a result of which those populations multiplied, differentiated, moved further afield, displaced one another, or became extinct. If we do take up this alternative point of view, we may indeed end by treating natural selection as a strictly 'causal' or mechanistic process; but this is only because, as a by-product of our chosen approach,

---

[1] See E. Lurie, *Louis Agassiz: A Life in Science* (Chicago, 1960), pp. 274–81 and 294–5.

[2] See, e.g., E. Mayr, *Animal Species and Evolution* (Cambridge, Mass., 1963).

our questions are restricted to causal processes and interactions.

Clearly, these alternative accounts of organic evolution are
closely related. Failing suitable niches, the genetic potential of
a given organic population will remain unexploited; while,
in the absence of suitable populations, the ecological demands
of a given niche will go unmet. Yet it would be equally unprofitable to ask, 'Do ecological niches exist only in virtue of
the current organic populations; or do those populations exist
only in virtue of the available niches?', and to ask the corresponding question about scientific ideas and institutions. In
either case, the question itself involves a false antithesis. Rather,
we must say that niches exist partly (though not solely) in
virtue of populations, and that populations exist partly
(but not solely) in virtue of niches. Once organic evolution is
under way, populations and niches have a complex, reciprocal
relationship; while current populations can develop into new
species, only where appropriate fresh niches are available,
the character of those very niches is powerfully affected by the
character of other coexisting populations of organisms. Furthermore: only those ecological niches have an evolutionary
significance which some existing population of variant organisms
was available to occupy, or fail to occupy, and only those
genealogies and mutations have an evolutionary relevance
which enabled, or might have enabled, the population concerned to meet the demands of an actual niche.

In saying this, of course, we are making no specific empirical
assertions, but simply analysing the general forms of connection
that are operative in the context of organic evolution. All
such terms as 'ecological demand' and 'niche', 'adaptiveness',
and 'success', are in fact correlative, and interdefined. The
Darwinian theory does not give us a way of discovering obligatory or universal criteria of evolutionary 'merit', 'success',
or 'superiority', independently of actual cases; still less can we
make it the basis for a universal definition, covering 'values'
of other kinds also. But we can quite properly use it to show
what occasions organic evolution provides for talking about
'success', 'superiority', and 'advantages' at all; and so what
kinds of ecological judgements and comparisons the ecological
context provides for. So we can see (for instance) that there

is no occasion to speak, in any absolute terms, of a 'perfect' organism, or to describe one species as 'better' than another without any qualifications.[1] Questions about evolutionary 'adaptiveness' and 'superiority' must, in fact, be understood in relation to some specifiable range of physical and biological environments. Within the actual history of organic species, there is no occasion for debating the 'merits' of different organic populations in bald and general terms; such completely general questions are—ecologically speaking—inoperative.

The same is true in the case of 'intellectual ecology'. What makes it worthwhile to extend ecological terminology from organic to intellectual evolution is, simply, the extensive parallels between the ecological account of organic change and the disciplinary account of intellectual development. Within intellectual history, any actual problem-situation creates a certain range of opportunities for intellectual innovation. The nature of these opportunities depends, of course, as much on the character of other coexisting ideas as it does on purely 'external' features of the social or physical situation. We may study the resulting conceptual changes either *qua* processes, by looking to see simply what historical course they took, or alternatively *qua* achievements, by asking how far those changes exploited the intellectual opportunities in the current situation. As in the organic case, the relationship between concepts and opportunities is a complex and reciprocal one. Earlier populations of scientific concepts can differentiate in such a way that they generate novel disciplines (say) only where suitable intellectual opportunities present themselves; meanwhile, the character of those opportunities is powerfully affected by the other conceptual populations already in existence, . . . and so on. The disciplinary account of scientific change accordingly studies the opportunities to be exploited in any problem-situation, analyses the demands created by those

---

[1] I say this, despite the arguments put forward by Julian Huxley in his book, *Evolution, the Modern Synthesis* (London, 1942), esp. Chapter 10 on 'Evolutionary Progress'. Certainly, Huxley succeeds in showing that the human species is remarkable, not just in being well-adapted to the particular 'niches' within which it originally developed, but in having subsequently made itself capable of surviving within a much wider range of potential environments; but this qualification gives us much less than we need to define any *absolute or universal* criterion of 'evolutionary merit'.

opportunities, and appraises the achievements resulting from the conceptual changes by which scientists actually responded to those demands.

Once again, to say this is not to prejudge the actual empirical relations between specific conceptual changes and the demands of particular niches, but is simply to analyse the general forms of relationship that are operative in the context of conceptual evolution. Here again, all such terms as 'merits' and 'superiority' are correlative and interdefined. They do not give us any way of establishing obligatory or universal criteria for judging the rational merits of actual conceptual changes, independently of actual cases; still less, for prejudging the philosophical questions that will subsequently arise about 'rationality', in any more general sense. On the contrary, we can now see why it is so out of place to speak, in any absolute terms, of a 'perfect' concept or theoretical system; or to describe one set of intellectual procedures as 'better' than another without any qualifications. Questions of this kind are operative only in relation to the demands of particular situations and problems. The rational issues confronting us in any particular situation reflect the specific problems arising in the course of our current intellectual activities; and, above all, the operative criteria of judgement in any rational enterprise depend on the practical occasions for choice, and internal self-criticism, actually existing within the activities of the enterprise.

Before Darwin demonstrated how the functional character of organic structures and behaviour might be explained as an outcome of their historical origins, it was customary to think of both organs and organisms as the product of Divine Craftsmanship. Each separate feature of organic life was then interpreted as a separate and deliberate Creation, manifesting unlimited Foresight and Wisdom and 'perfectly proportioned' to the 'tasks' for which the Almighty had designed it.[1] Darwin's achievement was to prove that, in practice, this functionality is never final, complete, or absolute, but that the merits of different organs and organisms are intelligible, only when we consider them as working together within a specific situation.

---

[1] On the providentialist background to the pre-evolutionary debate, see L. Eiseley, *Darwin's Century* (New York and London, 1958), and C. C. Gillispie, *Genesis and Geology* (Cambridge, Mass., 1951).

The present enquiries lead us to a similar conclusion. Any attempt to define a 'perfect' theory or 'uniquely meritorious' conceptual system lands us with questions which are at once too general, and too abstract, to be relevant to the actual, self-critical problem-solving procedures of any working discipline—questions that are no more relevant to the conceptual evolution of actual rational enterprises than the older questions about 'perfect' organs and organisms were to our historical understanding of organic speciation.

If we come down from our philosophical Olympus into the actual arena of scientific problem-solving, however, we can see how the 'rational considerations' relevant to the solution of conceptual problems are intelligible, as related to the specific details of particular problems; and how the character of these problems is related, in turn, to the longer-term development of the discipline within which they are currently 'problematic'. If we take the trouble to understand exactly, and in detail, what intellectual goals the men concerned were aiming at, in this or that discipline, and how that choice of goals came to face them with this specific conceptual problem, we shall then—but only then—be in a position to understand what *counted* for them as an intellectual 'achievement' or theoretical 'improvement', and how far—in that particular problem-situation—they were *justified* in applying the principles of judgement and criteria of choice they did.

CHAPTER 5   Interlude:
            Evolution and the
            Human Sciences

THE suggestion that cultural and intellectual change should be accounted for in evolutionary terms has had a long and chequered history. From the time of Charles Darwin on, there have been recurrent attempts to extend ideas from the *Origin of Species* to social or political, cultural or intellectual development. So we find Thomas Henry Huxley, in the course of a commemorative lecture in 1880, giving Darwinian idioms a direct application to the history of science: 'The struggle for existence holds as much in the intellectual as in the physical world. A theory is a species of thinking, and its right to exist is coextensive with its power of resisting extinction by its rivals.'[1] But Huxley did little to show what real light the phrase 'struggle for existence' could throw on the processes of intellectual change. Much more recently, in November 1967, the contemporary biologist, Jacques Monod, declared in a lecture to the Collège de France: 'A transmittable idea constitutes an autonomous entity . . . capable of preserving itself, of growing, of gaining in complexity; and is therefore the object of a selective process, of which modern culture is the current but in every way evolving product.'[2] Monod, too, puts forward this notion merely as a programme or schema, and admits that the evolutionary theory of ideas for which he calls has never been worked out in detail: 'The laws governing this selection are necessarily very complex (but perhaps) we shall some day have . . . a Natural History of the Selection of Ideas.' Up to

[1] T. H. Huxley, 'The Coming of Age of the Origin of Species', an evening lecture at the Royal Institution, London, April 9, 1880; reprinted in *Science and Culture* (New York, 1893), Chapter 12, pp. 319 ff.

[2] J. Monod, 'From Biology to Ethics', *Occasional Papers of the Salk Institute of Biology*, no. 1 (La Jolla, 1969), p. 16.

now, however, the results of all such attempts have been so generally disappointing that anybody who proposes to revive this suggestion must demonstrate how he proposes to avoid the trap into which earlier forms of 'evolutionism' fell.

The century between Huxley and Monod saw several abortive attacks on the problem of generalizing the concept of evolution. Ernst Mach, for instance, was among Darwin's earliest continental admirers, and he was much taken by the possibility of extending Darwinian categories from the history of organic species into the history of thought. Unfortunately, his theory of intellectual evolution started from the wrong end of the problem. If he had treated Darwin's account of organic speciation simply as one special case of a more general pattern of historical change—the pattern of 'variation and selective perpetuation'—and had looked for corresponding patterns with different parameters to explain the development of men's intellectual disciplines, all might have been well. Instead, Mach chose to emphasise just those features of organic evolution which throw least light upon intellectual change. The 'disciplinary missions' of natural sciences, he argued, were simply one aspect of the same broader 'historical mission' that underlay organic evolution itself:

Expressed very briefly, the task of scientific knowledge now appears as: the adaptation of ideas to facts and the adaptation of ideas to one another. Every favourable biological process is an event of self-preservation, and as such is also a process of adaptation . . . All favourable cognitive processes are special cases, or parts, of biologically advantageous processes . . . The cognitive process may display the most varied qualities: we characterize it in the first place as biological, and as economic.[1]

Yet, as Max Planck insisted in reply, what differentiates the historical development of scientific knowledge from the historical development of organic species is, precisely, its particular goals or missions. The tasks of scientific disciplines are intellectual, not biological or economic; and, except in the most flatulently extended sense of the words, the scientist

---

[1] E. Mach, 'My Scientific Theory of Knowledge and its Reception by my Contemporaries', in *Physical Reality*, ed. S. Toulmin (New York, 1970), pp. 30–1.

cannot be said to value his ideas on account of their 'biological' or 'economic' advantages.[1]

Mach's account of scientific evolution shipwrecked chiefly because he equated the intellectual selection-criteria of science with the quite different criteria operative in organic change and economic development: viz., differential reproduction rate and productive efficiency. This equation misled him into believing that he must put his account of scientific knowledge on a 'biologico-economic' basis, and at the same time distracted him from the possibility of giving a more general—and more valid—account of intellectual evolution in terms of other, more directly relevant criteria. The Social Darwinists made a similar mistake. Having recognised that something useful might be learned from a Darwinist analysis of sociopolitical change, they misguidedly imported the whole range of concepts and criteria developed for explaining biological evolution into their discussion of social affairs.[2] The resulting sequence of arguments and slogans was both intellectually confused and politically obnoxious. Phrases like 'survival of the fittest' and 'evolutionary success' were given a sociopolitical application in a way that blurred the crucial differences between organic species and human races, nations, or classes, confused ecological dominance with economic domination, and ignored the unanswered question, what truly gives unity, continuity, and common interests to a human community or society. So, instead of being a fruitful new source of explanatory ideas, the Social Darwinism of the 1890s and 1900s ended up by generating a pack of pernicious over-simplifications.

The most frequent and influential mistake, however, has been at once more fundamental than those of Mach and the Social Darwinists and also far cruder. It has resulted from confusing the 'evolutionary' approach of Charles Darwin, on the one hand, with the 'evolutionistic' ideas of Herbert Spencer and Lamarck, on the other: that is, from reading the Darwinian schema of explanation as entailing doctrines about the *overall* direction of organic (and even of cosmic) development like

[1] See Planck's essay on 'Mach's Theory of Physical Knowledge', originally published in *Physikalische Zeitschrift*, 11 (1910), 1186–90; *Physical Reality*, pp. 44–52.

[2] See, e.g., Richard Hofstadter, *Social Darwinism in American Thought, 1860–1915* (Philadelphia, 1944).

those of such historicist philosophers and social theorists as Lamarck and Marx, Spencer and Teilhard. Some writers on sociology, anthropology, and linguistics have found the conception of Evolution as a doctrine of Cosmic Progress—revealing a universal and irreversible direction of historical development in the natural and human worlds—a vastly appealing one. Others have found its apparent implications unacceptable. For or against, this conception has confused almost all evolutionary arguments in the human sciences for the last 100 years and more; and, in the resulting dispute, the true character of the Darwinian schema—which accounts, rather, for *local* changes in organic populations, by relating them to their adaptive advantages—has been forgotten.

The most recent and revealing illustration of this confusion, for our purposes, comes from T. S. Kuhn's very latest presentation of his views about scientific change. Faced with the charge of historical relativism, Kuhn has now retreated finally from the paradoxes associated with his earlier revolutionary theories to a new position, which he describes as being an 'evolutionary' one. To quote his own statement of the new position:

In one sense of the term I may be a relativist; in a more essential one I am not. What I can hope to do here is separate the two. It must already be clear that my view of scientific development is fundamentally evolutionary. Imagine, therefore, an evolutionary tree representing the development of the scientific specialities from their common origin in, say, primitive natural philosophy. Imagine, in addition, a line drawn up that tree from the base of the trunk to the tip of some limb without doubling back on itself. Any two theories found along this line are related to each other by descent. Now consider two such theories, each chosen from a point not too near its origin. I believe it would be easy to design a set of criteria—including maximum accuracy of predictions, degree of specialization, number (but not scope) of concrete problem solutions—which would enable any observer involved with neither theory to tell which was the older, which the descendant. For me, therefore, *scientific development is, like biological evolution, unidirectional and irreversible.* One scientific theory is not as good as another *for doing what scientists normally do.* In that sense I am not a relativist.[1]

---

[1] 'Reflections on my Critics', in *Criticism and the Growth of Knowledge*, ed. I. Lakatos and A. Musgrave (Cambridge, Eng., 1970), p. 264. (Italics added.)

About this re-statement, two things need to be said. To begin with, Kuhn's new position is 'evolutionary' only in an outmoded and unacceptable sense of that term. As evolutionary zoologists now understand the processes of organic speciation and change, evolution may be 'irreversible', but it is certainly not (in Kuhn's sense) *unidirectional*. The assumptions he makes about biological evolution are, in fact, quite unfounded. If we are presented with two unfamiliar organic forms, from different positions along any branch of the evolutionary tree, no sure procedures or criteria exist for telling which is the older, which is the descendant; an intestinal parasite of great morphological simplicity, for instance, may perfectly well be—and is very likely to be—descended from a more complex and highly-differentiated earlier species. To suppose that corresponding criteria could be designed in the case of intellectual 'evolution' is, therefore, not to reapply genuinely Darwinian ideas to the historical development of science; rather, it is to revert to an earlier, providentialist view of evolution, which it was Darwin's chief merit to have outgrown.

The supposition that the historical record of evolution can *by itself* reveal universal criteria of evolutionary 'advance' (degree of specialization, differentiation, etc.) was in fact the crucial element in pre-Darwinian, progressivist ideas about evolution which Herbert Spencer inherited from Lamarck, but which Darwin himself rightly abandoned. And Spencer's continued acceptance of this Lamarckian doctrine was—as we shall see—a chief source of the fatal flaw in late-nineteen-century attempts at constructing evolutionistic theories of social, cultural, and historical change: viz., the belief that History is not only Irreversible but also Unidirectional.[1] In contrast to Kuhn's account, our own descriptions of conceptual change as 'evolutionary' have implied only that the changes from one temporal cross-section to the next involve the selective perpetuation of conceptual variants. They have implied nothing to suggest that the 'evolutionary' changes in our concepts display any single long-term direction of change—still less, that it is their business to harmonise with a larger Cosmic Purpose. In this respect, we have profited from the experience of evolutionary

---

[1] See J. W. Burrow, *Evolution and Society: a Study in Victorian Social Theory* (Cambridge, Eng., 1966), esp. Chapter 6.

biologists, who have finally disentangled the 'populational' mode of explanation lying at the heart of Darwin's theory of speciation from the older shell of providentialist ideas with which the term 'evolution' had earlier been associated.

There may still (it is true) be some individual zoologists who believe that Evolution does have a larger, longer-term significance, and is the expression of some overall creative tendency in Nature,[1] but this belief no longer plays any collective, disciplinary part in the zoological explanation of speciation or ecological relationships. On the contrary, it survives merely as a last vestige of the 'natural theology' that played so large a part in Protestant thought from the days of Newton, Boyle and Ray[2] on; and it continues to draw strength from biology solely on the basis of 'virtue by association'. If the theory of organic evolution has a firmly established basis in the explanatory procedures of contemporary biology, by contrast, this comes not from its lingering providential overtones, but from the intimate co-operation of ecology, population dynamics, and the new systematics—all of which are concerned, rather, with the varied processes by which populations adapt to changes in their local environments.[3]

5.1: *Evolution and Cosmic Progress*

Looking back at Darwin's *Origin of Species* from a century later, we can see that his theories were built—from the very beginning—around the novel conception of organic species as modifiable populations, possessing not a 'specific essence' but a statistical distribution of properties; and of the 'peak' or

[1] For example: J. S. Huxley, *Evolution and Ethics 1893–1943* (London, 1947), C. H. Waddington, *Science and Ethics* (London, 1942), T. Dobzhansky, *Mankind Evolving* (New Haven, 1964); in this context, I do not mention Teilhard de Chardin, whose standing as a professional scientist—let alone a zoologist—has been very much in dispute.

[2] Consider, e.g., Newton's correspondence with Richard Bentley, Boyle's description of himself as a 'Christian Virtuoso', and John Ray's *Treatise of the Wisdom of God*; see also the essays of H. Metzger on the theological implications of Newtonian physics, *Attraction Universelle et Réligion Naturelle chez quelques Commentateurs Anglais de Newton* (Paris, 1938).

[3] See, R. A. Fisher, *The Genetical Theory of Natural Selection* (Oxford, 1930), J. B. S. Haldane, *The Causes of Evolution* (London, 1932), and Sewall Wright, 'Evolution in Mendelian Populations', *Genetics*, 16 (1931), 95–159; also the works of Ernst Mayr, as above, and T. Dobzhansky, *Genetics and the Origin of Species* (New York, 1937).

'mean' of these populations as shifting in the face of ecological changes, or in response to the colonization of new environments. When Darwin's book first appeared, the general intellectual debate was not ready to absorb this new populational mode of thought.[1] At that time, the term 'evolution' had other, quite different implications, being thought of as a naturalistic explanation of providential, progressive change in the organic world. In other spheres of thought—notably, in sociology and anthropology—the term retained this providentialist significance for much longer, and this is the sense which it had in the abortive 'evolutionism' of philosophy and the human sciences in the nineteenth century. In certain of these human sciences, indeed, the two senses of 'evolution'—populational and providentialist—have not been wholly disentangled even today.

If this disentanglement has taken so long, that is largely because the idea of evolution has had a much longer standing in such fields as linguistics and sociology than it has had in biology. This fact is often overlooked. Darwin's *Origin of Species* appeared only as late as 1859. If that had been the prime source of evolutionary ideas, we should expect theories of social or historical evolution to have made their appearance only in the 1860s and '70s, when Darwin's example had had a chance to make its wider influence felt. Yet the term 'evolution' had already been in circulation for a good half-century before 1859, and evolutionary speculations about historical and social change were a commonplace before Darwin had even begun his work on speciation. By the 1840s, Herbert Spencer was already developing a general account of Man's mental and social history, based on evolutionary notions learned from Lamarck;[2] while similar ideas had become

---

[1] See especially the impressive collection of essays, *Forerunners of Darwin: 1745–1859*, ed. Glass *et al.* (Baltimore, 1959), referred to below as *Forerunners*.

[2] As J. W. Burrow, *Evolution and Society* (Cambridge, Eng., 1966), makes clear, Herbert Spencer was substantially committed to the entire Lamarckian system of natural philosophy, which was based on the operation of two fundamental types of counteracting agencies in Nature, one constructive, the other destructive—a theory admirably analysed in C. C. Gillispie's essay on Lamarck in *Forerunners*. Derek Freeman of the Australian National University is also engaged in a useful reappraisal of nineteenth-century anthropological and sociological theories: see (e.g.) 'Social Anthropology and the Scientific Study of Human Behaviour', *Man*, 1 (1966), 330–40. He reaches substantially the same conclusions as myself about Herbert Spencer's Lamarckian background, as a factor responsible for misdirecting

familiar in France some ten years earlier. The novelist Honoré de Balzac actually planned his entire *Comédie Humaine* around an evolutionary cosmology,[1] in which his fictional characters were to serve as exemplars of different 'species' in an historically-developing social taxonomy: so we find him writing in a political journal as follows: 'L'échelle mystique de Jacob, l'échelle zoologique et l'échelle sociale; les sphères, les espèces et les classes; le mouvement ascendant de la Creation, la transformation des espèces et l'ambition des hommes, sont donc trois aspects différents d'une seule et même réalité.'[2] By the early 1830s, it was evidently safe to assume that French readers were generally familiar with the idea of 'evolution', equally, on the cosmic, biological, and social levels; and Balzac, who had learnt his Lamarckism from the botanist, Geoffroy de St. Hilaire, was by no means the first to link these three separate notions into a single unitary historical process.

Evolutionary categories originally entered Western thought, then, in the context of a comprehensive historicist system of natural and social philosophy; as a technical contribution to the biological theory of speciation, they proved their worth only much later. The progenitors of this overall historicist system, Lamarck and Herder, were both born in 1744, and their fundamental ideas took shape within an eighteenth-century framework. Both men set out to construct systematic historical cosmologies embracing changes of all kinds. Lamarck always regarded zoological evolution as a continuing expression of constructive organizing agencies previously operative in the pre-organic world on a chemical level.[3] Similarly, Herder spoke of the Chain of Human Cultures as continuing into the present era that same scale of Nature, or Great Chain of Being,

---

anthropologists and sociologists, by giving them false ideas about what an 'evolutionary' theory of human and social development would be like. See also the essay by P. B. Medawar on Spencer, in *The Art of the Soluble* (London, 1967).

[1] For a detailed analysis of the philosophical background to Balzac's life-work see P. Nykrog, *La Pensée de Balzac dans la Comédie Humaine* (Copenhagen, 1965); cf. also I. V. Guyon, *La Pensée Politique et Sociale de Balzac,* 2nd edn. (Paris, 1967), p. 416.

[2] See A. Maurois, *Prométhée, ou La Vie de Balzac* (Paris, 1965), p. 171; Eng. trans., (London and New York, 1965) p. 173.

[3] C. C. Gillispie in *Forerunners.*

by which lower forms of organic life had—in his view—successively come into existence.[1]

The evolutionary ideas circulating during the early nineteenth century were thus concerned, above all, with the overall direction of cosmic development, material, organic and human alike. They did nothing to provide a specifically zoological explanation of speciation; instead, they formed part of a sweeping progressivist description of cosmic history, as a single all-embracing process characterized by the appearance of progressively more 'advanced'—i.e., more complex, differentiated and specialized—substances, organisms, or social institutions. Far from being the prototype of evolutionary theorizing, Darwin's theory of speciation was a fairly late example; and it was one which broke with previous evolutionary thought by introducing a quite novel set of explanatory mechanisms to account for the zoological phenomena which were its specific concern. And, in the long run, Darwin's 'populational' explanation of speciation was to prove not merely independent of earlier providentialist ideas, but inconsistent with them.

To leave aside all question of mechanism for the moment: even on a descriptive level, the notion of 'evolution' has at every stage involved two separate ideas. On the one hand, it assumes the *fact of descent;* on the other, the *doctrine of progression.* To begin with, the occurrence of 'evolutionary' change was a matter for dispute, both in zoology and in human history, simply as a question of fact. For a long time, there were no conclusive arguments to decide between the hypothesis of evolutionary descent and the alternative hypothesis of independent creation. Given the limited time-scale of history, the hypothesis of descent long appeared the more implausible of the two. It seemed much more likely that different material substances, species, and societies had come into existence separately, and had preserved their present form throughout that existence. In this particular respect, there was one major difference even between Lamarck and Herder. Herder was the first man to develop a comprehensive progressivist cosmology, embracing the successive appearance of 'higher' forms of matter and life; yet he himself never accepted the idea of

[1] On Herder, see the essays by A. O. Lovejoy in *Forerunners,* pp. 207–21; and in his collection, *Essays in the History of Ideas* (Baltimore, 1948), Chapter 9.

descent. For Herder, the history of the world had been a creative process by which Mother Nature had brought progressively more advanced forms of being into existence independently and in succession: first inorganic, then organic, then rational, spiritual, and social.[1]

The first distinctive feature of any fully evolutionary theory was, therefore, its commitment to descent as a fact, and its rejection of the hypothesis of independent origins. In zoology, this meant thinking of different species as connected in a single genealogical tree, instead of as having originated independently. In social history, it meant treating different societies, cultures and languages as linked back by genealogical connections to some original 'birthplace of civilization'. At this point, Lamarck (like Erasmus Darwin) took one step beyond Herder and towards our twentieth-century view.[2] For he associated Herder's doctrine of progression with genealogical descent, and he insisted on something which men had previously dismissed as inconceivable—viz., that 'higher' forms of organic being could arise by direct descent from 'lower', without the necessity for an independent creation. In their joint belief in the creative direction of cosmic advance, however, Lamarck's evolutionism and Herder's progressivism were at one. And for many people, from the time of Lamarck on, the main charm of Evolution has sprung from its progressivist associations—that is, from the belief that Evolution reveals clues to the direction in which the entire Universe has developed in the past, and will presumably continue to develop in the future.

So understood, the idea of evolution rapidly became bound up with philosophical historicism, and shared its prophetic ambitions. During the thirty or forty years before Darwin's theories appeared, the pattern of evolutionary social theory crystallized around this doctrine of historical progression; a belief in this unidirectional picture of long-term historical development united men as different in other ways as Auguste

[1] See Lovejoy in *Forerunners*.

[2] The ideas of Lamarck have been sadly neglected by scholars: the weaknesses of his views about the *mechanism* of organic evolution seem to have obscured the wider originality of his approach. But see J. S. Wilkie, 'Buffon, Lamarck and Darwin', in *Darwin's Biological Work*, ed. P. R. Bell (Cambridge, Eng., 1959).

Comte, Karl Marx, and Herbert Spencer. Within the whole German tradition of historicism—both its Hegelian or idealist, and its Marxist or materialist forms—great stress was also laid on the Rationality of that Historical Process; and the idea of 'progression' was linked with the idea that History advanced in accordance with an inherent 'dialectic'.

That emphasis on the 'rationality' of historical change need again have done no harm, if it had been interpreted in ecological terms, as concerned with questions of 'adaptedness' or 'adequacy'. Unfortunately, the historicists interpreted the 'dialectic' in a stronger, teleological manner, which Hegel took over from Herder, and Marx and Engels inherited from Hegel and Lamarck jointly.[1] This interpretation gave rise to grave misunderstandings: for it implied that all socio-political changes take place, in some mysterious way, as the 'conclusions' of a Cosmic Argument, which unfolds 'logical implications' operative throughout the whole History of Society—and, perhaps, of Nature also. In practice, of course, we can judge the 'rational adequacy' of socio-political changes to the needs of particular problem-situations only with an eye to specific cases, as they occur; ideological principles of a totally general or cosmic kind give us no precise criteria of judgement. Whether in physical science or industrial organization, it is the particular changes by which men seek to deal with their current problems that are more or less 'rational', not some overall teleology, or 'direction', implicit throughout History-as-a-Whole.

In this respect, social and intellectual changes alike resemble organic changes. When an 'internalist' historian of science, for example, reconstructs the genealogy of certain scientific ideas, he may choose to expound the successive stages in their conceptual evolution as a sequence of steps in an explicit argument: detaching the particular conceptual problems that men succeeded in dealing with at each stage from the more general backcloth of scientific ignorance and confusion. But he can

---

[1] Marxist scholars disagree about the extent of Marx's responsibility for the structure and content of Engels's *Dialectics of Nature* (1873–86); the answer apparently depends upon how far you wish to regard Marx as a systematic and comprehensive philosopher—with his own philosophy of nature and all—and how far the obvious weaknesses of the book tempt you to unload entire responsibility for it on to Friedrich Engels, as a mere scientific popularizer in the same class as Haeckel!

do so only as a matter of retrospective judgement; the argument so reconstructed displays the disciplinary character of the resulting achievements, not the professional sequence of historical events whose outcome they were. Similarly, when a historian reconstructs a particular social or political episode, he may retrospectively characterize it as involving a dialectical sequence of problems and solutions, by dissecting out the needs that were successfully met in each period by new political or social procedures from the more general background of unresolved socio-political difficulties. But in this case, too, the appearance of 'formal necessity' arising from the dialectical mode of exposition is misleading. This appearance originates in our retrospective judgement on the fruitful outcome of the events concerned; it points to no intrinsic feature in the events themselves. The confidence we have in our retrospective appraisals imposes no retroactive 'compulsion' on the processes of social and conceptual change themselves. In social and political history, as in the history of scientific thought, the 'dialectic' lies primarily in our writing of history, rather than in the historical events about which we write.

We can properly speak of social and political institutions evolving in a more or less 'rational' manner, only if we consider the detailed ways in which such institutions develop—or fail to develop—in response to the specific requirements of changing historical situations. As for the specific forms these 'ecological demands' will take, in future situations,we can no more forecast these with confidence in the intellectual and social field than we can do in the field of organic evolution. Just because all the changing conditions that will in fact face future organic populations cannot be predicted, or even guessed at, the characteristics of later organic species cannot be said, in any real sense, to have been 'wholly implied in' those of the earlier species.[1] (Darwin himself had genuine reasons for shying away

---

[1] The impossibility of making categorical 'predictions' about the future emergence of novel organic species has been used, on the one hand, as an argument to discredit the 'scientific' standing of evolution-theory, by comparison with, e.g., celestial mechanics (cf. C. G. Hempel, *Aspects of Scientific Explanation*, p. 370); and, on the other hand, as an argument for breaking the logical empiricists' analytical connection between 'prediction' and 'explanation'. For this latter point of view, see M. Scriven, 'Explanation and Prediction in Evolutionary Theory', *Science*, 130 (1959), 477–82; and also my own argument in *Foresight*, Chapter 2.

from the term 'evolution', with its misleading implications of 'unfolding' and 'bringing hidden implications to light'.)[1] And the actual sequence of social and political events is repeatedly throwing up new constellations of circumstances and problems, whose character could never have been forecast in detail on the basis of the earlier historical record alone. As a result, men demonstrate the 'rationality' of their social and political procedures as much by their capacity to cope with unforeseeable problems as by their capacity to deal with problems that were already 'implicit' in what had gone before.

It is one of the ironies of European thought that historicist ideas about 'progressive evolution' in human affairs were worked out and debated at length, before Charles Darwin gave us a real understanding of historical development in the natural world. So a whole tradition of social and political philosophy had been created, by the time Darwin came on the scene, which has survived up till our own day. Many more moderate sociologists today who share the dissatisfaction of contemporary anarchists with static theories of the 'social system' have reacted against those theories, as a result, by reviving liberalized versions of Marxism.[2] To them, as to Collingwood in the 1940s, some kind of liberal Marxism seems to provide the only consistent basis for reintroducing historical categories into our theories of social and intellectual change.[3]

---

[1] Notice that the term 'evolution' plays very little part at all in Darwin's own writing: it is not used, for instance, in the first edition of *The Origin of Species*. The connection between Darwin's theory of 'natural selection' and traditional arguments about 'organic evolution' was made explicit only subsequently: as, for instance, in T. H. Huxley's article on 'Evolution in Biology' in *Encyclopaedia Britannica*, 9th edn. (1878).

[2] The most highly publicized of these neo-Marxist positions is that of Herbert Marcuse (*One-Dimensional Man*, etc.); but there is by now a very substantial body of serious neo-Marxist thought, especially in the German universities, which attempts to compromise between the historico-economic materialism of Marx's social theory and a post-Freudian approach to individual psychology. One real virtue of these newer styles of Marxist theory is their close attention to the role of such notions as 'consciousness' in the explanation of social and intellectual development. Although the rational dialectic of problem-solving is not *itself* a causal element in history, the 'awareness' which the problem-solvers have of the progress of their arguments can be. (See section 4.3 above.)

[3] Recall the discussion of Collingwood, above, in section 1.2. During the early 1940s, Collingwood's personal friends were much disturbed by rumours that he had in fact become a convinced Marxist.

Yet, in most respects, the attitudes of Hegel and Marx to human history were as providentialist and progressivist as the attitudes of Lamarck and Teilhard to organic evolution. While the Hegelians and the Marxists had a more profound feeling than their philosophical predecessors for the movement of history and the significance of historical change, they continued to view the March of History teleologically—as a single, upward-sweeping movement, with a consistent overall direction, as fundamental to its understanding as Divine Providence had been to the earlier Christian vision of history. In this historical providentialism, History-as-a-Whole became a 'logical' process whose 'inner rationality' had the same inescapability as the conclusion of a well-framed deduction. (As Hegel himself put it, 'Nothing else will come out from history but what was already there'.) So a genuine and original perception of the problem-solving element in social history, and of the rational adequacy of social innovations in meeting the needs of each new age, was refracted through the prism of traditional theology and misread as a recognition of the all-embracing Historical Dialectic, whose Rational Unfolding is expressed through the Logic of History.

At the time of Hegel's death, the young Charles Darwin had not yet joined H.M.S. *Beagle*, while the publication of the *Origin of Species* was more than 25 years away. As for Karl Marx, though he was inspired to dedicate *Das Kapital* to Darwin, neither he nor Engels understood much better than Hegel or Spencer how the Darwinian idea of 'variation and selective perpetuation' gave a new twist to our understanding of historical change. So we cannot hope to restate the historicists' insight into the 'rational justification' of social and political change in terms that are acceptable today, unless we first strip away from their theories that residual providentialism which they inherited from Herder and Lamarck. If we do this, we shall strip away also any temptation to follow them in their ideological misinterpretation of the historical dialectic— that is, the temptation to project the logical structure of our historical analysis back on to the facts, and to construe its dialectical 'necessity' as a species of 'causality' inherent in historical events themselves.

In contrast to all previous evolutionary theories, Darwin's populational theory of speciation accepted the hypothesis of descent, but was neutral about any supposed overall direction of historical development. All branches of the genealogical Tree of Life were zoologically on a par. Wherever one organic species gave rise to another successor-species, the resulting change called for a separate explanation in 'adaptive' terms; and it was quite arbitrary to pick out one particular line of descent within that genealogical tree—e.g. the succession of changes resulting in the appearance of the human species— as defining any overall 'direction'. For that would mean discriminating against the other genealogical relationships between organic species, in a way that had nothing to do with the explanatory goals of biology. So far as the Darwinian mode of explanation went, all evolutionary changes must be accepted for what they were, in relation to their own ecological contexts. The Darwinian theory supported no absolute scale of 'adaptedness'. Zoologically speaking, questions of 'adaptive merit' had a determinate meaning, only in relation to the demands of some specific range of niches or situations.[1] Darwin's populational theory thus sets aside all questions about the overall direction of cosmic change as irrelevant to the problem in hand. Just as Montaigne had criticized sixteenth-century attempts to read the Divine Will for Man in the Heavens as totally presumptuous, so the populational form of evolution theory treats the idea of cosmic teleology with scepticism, as an excuse for human wish-fulfilment to which biology lends no genuine support. Within the Darwinist tradition, Evolution is no longer a single, all-embracing cosmic process; it has become 'universal' only in the technical sense, of providing a general form of explanation applicable equally to all the phenomena of organic speciation.

So, since 1860, the evolutionary ideas of technical zoologists have significantly diverged from the evolutionistic ideas of

---

[1] The most elaborate attempt to base a 'cosmic philosophy' on evolutionary zoology is that of Julian Huxley, in his *Evolutionary Ethics*: I discussed this attempt at length in my own first published paper on 'World-Stuff and Nonsense', *Cambridge Journal* (1948). The tone of that essay was excessively sharp, but the general line of criticism still appears to me sound. See also my essay, 'Contemporary Scientific Mythology', in *Metaphysical Beliefs*, ed. A. MacIntyre (London, 1957).

social theorists and natural theologians. Herbert Spencer initially applauded Darwin's work, as providing Lamarck's hypothesis of descent with the explanatory underpinning it had lacked for so long. Yet, at a deeper level, the ideas of the two men were thoroughly inconsistent. The Darwinian process of 'variation and natural selection' might, in some cases, lead to greater differentiation and heterogeneity, but it was equally capable of producing the opposite result. To Darwin, this fact was of no special importance. He was interested only in finding a general explanation for the organic changes that in fact occur, not in demonstrating that the March of Evolution takes place always Onwards and Upwards. Properly speaking, therefore, Herbert Spencer was a Social Lamarckist, not a Social Darwinist. Like Herder and Hegel, Comte and Marx, he was interested in the detailed mechanisms of historical change only because—in his view—they helped to reveal the grand sweep of Human Progress.

Subsequent enthusiasts for the idea of 'social evolution' have all too frequently embraced Spencer's approach precisely because of these providentialist misconceptions. For their purposes, any detailed explanation of historical change which fails to guarantee the Upward March of History—whether on the organic or the socio-cultural level—is unacceptable on that account alone; and in due course Spencer's successors realized that Darwin's populational theories shared this defect. (Recall Leibniz's similar assault on Newton's theory of planetary dynamics, as failing to guarantee the providential stability of the Solar System.)[1] As a result, Darwin's ideas came under attack not only from Christian theologians, who saw him as denying the Hand of God in the Creation of Nature, but also from agnostics like Samuel Butler[2] and socialists like George Bernard Shaw,[3] who resented the failure of Darwinism

---

[1] See the opening paper in the Leibniz–Clarke correspondence, in which Leibniz is heavily sarcastic about the need for Newton's planetary theory to leave it open for God to intervene from time to time in the affairs of his Creation, and to readjust the paths of the planets when they are beginning to go astray.

[2] See Samuel Butler, *Life and Habit* (London, 1877), *Evolution, Old and New* (London, 1879), and *Unconscious Memory* (London, 1880).

[3] G. B. Shaw, *Back to Methuselah* (London, 1921), Introduction; see Standard Edition (London, 1932), pp. lii–liii.

to support their own belief in the inherent teleology of the natural world. Shaw's own intellectual ancestry was, of course, 'evolutionist', in the older, providentialist tradition, leading back to Lamarck by way of Marx and Engels. In a savage attack, we find him denouncing the idea of natural selection as not merely unacceptable, but contemptible: 'If it could be proved that the whole universe had been produced by such Selection, only fools and rascals could bear to live.'[1] The same Lamarckian tradition has been influential among other subsequent Marxist thinkers. If, for instance, the ideological authorities in the Russia of the 1950s took the side of Lysenko against the neo-Darwinists, this was no accident of politics. The ideas of Michurin and Lysenko were, in every respect, more congenial to Marxist ideas about the long-term direction of historical development than the more workaday neo-Darwinist analyses of selection-mechanisms and populational change.[2]

The confusions between evolutionary biology and evolutionistic sociology are not even confined to people ignorant of contemporary zoology. In our own time, for instance, Julian Huxley has argued once again for the idea which—as we saw— is at least as old as Balzac: viz., that pre-biotic, organic, and social history share a single consistent direction, which provides the basis for an 'evolutionary ethics', or 'religion without revelation'. Similarly, the philosophical palaeontologist, Teilhard de Chardin, based the system of natural theology expounded in his book, *The Phenomenon of Man*, on evolutionary ideas which owed nothing to Darwin, but whose intellectual ancestry went back through such men as Blondel, Bergson, and Balzac to the older Lamarckian tradition in French philosophy.[3] (However much Teilhard may have tried in his later years to turn himself into a good neo-Darwinist, he never escaped from the consequences of his intellectual upbringing; and the current *rapprochement* between the modernist followers of Teilhard and French neo-Marxist writers like Roger Garaudy is, as a result, the marriage of two schools of thought

---

[1] Ibid.

[2] See Z. Medvedev, *The Rise and Fall of T. D. Lysenko* (New York, 1969).

[3] See my own essay on Teilhard in *Commentary*, 39 (March 1965).

both of which are fundamentally pre-Darwinian and providentialist.)[1]

If we place Kuhn's 'evolutionary' account of scientific change alongside the general body of evolutionist ideas current in the human sciences right up to the present day—not just in social and political theory, but also in anthropology, linguistics, and elsewhere—we can say, without grave injustice, that these speculations have remained, for better and for worse, within the providentialist tradition of Lamarck, Spencer, and Marx. That tradition has helped to give evolutionist sociologists and anthropologists their very proper respect for historical development as a fact, but at the same time it has exposed the whole idea of socio-cultural evolution to damaging objections. Association with a belief in the inevitability of social 'advance', and the patronizing implications for our attitudes to more 'primitive' peoples and cultures, have brought the providentialist overtones of evolutionary sociology into discredit. Meanwhile, the very possibility of applying Darwin's own populational modes of explanation in the human sciences has been largely overlooked, and the full implications of the populational approach have never yet been worked out.[2]

That, of course, is the point from which our own analysis began. Instead of speculating about any universal and irreversible direction of conceptual development, we have tried merely

[1] It is interesting to notice that even Margaret Mead, who dedicates her Terry Lectures at Yale on *Continuities in Cultural Evolution* (New Haven, 1964) to Charles Darwin's grand-daughter, Nora Barlow, is nevertheless tempted to assume a scheme of evolutionary 'phases' and 'levels' which has no counterpart in regular neo-Darwinist zoology. On the other hand, such a scheme *is* characteristic of the 'evolutionistic' ideas about sociology inaugurated by Herbert Spencer.

[2] One early attempt to apply the idea of 'selection-principles' in epistemology is to be found in G. Simmel, 'Ueber eine Beziehung der Selectionslehre zur Erkenntnistheorie', *Archiv für systematische Philosophie*, I (1895), 34–45. More recently, the Darwinian pattern of explanation has been used in philosophy of science, with great sensitivity, in G. Holton, 'Scientific Research and Scholarship', *Daedalus*, 92 (Spring 1962), 362–99; also G. Holton and D. H. D. Roller, *Foundations of Modern Physical Science* (Reading, Mass., 1958), Chapter 14, esp. sec. 14.3. The most methodical and fully-worked-out attack on the problem of generalizing the Darwinian notions of variation and selection that I know of is by Donald T. Campbell, who has also collected references to many earlier forays in this direction, by Baldwin, Simmel, Konrad Lorenz and others. See (e.g.) D. T. Campbell, 'Variation and Selective Retention in Socio-Cultural Evolution', in *Social Change in Developing Areas: a Reinterpretation of Evolutionary Theory*, ed. H. R. Baringer *et al.*

to show how the process of 'variation and selective perpetuation' helps to explain the transformations of conceptual populations; and so to restate, in a more tractable form, those questions about the rationality of conceptual change which Collingwood and Kuhn left mysterious. In doing so, we have been able to present the structure of the processes by which historical populations develop in an entirely general form. There is—it appears—no reason for restricting populational analysis to problems about organic species and scientific disciplines. The notion of a 'historical entity', comprising successive sets of constituent elements which maintain a recognizable unity and continuity despite their changing content, can be discussed in abstraction from any specific application to (say) evolutionary zoology or the history of science. The populational approach accordingly provides an entirely general mode of historical explanation; and one that is a healthy antidote both to excessively systematic and ahistorical analyses of developing systems, and also to excessively sweeping 'evolutionist' doctrines.

A clear-headed application of this 'populational' model, however, demands certain precautions. In moving from field to field, we should take care to avoid Mach's fallacy, and remember that the only feature common to all populational changes is, precisely, the general form of this dual process of variation and selection. And we should notice, also, that the twin sub-processes of variation and selection can be related in either of two quite different ways. They may take place quite independently, so that the factors responsible for the selective perpetuation of variants are entirely unrelated to those responsible for the original generation of those same variants. Or, alternatively, they may involve related sets of factors, so that the novel variants entering the relevant pool are already pre-selected for characteristics bearing directly on the requirements for selective perpetuation. To mark this difference, we may say that in the latter case

(Cambridge, Mass., 1963), pp. 19–49; 'Natural Selection as an Epistemological Model', in *A Handbook of Methods in Cultural Anthropology*, ed. R. Novell and R. Cohen (New York, 1970); and 'Evolutionary Epistemology', in *The Philosophy of K. R. Popper*, ed. P. A. Schilpp (forthcoming). Cf. also Edgar S. Dunn jr, *Economic & Social Development: a Process of Social Learning*, pt. III ((Baltimore, Md., 1971).

variation and selection are 'coupled', in the former case 'decoupled'. On these alternative assumptions, the balance between variation and selection can be very different, and the unity and coherence of the resulting 'historical species' is maintained in different ways. Within a 'coupled' evolution, the *rate* of variation may be comparatively high, without the historical species thereby losing its compactness, for the coupling itself will tend to offset the effects of the unusually rapid variation. In a 'decoupled' evolution, on the other hand, the *range* of variation can be unlimited, since the historical species is preserved by a selection process which is severe in the treatment of all extreme novelties, except where these turn out to be 'pre-adapted' to a new niche.

This distinction, between 'coupled' and 'decoupled' processes, allows us both to understand and to answer the objections which some evolutionary zoologists have raised to applying the idea of evolution anywhere outside biology. Evidently enough, the evolution of organic species and the evolution of intellectual disciplines exemplify the formal pattern of populational explanation in rather different ways. Though conceptual change and organic speciation both result from the selective perpetuation of variants, the two processes are in other respects quite dissimilar. While conceptual variation and intellectual selection are coupled, for instance, genetic mutation and ecological selection are decoupled; and this difference has struck many biologists as so supremely important that they reject any argument based on other broader similarities between the two processes.

In a recent lecture, for instance, Medawar rightly criticized Herbert Spencer's views about social and cultural evolution, as having missed the point of Darwin's theories;[1] and elsewhere Medawar has several times attacked providentialist interpretations of evolutionary change.[2] If he does so, that is for understandable reasons. For he fears that a revival of evolutionist theories in the human field could have damaging consequences for evolutionary biology also. On the margins of contemporary biology, indeed, there are still plenty of writers who are as yet unreconciled to the decoupling of

---

[1] P. B. Medawar, 'Herbert Spencer and the Law of General Evolution', in *The Art of the Soluble* (London, 1967), pp. 40ff.   [2] Ibid., p. 58.

genetic mutation from ecological selection, which is the key-point of contemporary neo-Darwinism. And if people began taking theories of socio-cultural evolution seriously again (Medawar suspects) that fact itself might help to revive 'orthogenetic' speculations, which seek to 'couple' variation and selection, in biology also;[1] in which case all the old technical battles on which neo-Darwinism rests would have to be refought once again.

The charms of zoological 'orthogenesis' are, in fact, extraordinarily persistent. Although the ecological selection processes are quite enough, by themselves, to maintain the 'teleonomic' character of organic change, supporters of orthogenesis continue to recoil from the idea that zoological evolution is a 'decoupled' process, on the ground that this will make it a 'random' or 'blind' affair. Yet to suppose that mutation and selection might in fact be 'coupled' would have even more curious consequences. For this implies that the gamete has some clairvoyant capacity to mutate, preferentially, in directions pre-adapted to the novel ecological demands which the resulting adult organisms are going to encounter at some later time. And Medawar sees the only guarantee against such paradoxes in ignoring the pre-Darwinian history of the term 'evolution', restricting its use to zoology alone, and banishing it from all discussions of social, cultural, or intellectual change. This prescription is unnecessarily drastic. Given the whole range of historical precedents, the proper course is to treat the word 'evolution' as a general term, covering all historical processes in which a compact but changing 'population' is represented by successive sets of elements related by descent. On this definition, organic change, cultural change, social, conceptual, and linguistic change will be so many different varieties of historical 'evolution', all of which involve genealogical relationships between species, cultures, societies, and so on. Subsequently, we can go on to distinguish between the 'coupled' and 'decoupled' kinds of evolution; and so to recognize that neo-Darwinist zoology gives us first and foremost, not a general

[1] The writings of Arthur Koestler, esp. *The Ghost in the Machine* (London and New York, 1967), Marjorie Grene, Gertrude Himmelfarb, and Michael Polanyi come particularly to mind, as drawn towards some form of 'orthogenesis' in evolution.

account of evolution as such, but a well-established theory about the decoupled mechanisms involved in the special case of organic evolution.

This definition will leave us free to employ populational explanations in the discussion of human affairs, without getting ourselves entangled all over again in the progressivism of Lamarck, Spencer, and Engels. Indeed, in the discussion of human affairs, as much as in zoology, it is more than time that the providentialist interpretation of evolution was laid aside. Instead, discussions of social, cultural, and intellectual change should be paying proper attention to the populational mechanisms by which representative sets of institutions, practices and concepts are transformed in the transition from one temporal cross-section to the next; and to the criteria for deciding whether, having regard to the 'demands' of the relevant situation, these changes took place in a more or less 'well-adapted' manner.

5.2: *Languages and Societies as Historical Entities*

In our proper enthusiasm for zoology as an empirical science, therefore, we must not underestimate the profundity and generality of Darwin's conceptual achievement. For he gave us an insight into the character of historical entities, and the populational processes by which they evolve, which deserves to be illustrated in other fields of enquiry. By adopting a populational approach to organic species, Darwin overcame difficulties which had been inescapable, so long as zoology was committed to the older typological (or 'essentialist') concept of a species; and parallel difficulties have made themselves felt in other fields of study also.[1] In sociology or anthropology, political theory or linguistics, we may be able to distinguish between separate nations, cultures, societies, and/or languages (as we can between organic populations or intellectual disciplines) with sufficient precision for practical purposes, for just so long as we confine ourselves to 'synchronic' issues. But the moment we raise 'diachronic' issues, and set out to produce a properly historical analysis of the relevant changes, we must face the underlying question, 'How do distinct "historical entities" of the required kind come to exist at all?' The first demand on an

[1] See, e.g., M. T. Ghiselin, *The Triumph of the Darwinian Method* (Berkeley, 1969), Chapter 10.

adequately historical theory of political or cultural, social or linguistic change is thus to explain how specific and distinct nations (cultures, societies, languages) come into existence, and subsequently maintain their specificity, despite historical diversity and change. As R. G. Collingwood put the point: 'What was that Eternal City—that continuing "Rome"—whose essential life-history Livy saw himself as writing?'[1]

To begin with a comparatively straightforward example: on what conditions can we regard a single, specific language as maintaining its identity, despite the geographical distribution of varying dialects, and the historical introduction of novel usages, speech-habits and syntactic variants? (In what sense was the language in which Chaucer wrote 'English'?) On what conditions, likewise, can we claim continuing identity for an individual nation, culture, or society, in the face of synchronic and diachronic diversity in the constituent communities, customs, or institutions? In each case, the traditional debate has too often lapsed into a barren opposition between nominalism and realism, owing to the implicit assumption that any authentic 'entity' depends for its continuing 'existence' on its possession of some historically-unchanging 'essence'. Either (it has seemed) there are certain enduring and universal 'core' features shared —despite their historical or geographical diversity—by all exemplars of 'Rome' or 'Judaism', 'Soviet society' or 'the English language'; or else our application of a common name to such diverse exemplars must be purely arbitrary, lacking any real foundation in the nature of things.

Darwin's populational schema puts us in a position to break this philosophical deadlock. An organic species is not a permanent entity, defined by unchanging 'essential' characteristics; yet neither are its representative exemplars associated merely by our own arbitrary verbal decisions. And the same is true, equally, of a society, a culture, or a language. Rather, each is a historical entity: i.e. an individual whose component elements are at all times subject to diversification, but one within which

---

[1] See R. G. Collingwood, *The Idea of History* (Oxford, 1946), esp. pp. 43–4. For Livy, Collingwood writes, 'Rome is a substance, changeless and eternal'. He continues, 'Rome is described as "the eternal city". Why is Rome so-called? Because people still think of Rome as Livy thought of her: substantialistically, non-historically . . .'

selective factors are continually pruning out 'ill-adapted' innovations. How do we recognize the specific continuity and character of such a historical entity? The operative questions will now have to do, not with the 'functionality' of the current elements, regarded as components in a systematically structured entity whose existence is taken for granted; instead, they will have to do with the 'adaptive balance' between innovative and selective factors, without which no well-defined or clearly recognizable—in a word, no 'specific'—society, culture, or language would have existed in the first place. And, if the operative questions in sociology or linguistics are indeed of this form, we can carry over into those fields of enquiry some of the more sophisticated concepts and methods which have already proved their worth in the populational analysis of organic evolution.

To illustrate the virtues of this approach, we may look first at the field of linguistics.[1] Suppose that we take, as the component 'elements' of a particular language, the speech-habits of the different men or groups who use it; we may then follow out the changes that affect those speech-habits over a period of time. Having done so, we shall at once be faced by the question: 'How does any such language develop in such a way as to maintain itself as a recognizable or specific "entity" at all?' A language is a Protean thing: it is subject to a synchronic diversity in its local dialects and specialized usages, while a continual influx of neologisms and syntactical novelties leads to diachronic diversity also. So how is it that, in the linguistic as in the organic field, we find ourselves presented not with a totally unstructured continuum of speech-forms and dialects, but with distinct languages to which we can give separate names, such as English or Bengali, Xhosa or Yiddish?

Questions of this form become straightforward enough, if we treat a language as a historical entity, defined by an 'envelope' which embraces the different exemplars included in its temporal

---

[1] For a clear application of evolutionary and ecological concepts to linguistics, see particularly Charles A. Ferguson, in Ferguson and Bateson, *An Introduction to Linguistics for Non-Linguists* (duplicated, 1966); also the many works on the evolution of language by Joseph H. Greenberg and others. Cf. D. Hymes, *Language in Culture and Society* (New York, 1964), esp, pp. 82–5.

evolution; and if we make it a condition for acknowledging the continued 'unity' of the language, at any time, that members of the various speech-communities employing it should be capable of making themselves sufficiently intelligible to one another. (This corresponds to the requirement of natural interfertility, which plays a corresponding part in the populational definition of an organic 'species'.)[1] This done, we can frame some worthwhile hypotheses about the historical development of actual languages. Thus, we can extend to languages the arguments about hybridization and fragmentation touched on earlier in the case of intellectual disciplines. It is a historical commonplace (e.g.) that, once the chances of political and social history had fragmented the total population of Latin-speakers into geographically remote communities, whose speech-forms were no longer mutually intelligible, the way was opened for the evolution of new, and specifically distinct, 'Romance languages'. The moment the scholastic authority of the Roman Church, and the use of Latin as the language of education and administration, ceased to dominate the vernacular speech of the former Imperial territories, this possibility became an actuality. From then on, one could speak of (say) Portuguese, French, and Roumanian as becoming separate and specific languages in their own right. These new languages were (we might say) so many independent 'successor-species' to Latin.[2]

This, however, is only the first step. An analysis of linguistic evolution immediately leads, also, to questions about the sources and rates of linguistic variation, and about the criteria and selection mechanisms operative in this process. As to variation: in linguistic as much as intellectual affairs, variations appear naturally with or without any particular stimulus, though there are factors which can accelerate their occurrence. They appear naturally, simply because the transmission of a language from one generation to the next is never 100 per cent exact, so that the 'standard' speech of a younger generation departs in

[1] See, e.g., A. J. Cain, *Animal Species and their Evolution* (London, 1954), pp. 113-14, 141-3.

[2] The study of branching genealogies in the Romance languages was already well begun before evolutionary ideas got a serious foothold in zoology, and so considerably antedated the work of Lamarck and Darwin. Arguably, indeed, linguistics was the first subject in which the idea of 'evolution' found effective application.

certain respects from that of its immediate predecessor. This natural linguistic variability, or 'drift', is in some cases reinforced by changes in the conditions of life, by importations from other languages, by the deliberate coining of novel usages, and even—occasionally—by deliberate linguistic 'reforms'.[1] Just how far these further influences will serve to increase the rate of linguistic variation depends, in turn, upon considerations of many kinds: on the channels and level of education of the speech-community, on the circulation and degree of acceptance of newspapers, radio, and television, on the patterns of urban and rural life, and so on. In these respects, much of the basic empirical information still remains to be collected.

As to selection: once again, this is the crucial process for explaining the current 'direction' of linguistic evolution. At a first and coarsest level, the requirement of sheer intelligibility acts as an initial sieve. Recall how conceptual variants must achieve the standing of 'possibilities', before they can enter the real competition for incorporation into the 'established' body of a science. Similarly, a novel linguistic variant must achieve intelligible currency, at least within some localized 'in-group', e.g., professional-class teenagers in the Chicago public school system: otherwise, it does not even become a 'possible' candidate for eventual incorporation into English-at-large. Beyond this point, the procedures by which novel usages are sifted vary greatly from one linguistic community to another, and different languages evolve in correspondingly different ways. In some cultures, prime authority for approving linguistic novelties may be vested—at any rate officially—in some scholarly body, like the Académie Française or the Luso-Brazilian Academy. Alternatively, as with Arabic, it may be exercised by comparison with a supremely authoritative corpus of sacred writings.[2] Elsewhere, again, this authority may be diffused more widely and

---

[1] The term 'drift' is here being used in its normal evolutionary sense; but it also has a technical sense within theoretical linguistics, where it refers chiefly to rather longer-term changes in the character of the language. Clear examples of modern European languages in which there have been deliberate linguistic reforms are Dutch, Greek, and Norwegian.

[2] Although the Koran is itself written in slightly archaic Arabic, the ruling standard for classical Arabic remains a body of sacred writings, in which the Koran plays the central part, though modified by certain authoritative later commentaries.

informally: the determination of what is to count as 'correct English' (say) rests with a much less well-defined élite, relying on much less explicit standards and criteria.

These differences of reference-group and selection-criteria produce distinctively different consequences. At one extreme, there are a few highly specific languages which, for historical reasons, have evolved in isolation from their nearest cognates, e.g., Finnish, Magyar, and Basque. At the other extreme: given a large and stable agricultural area occupied by communities of common ancestry, but lacking any major centre of political power or scholastic authority, the linguistic selection process may not be severe enough to shape the current range of vernaculars into a specific common language. Instead, a continuous gradient may develop, along which neighbouring dialects are mutually intelligible, but more remote ones become too dissimilar to appear, by normal standards, as alternative dialects of the same language. (Such taxonomic gradients are familiar in botany, as with the wild tulips of Iran, in cases where the lack of an effective forum of selective competition leads to a continuous spectrum of varieties, within which clear specific boundaries can no longer be established.)[1] By the severest standards, indeed, we might even be tempted to say that the former class of 'isolated' languages were the only *fully* 'specific' ones.

Between these extremes, there lies a range of cases in which 'standard' or 'correct' speech-forms are maintained over larger or smaller linguistic territories, by different reference-groups applying different criteria. In this respect, the family of Romance languages can be contrasted with the family of Arabic vernaculars. What we know as 'classical Arabic' is the product of a strong, explicit, and deliberate selection process. The literate, scholarly *Ulema* have been responsible for exercising a collective censorship over linguistic innovations, using the Islamic scriptures both as their model and as their source of authority. For public affairs as well as religious purposes, this 'classical' language is superimposed on a geographical continuum of vernaculars, ranging from the Arabic dialects of Morocco in the West to those of Muscat in the East—to say nothing of hybrid dialects and languages spoken outside the contiguous

[1] See A. D. Hall, *The Genus Tulipa* (London, 1940), p. 32.

core-area of Islam, such as Swahili and Indonesian. Classical Arabic remains, as a result, a highly specific language, in which the selection procedures are unusually rigorous and the rate of variation is kept artificially low. By contrast, the 'standardized' vernacular forms of Arabic current in localized areas and culture-groups show a far lesser degree of unity and continuity than that of the classical language.

The present-day Romance languages are an indication of what the Arabic vernaculars would become, if the dominant influence of the classical form were removed. Under Imperial Rome, the Latin language played much the same part that classical Arabic plays today; and to a lesser extent this remained the case under the early medieval Church. But by now the 'forum of competition' has shifted from the imperial to the national level. In this case, as with Arabic, however, the divisions between specific 'languages' can still be drawn with complete clarity, only on a literary or 'authoritative' plane. The Romance dialects of Galicia and Minho, of Catalonia and Roussillon, of Piedmont and Haute Savoie resemble one another more closely than each of them resembles literary Spanish or Portuguese, French, or Italian. Literary authority apart, vernacular usage within the overall Romance-language territory displays just the 'linguistic gradients' we referred to earlier. If there were now to be marked shifts in the political organization of the region, those vernaculars might well provide the raw material for the evolution of other fresh Romance languages. A political reorganization of the Western Mediterranean basin, for example, could result in the speech-habits of Marseilles and Perpignan, Majorca and Barcelona converging on to a new Provençal–Catalonian norm. Under these circumstances, the establishment of new cultural authorities and other selective mechanisms might turn the word *Marseillaise* into the name, not just of a dialect and a national anthem, but also of a specific language.

(The relationship between classical and vernacular Arabic, literary French and *patois*, *Hochdeutsch* and Swiss-German dialects, are examples of the linguistic phenomenon known as *diglossia*.[1] Inhabitants of some linguistic territories grow up

---

[1] See the paper by C. A. Ferguson, 'Diglossia', *Word*, 15 (1959), 325–40: reprinted in Hymes, op. cit. pp. 429ff.

'bilingual', as between the standardized or official form of their native language and some particular local dialect form. As our model would suggest, the 'high' or official forms are in such case both more uniform and more stable, since they are subject to more severe monitoring and correction than the colloquial or 'low' forms; and there are even arguments for saying that specific languages are distinguishable only on the 'high' level.)

From the taxonomic standpoint, all of this is just what one should expect, if the evolution of languages follows a populational pattern; and the pattern can be taken further. Suppose, for example, that we use natural interfertility as our test for the specific distinctness of coexisting organic populations with overlapping territories; under certain circumstances, we shall then be confronted with what is known as a 'ring species'. For example, the habitats of these populations may lie (say) all the way around some natural obstacle—a mountain-range, an ocean, or a desert—and two separate sub-populations may begin by spreading round this obstacle from either side, eventually colonizing the far side from both directions. During this progressive colonization, natural variation sometimes proceeds at such a rate that, when they meet again on the far side of the obstacle, the successor-populations are no longer crossable.[1] In terms of our original criterion, the resulting situation is highly paradoxical. At every other point around the ring of habitats, neighbouring populations cross successfully, and are acceptable as varieties of a common species; yet, in the area of overlap, the sub-populations colonizing the habitat from different directions no longer cross, and give the appearance of belonging to different species.

If we substitute 'intelligibility' for 'interfertility', a similar linguistic phenomenon would be immediately explicable. Suppose that the island of Ceylon (say) had been colonized independently by two groups of Turkish-speakers, one from Eastern Siberia, the other from Istanbul. These might find each other's languages totally unintelligible, even though there is a linguistic gradient across Central Asia by which their two language-forms are continuously connected. Other refinements in taxonomic theory could also find a similar application in linguistics: the concept of a 'polytypic' species, for instance,

[1] Cain, op. cit., pp. 141ff.

might be used to throw light on the multifarious forms of contemporary English spoken in (e.g.) the United Kingdom and North America, India and the Caribbean; the concept of a 'paleo-species' would clarify the problems arising (say) over the relations between Chaucerian, Elizabethan, and Modern English; and so on.

To broaden the discussion, let us leave linguistics for sociology and anthropology. In both these human sciences, the opposition to evolutionary theories in the early decades of the present century was a legitimate protest against the 'progressivist' assumptions of Herbert Spencer and his successors. In place of Spencer's Lamarckian ideas, early twentieth-century sociologists and anthropologists analysed society and culture with physiological models in mind. In order to understand the 'structure' of a society, they claimed, one must recognize how—like bodily organs—its institutions co-operate 'functionally', to harmonize each other's purposes.[1] The true business of sociological theory was, then, to reveal the character of the 'social structure' or 'system'; and the outcome of this anti-historical move was an entire theoretical tradition concerned with static social structures and relations.[2] The assumption of 'systematicity' was fundamental to the whole resulting tradition, and was presupposed at the outset; as a result, it made the explanation of social change as hard as the explanation of conceptual change has been for 'systematic' philosophies of science. For how could the balance between institutions then be altered, except by 'dysfunctional' changes?

In recent years, there has been a widespread revolt against this tradition, in favour of romantic or revolutionary notions, and the resulting rejection of orthodox sociological theory might at first sight appear complete. Yet, on closer examination, the revolutionary doctrines of contemporary anarchism turn out to rest on the same over-systematic presuppositions as the theories they deny. Whereas academic sociologists have often accepted

---

[1] For a current critique of the ideas of men like Radcliffe Brown, see the papers of Derek Freeman (see above, p. 325, n. 2). See also M. Fortes, *Kinship and the Social Order* (Chicago, 1969), and my own essay, 'Rediscovering History', in *Encounter* (January, 1971), 53–64.

[2] Cf. T. Parsons, *The Structure of Social Action* (New York, 1937); new edn. (New York, 1949); and *The Social System* (Glencoe, Ill., 1951).

the notions of 'system' and 'functionality' without criticism—so taking a serious interest only in the functional successes of social institutions, and explaining away any tendency for co-existing institutions to frustrate each other's goals—their revolutionary opponents have retorted by denouncing 'The System' and all its works. As between the orthodox function-alists and the revolutionary anarchists, the only real point of difference is that the functionalists treat the existing institu-tions of any social system as working together for good, the anarchists see them as working together for ill.[1] Both parties presuppose a physiological model of society: one accepting harmonious interactions as the theoretical 'norm' for any intelligible 'social physiology', the other dismissing actual societies as totally cancerous and fit only for drastic political surgery.

Neither the academic theorists nor their anarchist critics, however, can give us an intelligible account of the occasions on which, and the manner in which, social institutions and relation-ships change, for the fundamental attitudes of both parties are as 'pre-Darwinian' as those of Herbert Spencer himself. Neither side, indeed, shows any sign of having learned the general historiographical lesson of Darwinism: that, in the development of historical entities or populations, it is not the *current* structure and relationships within that population which require to be explained as 'functional', but rather the *changes* taking place in them which require to be explained as 'adaptive'. Each of the coexisting institutions and customs making up a society or a culture will possess, at any particular time, its own internal forms and structures, which serve—in a quasi-physiological sense—to define the normal 'functioning' of each separate element. The relations between these different institutions or customs, on the other hand, will be—in this sense—neither 'functional' nor 'dysfunctional'. Each individual element in the society or culture will normally have come into existence at a different point in history, to meet some particular set of needs, problems, or 'demands'. And, just as we can legitimately raise questions about the 'functional' adaptedness or 'effective-ness' of a science, only when we are considering how changes in

[1] See, e.g., the essay on Herbert Marcuse by Maurice Cranston in *Encounter* (March, 1969), pp. 38ff.

350 Rational Enterprises and Their Evolution

the relevant concepts serve as responses to the current pattern of 'intellectual demands', so too with the institutions of a society and the customs of a culture.

The current institutions and customs are, thus, properly seen as forming largely unsystematic aggregates, or 'populations', whose elements work together, neither 'for good' entirely nor 'for ill' entirely, but simply as their joint histories dictate. Only when we consider the manner in which the current institutions or customs are changing, can we ask the operative question, 'What does this novel feature of society or culture achieve?'; and the answer will consist in showing how, of all the changes possible at that stage, this particular innovation came to win a place in the society or culture.[1] So the task of constructing an authentically historical sociology, anthropology, or political theory is again a Galilean rather than a Copernican task.

A society, then, is a 'historical entity', in the same sense as a scientific discipline or an organic species; and it is represented, at any given time, by a certain set of institutions, procedures, and relationships, whose historical origins and purposes were largely separate and independent. Throughout the development of a society, all these 'social elements' are subject to variation. But the selective modifications which result from the incorporation of adaptive novelties will normally have repercussions on a small part of the society only. What a populational sociology has to explain, is (i) the factors in any historical context favouring and/or hindering institutional and procedural innovation; (ii) the manner in which social or administrative innovations win acceptance, and establish themselves, within societies of different kinds; and (iii) the criteria by which one can legitimately judge how far any actual institutional change was genuinely 'adaptive', and effectively resolved outstanding social problems in a particular situation.

This means treating a society as a historical species developing more or less 'compactly', in response to changing historical demands.[2] Within any actual human community, the mere

---

[1] It is a pity that, even now, Lévi-Strauss sees his functional analyses as explaining the *structures* of cultural practices, rather than the factors responsible for *changes* in those practices.

[2] Cf. The writings of Lesley White and other more recent sociologists. Once we *do* allow ourselves to introduce genuinely 'evolutionary' categories into sociological

existence of any social institution, custom, or procedure will not be explicable in functional terms alone; since it may have survived a long sequence of historical changes, only because no better-adapted alternative was available, or because the institutional selection process was unusually conservative—so that it kept its place in spite of having lost its original significance, or adaptive advantage. The most profound measure of the comparative 'merits' of different societies will then be their respective degrees of 'adaptability': i.e. the scope they allow for social or institutional innovation, and the forums and procedures they recognize for the critical reform of current institutions and customs.

Such a populational analysis of social change is relevant, with appropriate changes, not merely to complete societies but also to their individual institutions. Within an institutional organization, the current conduct of affairs conforms—in the terminology of political theorists—to 'standard operating procedures' established during the foundation and subsequent history of the organization.[1] Formally speaking, its structure may be representable by a static 'organizational tree'; but, functionally speaking, its activities will be related less systematically, and represent the outcome of all those coexisting policies and procedures which have been established, for different purposes, at different times in the past. The collective outcome of these policies may in fact end by being inconsistent and self-frustrating. As with the occasional self-contradictions within a scientific theory, this fact is one familiar source of administrative problems calling for institutional reforms. (At this point, authentic analogies between the intellectual conduct of our rational enterprises and the practical conduct of our social

theory (as Catherine Bateson points out to me) all sorts of extra possibilities become available to us: e.g., explaining the adaptive advantages of leaving a substantial degree of 'redundancy', both in the structure of language and in the structure of our cultures.

[1] See, e.g., G. Lenski, *Human Societies* (New York, 1969), Chapters 3 and 4. Lenski explicitly bases his theory of social structure and change on the 'synthetic theory of evolution' of Ernst Mayr; and he goes on to use it to explain 'sociocultural continuity, innovation and extinction'. For the notion of 'standard operating procedures', as the component elements in the functioning bureaucratic institutions and agencies, consider the analyses by Richard Neustadt and his students in the Harvard School of Government. See, e.g., Neustadt's *Presidential Power* (New York, 1960) and *Alliance Politics* (New York, 1970).

enterprises invite the use of Marxist jargon about the 'inner contradictions' of social institutions.) We may then explain the historical development of a social institution, not as the result of a changing equilibrium between the internal 'laws' and 'forces' of a social 'system', but in terms of the adaptive evolution of its standard operating procedures.

This kind of theory of institutional development might even be used to bridge the gap between a populational analysis of historical change in entire societies and our own theory of conceptual change within rational enterprises. For there is now a continuous scale, leading from a single scientific discipline, at one extreme, to a complete society at the other. At every point on this scale, we are concerned with an aggregate, or population, whose elements comprise constellations of rule-conforming 'procedures'; and a comprehensive theory of historical change will have to show how the relevant constellations of procedures —whether conceptual, institutional or social—can be progressively modified in response to the ecological demands of successive problem-situations.

Like concepts, institutions find expression both through verbal appeals to certain symbolic words and principles, and also through practical conformity to certain established procedures. We could thus speak of (say) the Divine Right of Kings as a political 'concept' which, at one and the same time, symbolized the institution of Absolute Monarchy and found behavioural expression through an authoritarian pattern of political procedures; and likewise—for better or for worse— in the case of other social and political 'concepts', such as the Rule of Law, White Supremacy and the Dictatorship of the Proletariat. Just as the fully-fledged acceptance of a scientific concept (e.g. 'energy') commits one not merely to the symbolic employment of certain technical terms and forms of calculation, but to their practical use within the explanatory procedures of the science concerned, so too it can be helpful to describe current authoritative patterns of social, political, or judicial behaviour as the practical expressions of some collective 'concept'.

With these examples in mind, we could invert our earlier epigram, *Concepts are Micro-institutions*. The purpose of that epigram was, of course, to redirect attention away from the formal aspects of scientists' explanatory arguments and towards the behavioural aspects of their explanatory procedures. So, now, we might declare that *Institutions are Macro-concepts*. Like concepts, institutions find behavioural expression through changing constellations of 'standard operating procedures'. They preserve their self-identity through change by selective innovation in response to changing situations, in the name of collective social goals, and in this, too, they display an unremarked parallel to concepts and conceptual evolution. So regarded, indeed, the historical evolution of social institutions would simply become the process of conceptual evolution writ large.

In this chapter, I have tried to make explicit, in general terms, the broader implications of Darwin's historiographical method for the human sciences. But, in conclusion, this analysis puts us in a position to go still further. For, if carried through to the end, this historiographical transformation can justify us also in cutting off the traditional debate in metaphysics at its starting-point, by discrediting the problem which served as the opening gambit of the whole philosophical chess-game. That problem was the famous Parmenidean question about the One and the Many: 'How do the permanent, eternal objects of the world preserve their underlying identity, through all the surface appearances of change?'[1] If our present analysis is correct, that problem simply does not arise, in the form stated. Taking the question as asked, the only candid reply we can now give is, 'There are no such permanent, eternal objects.'

When Plato and Aristotle discussed the metaphysical theses of the pre-Socratic philosophers, they did not regard them as raising merely abstract questions, of a kind irrelevant to the programme of natural science. On the contrary, the pre-Socratics had set out to answer the question, 'What general kinds of things are there in the world?', not least as a methodological preliminary to constructing a rational science of nature. Any rational and coherent natural philosophy must be built around permanent entities, which preserved their identity despite the seeming flux of temporal change; and the task for their successors was to characterize these permanent entities in terms which were, at the same time, both general and also

[1] J. Burnet, *Early Greek Philosophy*, 4th edn. (London, 1930), Chapter 4.

true to life. Given Plato's commitment to geometry and planetary astronomy, the essential character of the permanent entities appeared to him—naturally enough—formal and mathematical. To a marine biologist and natural historian like Aristotle, on the other hand, it seemed equally clear that things-in-general must be classified on the same principles as animals, fishes and plants; so metaphysics became for him a kind of generalized taxonomy. Neither man challenged the validity of the pre-Socratic question, 'How is the identity of permanent entities preserved through change?': they merely gave fresh answers to the same old question.

The Greek metaphysicians' belief in Eternal Objects thus had two sources in their scientific view of the world, both of which have since been undercut. In its Platonist form, this belief was nourished by the success of mathematical physics (or 'natural philosophy') in building empirically-relevant theories around timeless concepts and universal principles. As expounded by Descartes, Newton, and even Maxwell, the tenseless theories of physics seemingly relied for their validity on the assumption that the World of Nature comprised certain eternal, changeless fundamental constituents—figures or particles, corpuscles or molecules—whose interactions gave rise to all physical phenomena.[1] By now, we view the source of this 'universality' or 'tenselessness' differently. Taken in isolation (as Hertz and Wittgenstein demonstrated) the general concepts and principles of physics tell us nothing directly about the empirical world;[2] their scope, empirical relevance, or applicability, depends on the auxiliary associations we subsequently make between abstract, theoretical categories and actual physical systems.[3] By our own time, indeed, the experience of working physicists suggests that no empirical system whatever can exemplify any theoretical category eternally or permanently.[4] (Even protons and neutrons do not have infinite half-lives.) So we must take care to distinguish the generality—or 'tenselessness'—of theoretical concepts and principles from the unbounded antiquity —or 'eternity'—formerly attributed to the ultimate constituents

[1] See *Architecture*, Chapters 8ff.

[2] Recall Wittgenstein, *Tractatus*, paras 6.34–6.36, as discussed above, pp. 172–3.

[3] Cf. the essay by Hilary Putnam in the collection, *The Philosophy of K. R. Popper*, ed. P. A. Schilpp (Forthcoming).

[4] Cf. the discussion by G. Chew, pp. 234–6, above.

of the world. We shall then recognize that employing 'universal' concepts and 'tenseless' principles in our theoretical physics in no way commits us to assuming the actual existence of permanent, unchanging objects in the empirical world.

In its other, Aristotelian form, the belief in Eternal Objects was linked equally closely to an interpretation of science—in this case, of zoology—which is by now discredited. The metaphysical doctrine of 'permanent essences' drew empirical support from the success of Aristotle's zoological theory of fixed species, which was its most convincing application to our actual experience of the world.[1] For Aristotle, the metaphysical permanence of entities-in-general and the historical fixity of organic species were all of a piece; the doctrine of fixed organic species simply exemplified, in the special sphere of biology, the permanent character of all 'rationally intelligible' entities. Conversely, Darwin demonstrated that Aristotle's most favoured examples failed to support, in actual fact, the metaphysical assumption on which orthodox Greek natural philosophy had been based. Species were not, after all, permanent entities; the earlier, 'typological' or 'essentialist' approach to taxonomy inherited from Aristotle misrepresented the long-term history of living things. In this respect, twentieth-century populational studies have merely reinforced the intellectual foundations of Darwin's original insights. However irrelevant the empirical details of Darwin's work may be to general philosophy, the abstract form of his explanatory schema has a much broader significance. So, when Darwin and his successors showed that the whole zoological concepts of 'species' must be reanalysed in populational terms, their demonstration knocked away the second scientific prop from under the traditional metaphysical debate.

What must our own position be now? If neither sub-atomic particles nor organic species exemplify the 'permanent entities' of Greek metaphysics, what else in the real world does so? To this question there is only one honest answer. Two hundred years of historical research have had their effect. Whether we

---

[1] As Ghiselin, op. cit., puts it: for Aristotle, 'if species change, they do not exist, for things that change cannot be defined [i.e. have no essence] and hence cannot exist . . . The Darwinian revolution thus depended upon the collapse of the Western intellectual tradition.'

turn to social or intellectual history, evolutionary zoology, historical geology or astronomy—whether we consider explanatory theories or star-clusters, societies or cultures, languages or disciplines, organic species or the Earth itself—the verdict is not Parmenidean but Heraclitean. As we now understand it, nothing in the empirical world possesses the permanent unchanging identity which all Greek natural philosophers (the Epicureans apart) presupposed in the ultimate elements of Nature. So, if we ourselves are to entertain metaphysical thoughts about the nature of things-in-general consistent with the rest of our late-twentieth-century ideas, we must explore the consequences of the modern, post-Darwinian or 'populational' approach, as applied not just to species, but to historical entities of all kinds. Confronted with the question, 'How do *permanent* entities preserve their identity through all their *apparent* changes?', we must simply deny the validity of the question itself. In its place, we must substitute the question, 'How do *historical* entities mantain their coherence and continuity, despite all the *real* changes they undergo?'

Darwin's straightforward populational schema may not, in all cases, give the correct answer to this question, but that schema certainly provides one of the legitimate forms of answer. The balance between variation and selection within a population of constituent elements is, evidently, one of the possible processes by which historical entities preserve their transient identity. The task of exploring the implications of this populational schema in completely general terms—regardless of whether it will eventually be applied to organic species, or languages, or intellectual disciplines—is one that philosophers should be prepared to take seriously now. In this respect, the implications of an evolutionary approach to our theoretical problems—in the sense of populational, rather than a progressivist approach—can take us beyond the limits of particular special sciences, and require us to reappraise our categories and patterns of analysis even on the most general philosophical level. Here again, our views about Human Understanding must keep in step with our views about the World which we have to understand; and our subsequent account of the concept of Reason, and the standards of rationality, must reflect the historically changing character of their interaction.

SECTION C   Rationality and
Human Adaptation

Introduction

THE positive results of these inquiries, so far, can be summarized in the following statement:

A collective human enterprise takes the form of a rationally developing 'discipline', in those cases where men's shared commitment to a sufficiently agreed set of ideals leads to the development of an isolable and self-defining repertory of procedures; and where those procedures are open to further modification, so as to deal with problems arising from the incomplete fulfilment of those disciplinary ideals.

The representative set of concepts or procedures which constitutes the current content of any enterprise develops in a disciplined manner—on this account—not because the collective aims of the enterprise by themselves provide a criterion of 'truth', or impose a single, uniquely 'correct' system of concepts. It does so, rather, because the disciplinary ideals determine an agreed direction of conceptual and procedural change, and so the criteria for selecting 'acceptable' variants. In the remaining chapters of this first part, we shall be exploring the wider relevance and implications of this definition.

To begin with: certain qualifications and restrictions are evident in the very terms of our definition. Nothing in it, for instance, limits the scope of 'disciplined' enquiry to the natural sciences. The phrase 'scientific discipline' is not a tautology; nor need the agreed goals or ideals of a collective human enterprise in all cases be explanatory ones. On the contrary, the existence of agreed, communal goals of (e.g.) a practical or judicial kind can create the basis for disciplinary activities of non-scientific kinds; to that extent, law or technology can provide fields for the rational development of improved collective procedures, quite as legitimately as science. So we may set aside at the outset the positivist thesis that other varieties of problem-solving, language- and concept-use, are tainted with emotiveness or subjectivity as compared with scientific ones.[1] And the first question here will be, how far our evolutionary analysis of conceptual populations and their development is relevant also to such non-scientific disciplines.

In the second place: our definition limits the scope of the present analysis to activities in which the collective procedures, methods, and concepts develop in accordance with a sufficiently agreed set of collective ideals. This implies restrictions of two kinds. Even within the natural sciences, the explanatory ideals current in a particular field do not always have the agreed precision required if our disciplinary pattern is to apply exactly. Alongside fields like atomic physics and biochemistry in which there is a clear consensus about the intellectual goals, so that conceptual development can proceed in a 'compact' manner, there are other fields where the explanatory ideals are less clear or agreed, and the evolution of concepts is correspondingly 'diffuse'; while there are others again in which no substantial agreement has yet been arrived at about the goals at which scientific enquiry can reasonably aim, and which are— at any rate, to date—hardly more than 'would-be disciplines'. Furthermore, if we go entirely beyond the disciplinary activities of scientists, lawyers, and engineers, we shall find other spheres of human activity (e.g. the fine arts) in which the current goals and ideals are shared only in part by the practitioners concerned; and we may look and see how far the historical development of those partly disciplinary enterprises departs, as a result, from our straightforward evolutionary pattern.

Thirdly: we have defined 'disciplinary' activities as concerned with isolable and self-defining repertories of procedures, and this aspect of the definition involves an element of artificiality. Though scientific and technological enquiries may be pursued in a more or less self-contained and self-sufficient manner, this isolation is never in practice absolute. The coexisting enterprises of a milieu interact, both with one another and with the broader ideas and assumptions of the age. As well as studying the development of these parallel human enterprises separately, therefore, we shall do well to look also at the ways in which such wider interactions affect the evolution of their ideas, procedures, and techniques.

[1] See, e.g., A. J. Ayer, *Language, Truth and Logic* (London, 1936), p. 48, and H. Reichenbach, *The Rise of Scientific Philosophy* (Berkeley, 1951), *passim*.

Fourthly: we may look beyond the boundaries of our definition entirely, and consider what light this analysis throws on those spheres of human activity that are not disciplinary at all. If we are to understand fully how the notion of rationality applies to collective concept-use, we must pay attention not merely to those human activities and enterprises which are by nature 'disciplinable', but to collective concept-use more generally. This will mean going on, in due course, to see whether the same ecological relationships that we have illustrated in the case of scientific and other disciplines can be found in the case of these non-disciplinary activities also.

Notice, finally, that the theoretical pattern of 'disciplinary evolution' arrived at in the present analysis is, in one crucial respect, abstract and idealized. We have discussed the nature and historical development of the disciplines which are the product of our collective rational enterprises; we have done so, with the aim of throwing light on the communal use of concepts, and on the process of conceptual change; and we have deliberately framed the resulting account of concepts and disciplines in the most general terms possible. This being so, the question can—and must—be faced, just how far the resulting pattern is ever, in actual fact, exemplified by any real-life branch of science or learning. Does any real-life rational enterprise ever conform with absolute precision to our disciplinary ideal? Has any actual intellectual discipline, in other words, ever been perfectly 'compact'? We can answer these questions only by making explicit the empirical conditions on which our theoretical analysis can be expected to apply to real-life cases.

In itself, of course, the abstract character of our account is no basis for an objection. There are plenty of other fields where similar idealizations are employed. In any event, when we set out to apply such an idealized term, the operative question is not whether actual instances exist in reality to be designated by the new term. Rather, the question is, on what conditions the new idealized concept has any empirical relevance, and how it succeeds in throwing light, both on those situations to which it directly applies, and on those to which it does not. No actual object exemplifies with unlimited precision the mathematician's specification for a 'geometrical point' or a 'Euclidean straight line';[1] no real-life material system fully answers to the physical definition of a 'rigid body' or 'ideal gas'; no human being conforms unfailingly to the judicial ideal of a 'reasonable man'; nor are there any absolutely perfect instances of the economist's 'free market'. Yet this leaves the explanatory significance of such idealized concepts unaffected and undiminished. Theoretically speaking, indeed, they can be just as revealing in negative as in positive cases. When we consider, for instance, why carbon dioxide departs from the physicist's ideal of a 'perfect gas' more strikingly than oxygen, or when we discuss the conditions on which the workings of an oligopoly can approximate to those of a 'free' market, it is the explanatory fruits of the concepts that matter, not their exemplifications.

Having expounded the disciplinary model itself, then, we must now show (in Chapter 6) to what extent, and on what conditions, the present pattern of conceptual change is applicable to real-life examples. This does not, of course, mean presenting a formal mathematical proof that the model is logically adequate and consistent, as we should have to do if we were exploring the entailments of some previously-agreed set of axioms and postulates. Nor does it mean compiling a list of 'confirming instances' to which our account applies perfectly, and proving that all supposed 'counter-examples' are invalid, as we should have to do if we were concerned with some straightforward empirical generalization.[2] The significance of our account must instead be demonstrated progressively, as we go along: partly, in a positive way—by showing how it helps us to understand the development of those enterprises to which it applies with sufficient accuracy—and partly in a negative way, by explaining in some representative cases why the model fails to apply to activities which are not the concern of 'compact' disciplines.

[1] The idealization of geometrical concepts has been a source of philosophical perplexity, too. See, e.g., G. Berkeley, *Commonplace Book*, ed. G. A. Johnston (London, 1930), p. 41, para. 358: 'Memo: to ask the mathematicians about their *point*, whether something or nothing, and how related to the *Minimum Sensibile*.' Berkeley was much puzzled about the *sensibilia* of cheese-mites, which had recently been revealed by microscopists and existed at the limit of human vision.

[2] Even Karl Popper exaggerates the *closeness* of the relationships between theoretical statements in science and the empirical observations which illustrate them, and which they are used to explain. See, e.g., Chapter 6, 'Testability', of *The Logic of Scientific Discovery* (London, 1959).

This done, we must return (in Chapter 7) to a topic we set aside at the beginning of our exposition. Up to this point, we have been concentrating on those spheres of experience in which men's patterns of thought, action, and language are most clearly historico-cultural variables. But our analysis will not have been wholly successful, unless we do what we set out to do at the outset: namely, to explain not only (i) why those of men's concepts and patterns of thought which are local and variable change as they do, but also (ii) how those that give the appearance of invariance and universality come to be so stable and widespread.

CHAPTER 6

The Variety of
Rational Enterprises

6.1: *Non-Scientific Disciplines*

THE crucial element in a collective discipline (we have argued)
is the recognition of a sufficiently agreed goal or ideal, in terms
of which common outstanding problems can be identified.
Where this common goal is an explanatory one, the discipline
is a scientific one; but men's common efforts can equally be
directed towards technical or judicial ideals, and the concepts
of law or technology are therefore subject to a comparable
rational development. So we may first show, briefly, how our
disciplinary model might be applied in more detail to rational
enterprises outside the sphere of natural science. The area to
which this analysis can be extended most straightforwardly,
and with least strain, is that of the crafts and technologies.
These have developed to serve common human needs for more
effective and useful goods or materials, equipment or services;
they are concerned with the problems which arise in designing,
manufacturing, and distributing such goods or services; and their
historical development can be described in terms closely
parallel to those that apply to scientific disciplines.

Once again, each particular craft or technology has its own
characteristic constellation of ideals and ambitions, to which
anyone who takes it up as a professional career thereby commits
himself. In this respect, iron-working is a disciplined, professional activity as much as crystallography, medicine as much as
physiology, electronic engineering as much as atomic physics.
In the nature of the case, the collective ideals governing
technological development are explanatory neither in intention
nor in effect; instead, they are practical—being directed at
improving the techniques for producing and distributing
materials, vehicles, communications devices, information, or
whatever. Correspondingly, the 'transmit' of a historically

developing technology comprises not a changing population of theories and concepts, but a changing population of recipes and designs, techniques, manufacturing-processes, and other practical procedures. This apart, the specialized professional crafts and technologies have as much claim as the sciences to be described as 'disciplines', and they share the same patterns of historical change. In each case, we are concerned with the historical evolution of a 'transmit' or population of procedures. In each case, the apprentice qualifies for his profession by accepting its collective ideals and mastering its current procedures. In each case, the index of his proficiency is given by a demonstrated grasp of these ideals and procedures and of the uses to which they can be put. In each case, the current procedures are themselves subject to a critical process of innovation and selection, in the light of the disciplinary ideals. And, in each case, the overall process of disciplinary evolution is a locus for the operation of both 'causal factors' and 'rational considerations'.

As in natural science, the practical ideals of the different technologies are chosen so as to divide up the entire area of industrial arts and crafts into sub-areas, each of which belongs to a recognizably specific and self-contained discipline or sub-discipline. As in science, these technological disciplines and sub-disciplines preserve enough unity and continuity to warrant 'specific' status, without any universal, unchanging 'essence' being evident in their content, methods or ideals. Thus, for professional purposes, the broad field of electronic engineering is sub-divided into a number of sub-fields, concerned with (e.g.) cryogenics, micro-miniaturization, coherent-wave, and printed-circuit techniques. Each specialized sub-discipline defines its own boundaries and direction of development in terms of a correspondingly specialized goal: the least possible 'noise' or the greatest possible compactness, the highest possible reliability, directionality, or security.[1] And, in any sub-field, the established repertory of practical knowledge—commonly referred to as 'the state of the art'—comprises the most advanced set of techniques whose proven reliability is currently accepted by professional practitioners of a discipline. So the

[1] On selection-criteria in technology, see H. Skolimowski, 'The Structure of Thinking in Technology', *Technology and Culture*, Vol. 7, no. 3 (Summer, 1966).

current content of a technological sub-discipline 'represents', once again, a single temporal cross-section from the entire, historically developing technology. The collective task of systematically improving this repertory is the thread around which the corresponding technological profession crystallizes; and this commonly comprises an international circle of expert devotees—not unlike the 'invisible colleges' of the natural sciences—whose members are in close personal contact, study each other's work, and engage in a respectful but competitive rivalry.

Let us here hint at some of the parallels in historical development between the practical procedures of technology and the explanatory procedures of the natural sciences. (1) Just as a science comprises theories and concepts having different degrees of generality, so in a technology we can distinguish specific, detailed applications of technical procedures to particular problems from more general methods and techniques, and these again from the most general and fundamental principles of design practice current in the field concerned. (These most fundamental principles are, in effect, 'constitutive of' the relevant technologies: the notion of a 'circuit' is as much a defining feature of certain branches of electronics, as we know them, as the concept of 'inertia' is of dynamics.) During the last eighty years, for example, confectionery manufacturers have developed many novel and ingenious machines for producing wafer-biscuits; yet all these designs are recognizably variations on certain fundamental design principles—one might even say, on a general technological 'concept'—originally worked out in the 1880s.[1] Correspondingly, a radically novel technical innovation will frequently serve, in technology as in science, as the focus for a brand-new sub-discipline: as with 'cryogenics', the art of producing electronic circuits which operate at ultra-low temperatures, as as to take advantage of the phenomenon of superconductivity.

(2) The 'state of the art' in a technology develops, likewise, by the selective perpetuation of procedural innovations. For the current repertory of designs and procedures never wholly fulfils the collective ambitions of the discipline in question. The

[1] Augustus Muir, *The History of Baker Perkins* (Cambridge, Eng., 1968), pp. 37–8.

gap between the current disciplinary ideals and the most advanced repertory of techniques at present available defines a reservoir of technical problems; alongside the established state of the art, there is a pool of tentative technological variants or 'possibilities', currently under consideration; these variants are, in due course, judged with an eye to the 'selection criteria' of the discipline; and, as a result, some of the variants from this pool are incorporated into the art. In a technology as in a science, the disciplinary coherence of the 'historical population' is thus maintained by a balance between variation and selection. The current repertory then comprises all those techniques that have survived the selective professional judgements, the prejudices and tastes, memories and forgetfulnesses of the technologists concerned.

(3) In one significant respect, however, the process of technological innovation and selection differs markedly from that in science. In the natural sciences, the forum of competition and selection is predominantly 'internal' to the scientific profession. In technology, this is much less completely the case. In part, no doubt, technological innovations are appraised with an eye to 'expert' professional considerations, having to do (e.g.) with feasibility, efficiency, and ease of manufacture, all of which can be judged only by the relevant group of professional technologists. But, in addition, technical innovations are inevitably exposed to demands of other kinds also: in particular, those which arise in the patent-office, the market-place or the forum of public debate, rather than in the laboratory or on the test-bench. A worthwhile technological innovation, that is to say, must be not only technically feasible, effective, and suitable for regular production, but also original, competitive, and free from objectionable nuisances. This multiplicity of requirements does something to blur the collective disciplinary aims in the technological field, and makes the full single-mindedness of a science less easy to preserve. There can always, for instance, be legitimate differences of opinion about the relative priority of different demands; so that in a technology the question, what constitutes a 'success' or an 'advance', is one about which there is frequently less agreement than there is in science. Faced with that question, indeed, engineers and technologists from different backgrounds will display substantial

differences in outlook, temperament, and research strategy; and the resulting variety in habits of mind will give rise to correspondingly different 'schools' of engineering or technology.

The differences between these schools will commonly cut along national lines, like those which Duhem found between the scientific ideals of different national traditions. In one country, technologists will tend to be more academically minded, tackling their problems in a quasi-scientific spirit and valuing above all the esteem of their fellow-experts; elsewhere, they may tend to have a mere commercial orientation, and approach their problems with an eye equally—even primarily—to the market. In this way, one might define a professional and temperamental spectrum, ranging between the extremes of the purely academic and the purely commercial. A similar diversity is apparent also in the professional institutions of technology. Among technologists of any particular bent, different sub-fields will organize themselves in ways that tend to reinforce the original balance of emphasis of their members as between the 'expert' and 'commercial' aspects of their subjects. As elsewhere, such professional institutions will tend to operate —for good and for ill—as a conservative influence, acting 'in the name of' the ideals of the relevant discipline as locally conceived, and operating in ways that have the effect of preserving its original orientation and traditions.

(4) Given the parallels between scientific and technological change, we can go on to ask certain more detailed questions about the factors and considerations involved in the process of practical innovation and selection. What factors, for instance, affect most critically the absolute and relative rates of technological innovation in one branch of engineering or another, in one country or another, in one type of economy or another? What influences determine the dominant directions of technological experimentation within any milieu? By what standards are technological variants judged successful, and who does the selecting? In technology as in science, these questions fall along a spectrum. On the one hand, some aspects of this process are responsive predominantly to external, socio-economic factors, on the other there are aspects which depend predominantly on internal professional considerations.

As to the absolute pace of innovation, national differences depend for the most part on external, social and economic differences, having little to do with the technicalities of the disciplinary problems involved. As to the relative pace of innovation in different branches of technology, however, these will have to be explained in a more complex manner—partly by internal disciplinary considerations (e.g. the comparative 'ripeness' of the different problems), partly by the external influence of the market, and the social occasions for innovation in different fields. As to the selection procedures by which technical innovation are judged, in so far as these involve questions of feasibility and other such expert matters, they raise predominantly internal disciplinary issues; in so far as they involve economic and aesthetic elements, selection depends also on external issues of commercial promise and public interest. Turning to the professional aspects of technology, likewise, our earlier analysis suggests some fruitful lines of enquiry. How far (for example) do the channels of transmission and apprenticeship operate in the same manner in technology as in science? How far does the function of technical literature in engineering resemble, and differ from, that of scientific journals? Again, in what respects are the roles of authoritative textbooks in science comparable to those of manuals of standard practice in technology? How is magisterial or institutional authority distributed within a technological profession, and what kinds of politics (e.g. generational conflicts) are involved in the redistribution of this authority? And does the institutional structure of the technological professions tend, as in science, to lag behind the organizational needs of the most advanced disciplinary practices? In all these respects, the 'populational' character of technologies and technological change opens up worthwhile lines of historical and sociological investigations.

For our own purposes, however, these detailed historical and sociological issues are less significant than certain deeper philosophical questions: notably, those concerned with the 'rationality' of technological development. First and foremost, we can now demonstrate—and explain why it is—that questions of rationality do indeed arise about technological changes, as legitimately as they do about conceptual changes in science. In recent years, the philosophy of science has been cultivated intensively, but the philosophy of technology has been almost totally neglected: a dark, yawning void explored only by a few Marxists, some social and aesthetic philosophers, and Polish praxiologists from the school of Kotarbinski.[1] During this half-century, the logicist identification of 'scientific explanations' with 'explanatory arguments' has resulted in rationality being analysed in terms of the formal relations between the propositions of an explanatory 'system'; and, in the process, the procedural parallels between science and technology have been overlooked. Our own analysis leads to a quite different result. By rejecting any equation of rationality with logicality, and recognizing that rationality is a characteristic of human behaviour generally—not merely of logical relations and linguistic performances—we remove the objections to describing technological innovation as 'rational'. Instead, we can now see scientific rationality and technological rationality as being very closely related: both being concerned with the manner in which new experiences are mobilized to justify changes in our ideas and procedures.

What marks a scientist's work as rational is not his competence in the formal manipulation of established concepts and arguments: rather, it is his readiness to think up, explore and criticize new concepts, arguments and techniques of representation, as ways of tackling the outstanding problems of his science. So, in both science and technology, the operative questions of rationality arise over the justification of procedural changes; and they arise in similar ways, whether the procedures in question are 'scientific', i.e. explanatory or representational, or 'technological', i.e. practical or technical. The difference in end-products has nothing to do with the rationality of the changes themselves. In both cases, failures of rational judgement manifest themselves in (e.g.) resistance to procedural changes whose merits, practicability, and lack of 'side-effects' have been fully demonstrated, or in the premature acceptance of unproved innovations. And, in these respects, technological changes can clearly be 'premature' or 'overdue' just as much as scientific changes. The hindrances to rationality, too, are the same in technology as in science: e.g. institutional conservatism, the

[1] On Kotarbinski and his pupils, see H. Skolimowski, *Polish Analytical Philosophy* (London, 1967), Chapter IV.

interests of dominant individuals, reckless or overcautious management, and excessive rivalry between professional generations.

If we can only bring ourselves to accept David Hume's invitation to leave the philosophical study for the outside world of practical life, we shall find more similarities between the rationality of science and the rationality of law, technology, and other practical affairs, than are dreamt of in academic philosophy. These rational parallels cut deep. One crucial conclusion from our analysis of conceptual development in science, for instance, was that judgements about the rationality of conceptual changes are always comparative, rather than positive or superlative in form; and the same conclusion applies here. There is no such thing as an 'intrinsically rational' concept in science, and there is no such thing in technology (or law) as an 'intrinsically rational' engineering design (or legal procedure) either. What is 'rational' is the judgement that, of the available novelties, some given design, or procedure, is superior to existing ones in a sufficient range and/or combination of respects to be worth adopting; and this does not imply either that it is the only 'rational' procedure possible, or that it is the best which could coherently be conceived. In practical as in intellectual affairs, we must therefore take issue with Kant. Neither the rationality of theoretical concepts nor the rationality of practical procedures can be judged definitively, timelessly, or once-for-all. The Practical Reason like the Pure Reason must have an eye, not to the Good or the Best, and still less to the Only-Coherently-Conceivable, but rather to the Better; and, the rationality of collective human enterprises being what it is, this means always the Better-for-the-Time-Being.

Up to this point, we have been treating men's varied collective activities—in science, technology, and elsewhere—as the concerns of separate 'rational enterprises', each of which develops in its own independent and more or less 'disciplined' manner. But this is only a first step. While men's rational enterprises may be distinguishable, they are neither absolutely independent, nor do they develop entirely separately. So it is necessary to consider, also, how and at what points different activities and enterprises interact. For instance, can a discipline or sub-discipline of one kind (technological, say) ever be subordinated completely to one of another kind (e.g. scientific) so that its direction of development is strictly controlled by changes in that other field; or do different enterprises develop, rather, autonomously and in parallel? In either case, by what channels does one discipline borrow, and 'apply', ideas or procedures originating in another? Do the links between different disciplines operate on their pools of conceptual variants, on their current directions of variation, or on their selection criteria? ... We could quickly think up a dozen questions of this general sort. While no one doubts that scientific and technological disciplines interact and influence one another, for example, it is still unclear through exactly what channels this influence is exerted, and precisely how changes in one modify the rate and direction of development in the other.

Notoriously, there are several different views about these relationships. In the more industrialized countries, the conventional wisdom commonly regards technology itself as being simply 'applied science', the implication being that changes in technology are always preceded by, and dependent upon, corresponding changes in science. By a suitable choice of illustrations, we can lend a lot of force to this view; particularly if we focus our attention on the 'science-based' industries of the mid-twentieth century. (Where would radio and electronic engineering be, without the earlier physics of Maxwell and J. J. Thomson, cryogenics without the low-temperature research of Kammerlingh Onnes, or the modern pharmaceutical industry without the scientific work of men like Florey and Fleming?) According to a rather more sophisticated view, the fundamental interaction between science and technology takes place in the opposite direction, and science itself is—so to say—the 'ideology' of technology; on this view, changes in economic and technological demand antedate, and pose the crucial problems leading to, changes in scientific theory. (This thesis has been supported explicitly by Marxist historians of science such as J. D. Bernal and Boris Hessen, but is implicit in a weakened form also in, e.g., J. B. Conant's well-known remark, 'Science owes more to the Steam-engine than the Steam-engine owes to Science.'[1])

---

[1] For Bernal and Hessen see above, pp. 301–5; see also J. B. Conant, *Science and Common Sense* (New Haven, 1961).

Finally, midway between these extremes, there is an intermediate thesis, according to which the interaction between science and technology is a reciprocal one: i.e. the vigorous development of either encourages, yet is at the same time dependent on, the vigorous development of the other. On this account, a natural science and an associated technology are partners in a kind of historical gavotte, developing most effectively when their changes are harmoniously synchronized.[1]

Our own analysis has the merit of permitting us to go behind these three rival accounts, and to explain the actual channels of mutual influence, by showing what occasions of interaction the historical evolution of scientific and technological disciplines actually creates and provides. The first step is, once again, to distinguish between 'variation' and 'selection'—i.e. between the factors and considerations responsible for the appearance of novel variants within a disciplinary pool, and those responsible for establishing certain selected variants in the repertory of a discipline. We may use as our illustration the familiar and much-discussed example of the interaction between theoretical thermodynamics and steam-engine design in the early nineteenth century.

At any particular stage in the development of steam engines, the 'state of the art' leaves unsolved a reservoir of outstanding technical problems; these are concerned (e.g.) with methods of utilizing higher-pressure steam, preventing boiler-leaks or minimizing the loss of heat through the exhaust. At every stage, in consequence, engineers specializing in steam-engine design have approached novel lines of investigation and technical variants with an eye to their potential as providing solutions to these particular technical problems. Wherever a novel idea or suggestion has shown promise of yielding a technical improvement in one or another of these respects, it has become—from the standpoint of the technology—a candidate (or 'possibility') for incorporation into the disciplinary repertoire. The ultimate source of the new idea has then been quite unimportant. It may arise out of previous experience in the same discipline, it may be suggested by theoretical considerations in physics, or it may

---

[1] See the papers by D. J. de S. Price, in *Technology and Culture*, 6 (1965), 553ff, and in *Factors in the Transfer of Technology*, ed. W. H. Gruber and D. G. Marquis (Cambridge, Mass., 1969).

be prompted by further-fetched analogies: that does not matter. All that finally counts is that the innovation in question shall provide a genuine step forward, in moving towards the practical goals of the technology.

Conversely, if we look at the corresponding branch of physics —viz., thermodynamical theory—we find at every stage in its development a reservoir of outstanding problems, for which the currently established repertory of concepts provides no solution. At every stage, in consequence, physicists specializing in theoretical thermodynamics have approached novel lines of investigation and conceptual variants with an eye to their potential as providing solutions of those theoretical problems. Wherever a novel suggestion or idea has, in one respect or another, shown promise of yielding improved methods of theoretical representation or a better grasp of underlying mechanisms and processes, it has become—from the standpoint of thermodynamical science—a candidate (or 'possibility') for incorporation into the conceptual repertoire of thermodynamics. The ultimate source of this conceptual innovation, is, once again, the last thing of serious importance. It may be based on existing theories in some related branch of physics, it may be suggested by technological experience, or it may be prompted by some more remote comparisons: that does not matter. All that finally counts is that the variant in question should provide a genuine step forward, in moving towards the explanatory goals of thermodynamics, or of physics generally.

Consider the actual history of this interaction. On the one hand, it is clear that the French physicist Sadi Carnot found the behaviour and properties of steam engines highly suggestive when he began developing the theoretical arguments which eventually led to the famous Second Law of Thermodynamics; it was not for nothing that his treatise on thermodynamical theory was an essay *Sur la Puissance Motrice de Feu*.[1] Indeed, Carnot expounded his theoretical argument as applying to an idealized heat-engine, and in this way introduced the novel concept of a 'reversible cycle of operations', around which so

---

[1] Originally published in Paris in 1824; available in English translation as *Reflections on the Motive Power of Fire*, ed. E. Mendoza (New York, 1960); see also D. S. L. Cardwell, *Steam-Power in the Eighteenth Century* (London and New York, 1963).

much thermodynamical argument was subsequently to be constructed. Yet the scientific validity of Carnot's conclusions did not depend on this choice of technological illustrations. When his arguments were reformulated and extended by later physicists, such as Clapeyron, Helmholtz, and Boltzmann, these references to heat-engines dropped out of sight, and the considerations involved became more and more explicitly 'physical'.[1] While Carnot's source of inspiration may have lain in the field of engineering, the merits of his ideas were strictly scientific. Likewise, in the reverse direction: the theoretical analyses of men like Carnot and Rankine themselves became the starting-points for developing novel types of steam engine. So nineteenth-century thermodynamical theory eventually became a fruitful source of ideas for mechanical engineering. But, as before, the merits of the resulting designs had nothing specifically 'scientific' about them. Although the technical procedures of engineering gradually incorporated ideas and techniques of scientific inspiration, mechanical engineering itself remained—as much as before—a distinct craft or technology, rather than a subordinate 'application' of physical science.

In this sense, the problems, selection-criteria and historical genealogies of a natural science and its associated technologies are autonomous. Still, while these associated disciplines remain in principle distinct, they have tended more recently to become linked into close standing partnerships, by which one acts as a regular source of 'variants' or 'possibilities' for the other. Whether intellectual or practical, each discipline then provides occasions for reconsidering the problems of the other and suggests new lines of investigation for the other to explore. Efforts to improve steam-engine design, for instance, may spotlight weaknesses in current thermodynamical analysis and so prompt the investigation of ideas and models aimed at overcoming those theoretical weaknesses; meanwhile, the progressive refinement of thermodynamical theories draws attention,

---

[1] The tendency for thermodynamics to become progressively more abstract, especially in the hands of Boltzmann, was one of the chief points emphasized by Max Planck in his Leiden lecture on 'The Unity of the Physical World-Picture'; see *Physikalische Zeitschrift*, 10 (1909), 62–75; or for an up-to-date English translation, *Physical Reality*, ed. S. Toulmin (New York, 1970), pp. 3–27.

in return, to unsuspected sources of thermal inefficiency in actual heat-engines, and other such problems of design. Whatever its ultimate origin or inspiration, however, any novel variant must be judged, either as one theoretical concept in competition with other possible theoretical concepts, or else as one practical technique in competition with alternative practical techniques. Though either discipline may act as a standing source of variants for the other, the question whether those variants represent acceptable solutions is always a matter for the recipient, not for the source discipline.

In the light of this analysis, we can see why the characterization of technology as 'applied science' should not be taken too seriously. Such terms as 'apply', 'applied' and 'application', are in fact ambiguous and potentially misleading. We can speak, for instance, of Isaac Newton 'applying' his general Laws of Motion and Gravitation to the special case of planetary motion, and so explaining Kepler's Laws as one particular 'application' of his universal theory: in this sense, Newton's dynamical principles provided the intellectual underpinning for planetary kinematics, and led observational astronomers to restate their very questions.[1] But, in such cases as this, both the 'theory' and its 'application' are elements in the same scientific discipline, and the relationship between them is correspondingly close. Similarly, we may speak of some particular technical design as 'applying' more general engineering principles: of the F-111 aircraft (say) as an 'application' of the variable-geometry principle in wing-design; and, in this case, both the principle and its application are elements in the same technological discipline. If we consider the cross-connections between a natural science and its technological partner, by contrast, we find that these are nothing like so close. Neither partner can directly underpin the other, or be the 'source' of its ideas, apart from the specific criteria imposed by the standing ambitions of the 'recipient' discipline. At most, a scientific theory can explain, in retrospect, why certain practical enquiries were for so long without result; and certain technological successes can add a quantum of further conviction to the scientific innovations which suggested them in the first place. Either way, the rela-

---

[1] On this point, see the useful discussion in N. R. Hanson, *Patterns of Discovery* (Cambridge, Eng., 1953), Chapters 1-4.

tionship is both symmetrical and indirect. It is just as appropriate to do as Carnot himself did, calling the thermodynamic analysis of entropy an 'application' to physics of technological ideas about the motive-power of fire, as it is to speak of technical improvements in refrigerator design (say) as 'applications' of the scientific ideas developed by Carnot's successors.

Historically speaking, in fact, science and technology have developed hitherto independently and in parallel. As a result, different cultures have treated the relationship between them very differently. Some civilizations—e.g. that of Classical Greece—have developed strong traditions in pure mathematics and physics, without doing much to pursue their implications for mechanical engineering.[1] Others have been technologically fertile, despite having little effective mathematical or physical theory—as in China.[2] The technologies and practical arts have, likewise, had quite different historical origins from the natural sciences with which they are nowadays associated, and in many cases the crafts in question have antedated the sciences by many centuries. Originally, for instance, computational astronomy was not 'applied astrophysics' but a branch of divination;[3] for several millennia, metallurgy and material-processing were flourishing crafts although as yet lacking any scientific affiliations;[4] while clinical medicine and scientific biochemistry have become closely linked only during the last short chapter in the long history of the medical arts.[5] If in recent years the development of the technical and industrial arts has seemingly been 'revolutionized' by science, we must not be

[1] The limited application of Greek physical ideas to engineering has been well described by Brumbaugh and Drachmann; see, e.g., R. S. Brumbaugh, *Ancient Greek Gadgets and Machines* (New York, 1966), and A. G. Drachmann, *The Mechanical Technology of Greek and Roman Antiquity* (Copenhagen, 1963).

[2] See Joseph Needham, *Science and Civilisation in China*, Vol. 4, Part 2 (Cambridge, Eng., 1965), Sec. 27 (2), pp. 10ff.

[3] See O. Neugebauer, *The Exact Sciences in Antiquity*: cf. p. 216, n. 1, above.

[4] This point becomes particularly apparent from the recent scholarly work of such men as Levey and Forbes; see, e.g., M. Levey, *Chemistry and Chemical Technology in Ancient Mesopotamia* (Amsterdam, 1959).

[5] Even in the late nineteenth century, medicine still remained an empirical art, and only recently has it begun to acquire a biological rationale. Recall, e.g., the remarks of T. H. Huxley in his essay, 'The Connection of the Biological Sciences with Medicine' (1881), printed in *Science and Culture* (New York, 1893), pp. 333–57: 'A scorner of physic once said that nature and disease may be compared to two men fighting, the doctor to a blind man with a club, who strikes into the *melée*, sometimes hitting the disease, and sometimes hitting nature.'

misled. This does not mean that the essential nature of technology has in any way changed; only that its contemporary partnership with science has accelerated the solution of technical problems which had previously been intractable.

Even so, our recent experience with such science-based industries as electronics and pharmaceuticals may be unrepresentative. Rather than giving rise to brand-new technologies and industries, science-based innovations more typically help an existing technology to solve its own previous problems more rapidly. For a long time, indeed, they may not even succeed in doing that much. At least until the time of Laplace, Newton's theories in planetary astronomy made only marginal contributions to the practical craft of astronomical computation.[1] Throughout the eighteenth century, the technical resources of pre-Newtonian computation still outran the scope of Newton's explanations; indeed, their practical superiority only underlined the scientific incompleteness of Newton's analysis, and posed problems which astrophysicists had to solve before they could hope to give computational astronomers any real help. So, between Newton and Laplace, the leading topic of discussion in planetary theory was the so-called 'inequalities' in the motion of the Moon and the planets—i.e., the points of apparent failure in Newton's scientific calculations. And, even today, the practical work of computational astronomers (e.g. on satellite orbits) still does more for theoretical physics than the other way around. Here again, the astronomer's craft preserves a disciplinary autonomy, and standards of precision and success, that go beyond the scope and authority of astrophysical science.

6.2: *Compact, Diffuse, and Would-be Disciplines*

Among all those human activities and enterprises which provide loci for rational choice—in which decisions are made, procedures followed, considerations taken into account, conclusions arrived at, new possibilities entertained, and 'reasons' given for the resulting conclusions or actions—we can here distinguish, at one level, between those which are 'disciplinable', and those whose concerns and concepts do not, in the nature of the case,

---

[1] Until the publication of Laplace's *Mécanique Céleste* (1799–1825), Newtonian dynamics gave only an explanation 'in principle' of the planetary motions, and made little substantive contribution to computational astronomy.

lend themselves to such 'disciplined' debate and improvement. Meanwhile, at another level, we can draw a second distinction, which holds within the class of disciplinable enterprises itself: between those which already have, and those which have not yet, achieved the disciplinary status at which they rightly aim. A rational enterprise whose conceptual repertory is exposed at every stage to critical reappraisal and modification by qualified judges, in the light of clearly recognized and agreed collective ideals, develops—we say—in a 'compact' manner, while one that conforms only loosely to those requirements develops 'diffusely'. As for those subjects which might in principle become fields for disciplinary cultivation, but in which effective disciplinary development has as yet scarcely begun, we shall speak of these as 'would-be disciplines'.

How, then, do diffuse and would-be disciplines differ from compact ones; and what conditions must be satisfied before a disciplinable activity can actually set out on the historical path of a genuine discipline? We can best answer these questions by recalling the conditions required for the development of a compact, well-structured discipline, and then considering the various possible ways in which a potential discipline may fail to satisfy them. A compact discipline, then, has five connected features. (1) The activities involved are organized around and directed towards a specific and realistic set of agreed collective ideals. (2) These collective ideals impose corresponding demands on all who commit themselves to the professional pursuit of the activities concerned. (3) The resulting discussions provide disciplinary loci for the production of 'reasons', in the context of justificatory arguments whose function is to show how far procedural innovations measure up to these collective demands, and so improve the current repertory of concepts or techniques. (4) For this purpose, professional forums are developed, within which recognized 'reason-producing' procedures are employed to justify the collective acceptance of novel procedures. (5) Finally, the same collective ideals determine the criteria of adequacy by appeal to which the arguments produced in support of those innovations are judged.

Now, there is no saying beforehand just what ideals or ambitions will prove realistic in any sphere. The fact that some range of intellectual, practical, or judicial activities (say) can be organized and handled by a particular repertory of disciplined methods has to be discovered and demonstrated in the course of human experience. By the present time, these conditions are substantially satisfied in the better-established physical and biological sciences, in the more mature technologies and in the better-conducted judicial systems; and, for much of the time at least, conceptual and procedural developments in these areas follow a reasonably 'compact' pattern. Conversely, diffuse and would-be disciplines may fail to satisfy these conditions, and so differ from fully-fledged, compact disciplines, in either or both of two main respects: methodological and institutional. In the former case, their shortcomings spring primarily from the absence of a clearly defined, generally agreed reservoir of disciplinary problems, so that conceptual innovations within them face no consistent critical tests and lack any continuing rational direction. In the latter case, the deficiencies spring primarily from the absence of a suitable professional organization, so that the disciplinary possibilities of the subject are not fully exploited, and the rational purposes of its practitioners are frustrated.

The lack of well-defined problems can have several kinds of origin, and can show up in several ways. Typically, it springs from the absence of a properly analysed disciplinary goal. The different men attempting to co-operate in launching a new science (say) may not merely disagree about their particular observations and interpretations, concepts and hypotheses: they may even lack common standards for deciding what constitutes a genuine problem, a valid explanation, or a sound theory. Whereas all atomic physicists, for instance, take for granted 'ideals of natural order' which are largely shared, and identify a successful theory of atomic structure by very similar tests, the various practitioners of a scientific 'would-be discipline' can presuppose no agreed aims, ideals, or standards. The immediate result of this lack is that theoretical debate in the field concerned becomes largely—and unintentionally—methodological or philosophical; inevitably, it is directed less at interpreting particular empirical findings than at debating the general acceptability (or unacceptability) of rival approaches, patterns of explanation, and standards of judgement.

In a scientific field that has not yet achieved its proper status as a discipline, accordingly, the theoretical debate can—at best—concentrate on the acknowledged methodological failings of the subject, in the attempt to analyse, and see the relations between, the alternative intellectual goals which it might be pursuing. So long as a would-be discipline remains in this preliminary, inchoate condition, no agreed family of fundamental concepts or constellation of basic presuppositions—no 'paradigm', in Lichtenberg's sense—can establish itself with authority. For, where different scientists are fixing their professional sights on quite different intellectual targets, no agreed set of criteria is available for judging new hypotheses or conceptual variants, and no single body of theory will serve all their purposes alike.

On the other hand, one of the best indications that a new science has arrived at a clear definition of its intellectual goals, and achieved a proper disciplinary status, is the eventual enthronement of an agreed set of fundamental concepts and selection-criteria. This rarely happens suddenly or all at once. Even Newtonian physics took more than a generation to get its authority accepted throughout the scientific circles of Europe, and meanwhile a running fight rumbled on over methodological and metaphysical issues, between rival schools of Cartesians, Leibnizians, and Newtonians.[1] Nor is such an enthronement ever irreversible. No single scientific theory can claim unqualified authority, or do justice to all legitimate insights whatever; as a result, complementary points of view frequently survive in the form of 'minority' traditions.[2] Still, the various indispensable steps in the creation of an effective discipline are correlative and interdependent, so that scientists must in due

---

[1] The classic documents are Samuel Clark's edition of Rohault's textbook of Cartesian physics (1967), with an extensive commentary written from the Newtonian point of view; and the subsequent correspondence between Clarke and Leibniz, first published in 1717 and recently re-edited in a useful edition by H. G. Alexander (Manchester, 1956).

[2] Perhaps the best example of such a 'minority tradition' is the anti-corpuscular interpretation of Newton's theories, first expounded at length in R. J. Boscovich's *Theory of Natural Philosophy* (Vienna, 1758; English trans., Cambridge, Mass., 1966) and transmitted by Priestley to Faraday, Maxwell, and the creators of twentieth-century wave mechanics.

course take them all of a piece. The establishment of agreed goals and strategies, around which the cumulative development of a well-structured science can proceed, requires a measure of agreement on all its constituent elements—its disciplinary goals, ideals of order and explanation, forms of theoretical problem, and criteria of judgement—at one and the same time.

The characteristic features of 'would-be disciplines' can best be illustrated, at the present time, by contrasting the state of the behavioural sciences with that of the physical sciences. Throughout the history of modern science, rival schools of physicists have seen their problems in different proportions, have evaluated their basic concepts somewhat differently, and have interpreted certain crucial phenomena in distinctively different ways; even in physics, that is to say, there has never been unanimity. Nevertheless, by disciplinary standards, these disagreements have remained marginal, and have commonly been limited to alternative intellectual strategies for promoting certain generally agreed goals. The current position in the behavioural sciences is very different. Whether we turn to professional psychologists for explanations of the behaviour of individual human beings, or to professional sociologists and anthropologists to account for the collective behaviour of groups, societies, or cultures, in each case we find a diversity of approaches of a kind unparalleled in physics. Instead of their being united by agreed conceptions of what a 'human science' should aim at, or can ideally hope to achieve, we find behavioural scientists split into parties, factions, or sects, which have not managed to hammer out a common set of disciplinary goals.

In the behavioural field, no generally agreed criteria yet exist for deciding when a human action is intelligible or unintelligible, or what types of conduct constitute genuine 'phenomena', and so pose theoretical problems for psychology or sociology at all. Still less do behavioural scientists with different orientations share agreed standards for recognizing when a new theory of human behaviour has provided a 'complete explanation' of the relevant modes of conduct; or for recognizing what kinds of regularity or mechanism will make human behaviour intelligible; or for judging when a novel conceptual variant has the merits needed for incorporation into an established body of general theory. All this being so, the preconditions do not yet apparently exist, in the fields of individual and collective human behaviour, for establishing a 'compact' scientific discipline, possessing a definite strategy and an authoritative body of current theory.

As between a Skinnerian behaviourist and a Gestalt psychologist, a transformational linguist and a cultural anthropologist, a Freudian psycho-analyst and an existential psychologist, a clinical neurologist and a student of human ethology, differences in explanation, concepts, and approach spring less from factual disagreements about the observed circumstances of particular types of behaviour than from inconsistent conceptions of the form that 'explanations' of human behaviour should take in the first place. An existential psychologist and a transformational linguist, for instance, will concur in rejecting the Skinnerian's conclusions: ignoring his experiments on the grounds, not that they are ill-conducted, but that they are devoid of theoretical significance—for how can human behaviour be 'fully explained', or made 'completely intelligible', by pointing only to empirical correlations or regularities of kinds found equally in the behaviour of rats or pigeons?[1] In return, the Skinnerian will dispute their conclusions, not merely attacking the details of his rivals' explanations, but denouncing their appeals to 'mental entities' as a lapse into groundless (and probably meaningless) metaphysical speculation.[2] Next, however, the existential psychologist will turn on his former ally, the transformational linguist, and criticize him for 'over-intellectualizing' human behaviour: for, surely, the occurrence of grammatical patterns in linguistic behaviour is not self-explanatory, but is rather a 'phenomenon' requiring to be explained, in turn, as the product of profound motivational needs? . . . And so the factional splitting will continue, justified less by agreed empirical demonstrations than by philosophical appeals to general methodological considerations.[3]

In many important respects, therefore, the current debate

---

[1] See, e.g., N. Chomsky, review of B. F. Skinner's *Verbal Behavior*, in *Language*, 35 (1959), 26–58.

[2] See, e.g., W. C. Holz and N. H. Azrin, 'Conditioning Human Verbal Behavior', in *Operant Behavior*, ed. W. K. Honig (New York, 1966).

[3] The metaphysical element involved in current theoretical debates about the methodology of psychology is brought out in Charles Taylor, *The Explanation of Behaviour* (London and New York, 1964).

about the explanation of human behaviour lacks focus and definition. There is no consensus even about those analytical and pre-scientific issues on which the construction of a 'disciplined' science of human behaviour depends: e.g., about what kinds of thing are puzzling about human behaviour at all, what tests a fruitful new psychological concept must pass, or what a comprehensive theory of human behaviour should seek to achieve. So long as this is the case, it can be no surprise that, on a more substantive level, psychology still lacks an agreed body of theoretical concepts, or that one psychologist's great discovery often strikes other psychologists as a waste of energy and breath. This is not to suggest that the analytical clarification required on the disciplinary or methodological level could possibly be carried out, in advance of more detailed substantive investigations into specific areas of conduct. On the contrary: independent investigations into many different puzzling features of human behaviour—aphasia or neurosis, pattern recognition, personal commitment, or language-learning—may eventually be needed as preliminaries to the development of any more general explanatory theories. If this point is ever reached, however, it will happen, not as a result of any existing mode of psychological inquiry establishing itself monopolistically, as 'the one and only sound science' of human behaviour. Instead, it will do so because, after analysing and comparing the respective goals and successes of different modes of explanation, psychologists have come to see their mutual relations and relevance in clearer proportion.[1]

If this is true in the case of psychology, it is true, with variations, of sociology also. But here we can usefully make one further point. During the last 100 years, conceptual development in the physical sciences has displayed some striking changes of direction; yet, taken overall, it has nevertheless been progressive and cumulative. At each stage, the physico-chemical ideas of one generation have tended to move progressively further away from those of its predecessors. The fundamental concepts of Bohm and Chew, for example, may differ sharply from those of Heisenberg and Bohr; but, by comparison, the common ground shared by all four men makes the ideas of

[1] For a fuller discussion of this point, see my paper on 'Concepts and the Explanation of Human Behavior', in *Human Action*, ed. T. Mischel (New York, 1969).

Rutherford and Thomson—and still more those of Maxwell —look like ancient history. In the sciences of collective human behaviour, the pattern has been very different. Instead of being progressive or cumulative, theoretical development has gone through a series of pendulum-swings. As a result, the leading ideas current at any stage in the development of twentieth-century social science have tended to resemble those current two or three generations before, more than they have those of the immediately-previous generation.

Up to the 1890s, for instance, it was still an accepted commonplace that sociological and anthropological theory had an essentially historical dimension. But, around the turn of the century, a new generation of social and anthropological theorists began to develop very different aims and strategies for their sciences. Their new explanatory ideals were static and ahistorical, putting aside all earlier historical or evolutionary notions.[1] These men saw the goal of a social science as the discovery of a universal pattern of social 'groups' or 'systems', and the establishment of general 'laws', according to which the interconnected and interdependent 'elements' in such a social system interacted to produce their characteristic 'effects'. Disavowing any concern with historical change for its own sake, early twentieth-century anthropologists and sociologists saw themselves, rather, as the anatomists and physiologists of human culture and society. So, for a whole generation, the key categories in sociology and anthropology were such synchronic ones as 'structure', 'function' and 'interaction': historical terms like 'primitive' and 'advanced', 'differentiation' and 'evolution' were ruled out. Yet now, a further generation on, we find ourselves in a fresh situation. The conceptual pendulum has swung back; historical or diachronic categories are re-entering the sciences of collective human behaviour; and even so hardened a 'structuralist' as Talcott Parsons is prepared to entertain 'evolutionary' notions once more . . .[2]

Yet this, too, is a kind of historical development we might

---

[1] See the writings of L. H. Morgan, Radcliffe Brown, and Malinowski, who were the leaders of this movement in anthropology; also the retrospective discussion in M. Fortes, *Kinship and the Social Order* (Chicago, 1969).

[2] See, e.g., Talcott Parson's recent book on *Societies: Evolutionary and Comparative Perspectives* (Englewood Cliffs, 1966).

well expect to find in a would-be discipline. In atomic physics, conceptual change can take place cumulatively, just because the consensus over intellectual goals and selection-criteria imposes a corresponding continuity on the rational development of the discipline. In a would-be discipline, by contrast, no such state has yet been achieved. The methodological pendulum-swing from historico-evolutionary to formal-structural categories, and back again, was not the consequence of applying systematic, agreed procedures of conceptual variation and selective judgement. Rather, it represented a sequence of inconclusive switches in intellectual strategy—so many successive attempts to make theoretical progress from new directions, by adopting fresh strategies and ideals of explanation—in a field where the fundamental theoretical issues have remained at bottom intractable, for want of sufficient analysis.

If I have argued here that, at the level of general theory, psychology and sociology remain today 'would-be disciplines', I am not claiming any absolute or permanent contrast between the social and the physical sciences. On the contrary: I have merely been trying to diagnose certain special difficulties which face the theoretical sciences of human behaviour at the present time. In earlier centuries, physical theory too had the same inconclusive character; indeed, many of the methodological difficulties afflicting sociology and psychology today had counterparts in earlier physical science. It may yet turn out, for instance, that human behaviour in all its aspects whatever does not lend itself to explanation by a single, completely general and comprehensive theory; instead, the desire for a single, all-embracing 'scientific psychology' may itself prove to be a will-o'-the-wisp. Certainly, a similar will-o'-the-wisp had to be disregarded, before modern physics could become the discipline it now is; and the reasons why this was so throw some light on the contemporary state of behavioural science.

Aristotle (to recall) had set himself the task of explaining 'change' in entirely general terms. His intellectual goal was to devise, not several parallel but separate theories, each explaining a different type of 'change'—changes in colour, position, mass, chemical composition, or physiological function—but, rather, a single comprehensive theory capable of explaining changes of any kind whatever.[1] For some 2,000 years, philosophers and mathematicians followed Aristotle's example, with inconclusive results. They, too, discussed physical and physiological processes in an entirely general manner, seeking to present a common analytical pattern manifested alike—*mutatis mutandis*—in 'changes' of all kinds, however different these might appear on the surface. Even the fourteenth-century scholars at Paris and Oxford, whose work on 'rates of change' laid the analytical foundations for all later mechanics, saw their definitions and graphs as applicable not just to the rates of change and integrated effects of kinematical variables, but to all variable magnitudes whatever. Their formulae were intended to cover not merely the velocities and accelerations of moving bodies, but equally the warming-up and cooling-down of material objects, the intensification and dying-away of sounds, or the maturation and ageing of human beings.[2]

Later science has demonstrated its maturity, not least, in the fact that scientists have given up this prematurely general ambition. Instead, physicists and physiologists now believe that changes of different kinds must—to begin with, at least—be studied independently: i.e. that we shall do better, in these fields, by working our way towards more general concepts progressively, as we go along, rather than insisting on complete generality from the outset. And if behavioural scientists eventually reach a similar conclusion—deciding that 'human behaviour in general' (like 'change in general') represents too broad a domain to be encompassed within a single body of theory—that could again be a sign of maturity rather than defeatism. Until this preliminary point has been clarified, the deepest theoretical issues in the behavioural sciences may well remain on a purely analytical level, and appeals to experiment and observation may be premature. In the fourteenth century, the collection of extensive 'empirical data' would have been of little use to physics, since no clear theoretical questions had yet been formulated on which such observations could throw light;[3] and it would probably be a similar mistake for

---

[1] Cf. J. H. Randall, jr., *Aristotle* (New York, 1960).
[2] See M. Clagett, *The Science of Mechanics in the Middle Ages*, (Madison, 1959).
[3] See. H. Butterfield, *The Origins of Modern Science* (London, 1957).

behavioural scientists today to insist on showing themselves 'empirically-minded', at all costs. For, until they are in clearer agreement about their own legitimate intellectual ambitions, they too may be in little position to state any general theoretical questions, to which the accumulation of empirical observations can by itself provide acceptable answers.

Even the theoretical pendulum-swings characteristic of recent sociology have parallels in the history of the physical sciences. The provisional acceptance of one strategy in physics has never ruled out finally the coexistence of alternative, minority traditions: as (e.g.) with the long-term alternation between atomistic and continuum theories of matter. But what can reasonably be demanded of a fully-fledged, 'compact' discipline is that strategies should displace one another as the outcome of positive new explanatory achievements, rather than being the mask for a mere uncertainty about where to turn next. Within twentieth-century physics, atomistic and continuum strategies still coexist within the broader theoretical framework of quantum mechanics, and as a result their respective scopes and merits have become more apparent. This has come about in a way that a mature social science might with advantage copy; we can hardly suppose that all collective human behaviour will prove completely explicable either in evolutionary terms alone, or in formal-structural terms alone —either by a glorified abstraction from social history, or by a timeless 'social physiology.' The experience of physics points to a different outcome. Perhaps the real theoretical failing of the social sciences today lies in their need for a less restrictive body of concepts: one within which changes in social relations and functions are no longer dismissed as accidental 'dysfunctions', but in which synchronic structures and institutions are themselves described in terms that lend themselves, without paradox, to diachronic variation.

Quite apart from such disciplinary weaknesses, however, an immature science may remain a 'diffuse' or 'would-be discipline' as the result of institutional failings. True, a diffuse or would-be discipline will often be deficient in both respects at once; until its fundamental explanatory tasks have been clearly defined, neither an agreed set of disciplinary methods and concepts nor a common forum of scientific debate will be able to establish its authority. But the nature of these institutional failings are worth considering here separately. For, even though the men involved in a would-be science may have come to understand the nature of their proper disciplinary goals, problems and selection-criteria, the potential of their science may be unrealized because its professional organization falls short of the subject's current needs.

The institutional weaknesses of would-be disciplines can again show up in a variety of ways. The most typical indications are the failure of communication and the maldistribution of authority. The effective progress of the subject may, for instance, be handicapped, either because it lacks any authoritative reference-group of theoretical judges, or else because it is dominated by men whose professional power has outlived their intrinsic claims to authority. Or it may be held up because the literature of the subject is deficient: either because not enough critical and well-edited journals exist to sift and publish the available research reports, or because the flood of new research reports is overwhelming the capacity of the scientists concerned to absorb their colleagues' results, even with the help of 'abstracts'. Or again, a science may develop less effectively than it should, because its professional organization is fragmented by boundaries irrelevant to its proper intellectual purposes: as, for instance, between theoretical psychology in contemporary America and Russia,[1] between eighteenth-century chemistry as written in the French and German language,[2] or between nineteenth-century physiology as studied in research hospitals and in the biological faculties of universities.[3]

The case of Mendel illustrates well how disciplinary and professional weaknesses can reinforce one another. To start with, it involved a breakdown in communication. If, on its first

[1] In this case, some effective contact seems to be being established at last, particularly since the translation of A. R. Luria's major work, *Higher Cortical Functions in Man* (Moscow, 1962, English trans., New York, 1966). See also the duplicated translations circulated by the National Science Foundation, e.g., *Works of the Institute of Higher Nervous Activity: Pathophysiological Series*, Vol. 2 (Moscow, 1956: English trans., Washington, D.C., 1960).

[2] See, e.g., H. Metzger, *Newton, Stahl, Boerhaave* (Paris, 1930); and especially P. Duhem, *La Chimie: une Science Française* (Paris, 1916).

[3] Ben-David, op. cit. (see above, p. 216, n. 2).

publication, Mendel's classic paper had come to the general attention of his biological contemporaries, the significance of his work would surely have been appreciated some twenty years earlier. Auguste Weismann was particularly qualified to see how Mendel's results bore on the outstanding problems of theoretical biology; and, although biological theory then had no universally-agreed strategy, half a dozen other leading biologists shared enough of Weismann's conceptions to do the same. Unfortunately, Mendel was in regular correspondence only with Nägeli at Munich; and Nägeli failed to recognize the theoretical implications of his work. It was only under the spur of theoretical problems which were becoming both clearer and more urgent by the late 1890s that geneticists, at the end of the century, searched back through the literature and discovered in Mendel's paper the novel concepts and patterns of explanation their own new science needed.[1]

The case of Mendel throws light, also, on the factors which determine the exact moment when a new sub-discipline can crystallize out from a current amalgam of less compact scientific investigations. Both the slow reception of Mendel's ideas, and the absence before 1900 of any professional forum for genetics, were indications of the comparative diffuse intellectual state and professional organization of theoretical biology. The new science was able to develop cumulatively, and to generate its own institutions only because, by the union of Mendel's theory of heredity with Weismann's work on cell-division and reproduction, men like Bateson, de Vries, and Morgan could at last define a clear disciplinary goal for their science.[2]

The creation of authoritative reference-groups and journals has a particularly significant part to play in the maturation of a would-be discipline. The shortcomings of contemporary academic psychology show themselves, for instance, not only in its lack of well-defined intellectual goals, but also in its professional institutions. These too often speak for a loose confederation of proselytizing sects rather than for a well-established

---

[1] W. Bateson, *Mendel's Principles of Heredity* (Cambridge, England, 1909).

[2] The history of this period is only now beginning to be written, e.g., by Arthur Hughes, *A History of Cytology* (London and New York, 1959) and E. A. Carlson, *The Gene: a Critical History* (Philadelphia, 1966).

science. Few authoritative reference-groups have developed in theoretical psychology of the sort that exist in the physical sciences. Rival factions are separated less by a methodical and agreed sub-division of the outstanding problems than by radical differences of theoretical approach, and each group tends to operate under the magisterial dominance of its own 'chief' or 'high priest'. Meanwhile, the methodological partisanship of its journals and textbooks tends simply to aggravate the self-isolation of those sects. Although the literature of theoretical psychology appears at first sight very like that of physical science—the same profusion of journals, symposium reports and the like—on closer scrutiny, important differences become apparent between the two sets of journals and texts. All of the physical sciences aim at distinct, but co-ordinated explanatory goals; it is the complex variety of the resulting tasks that has imposed the current intellectual division of labour, and with it a division of authority between autonomous but interdependent sub-disciplines, each with its own reference groups and journals. In the psychological field, by contrast, there is no such co-ordination of goals. Rival factions work with different conceptions of what a 'behavioural science' should be, and organize their literature less on an agreed specialization of functions, than on conflicting claims to sovereign independence.

This contrast shows up clearly in the corresponding patterns of 'citations'. While different physical sub-disciplines interact only to a limited extent, this specialization has a pragmatic justification: the conceptual problems of nuclear, electronic, molecular, crystalline, and macro-molecular structure (say) each involve their own mathematical and technical difficulties. Yet, despite this specialization, there is a substantial amount of cross-citation between the specialized journals of the different sub-disciplines; and, on a broader level, certain common forums of discussion—e.g. the *Physical Review*, and the *Proceedings of the Royal Society* (Section A)—enjoy general respect and authority throughout the whole area. In academic psychology, we find a very different pattern. Alongside journals divided by domain or subject-matter, there are others which are distinguished chiefly by methodological approaches. The papers published in these journals—e.g. the *Journal of Experimental Animal Behavior*, or the *Journal of Verbal Learning and Verbal*

Behavior—cite a much more restricted range of other authors and journals: in the different branches of theoretical psychology, as a result, citations are quite unusually 'inbred'.[1] So whereas, in physical science, the profusion of sub-disciplines and journals exploits the quasi-economic advantages of intellectual specialization, the reverse is often the case in academic psychology. Instead, suspicions of methodological heresy prevent fruitful intellectual debate between members of different factions, and the proliferation of sectarian journals only delays further the differentiation of psychology into a set of genuinely compact disciplines.

The current literature and channels of communication may, indeed, play quite a subtle part in the historical development of a scientific discipline. Once a science has well-established goals and strategies, any factors restricting conceptual variation or distorting the procedures of critical judgement will commonly act as obstacles to conceptual evolution. (Within a fully fledged scientific discipline, ideological and national barriers exist only to be ignored.) But, at earlier stages in the maturation of a science, the situation can be more ambiguous. During its initial crystallization, for example, communication-barriers may actually help a new science to become 'viable', by giving the relevant forum of discussions the protective isolation they need. It has been pointed out (e.g.) that Charles Darwin was very weak both in physiology and in foreign languages. The quotations from French and German authors in the *Origin of Species* were apparently borrowed, in translation, from Lyell and other English writers; meanwhile, Darwin was largely isolated from French and German physiology, with their sophisticated techniques of microscopy and their strongly theoretical basis in physics and chemistry. All these factors made it easier for Darwin to develop the novel—and peculiarly *historiographical*—strategies needed by his theory of organic speciation: shielding his brand-new intellectual strategies against the hostile but irrelevant criticism that they would otherwise have encountered. These factors made it possible for

[1] See particularly the work of David Krantz on the evolution of scientific groups, with special reference to the case of Skinnerian psychology: e.g. 'Schools and Systems: the Mutual Isolation of Operant and Non-Operant Psychology as a Case-Study', *Journal for the History of the Behavioral Sciences*, July 1971. (A fuller report is in preparation.)

the potentialities of the new approach to be explored, to the point at which relevant new critical standards could be developed, and applied to the resulting theories.

When applied to genuinely 'disciplinable' activities and inquiries, therefore, our populational pattern of historical analysis can be given a normative as well as an explanatory interpretation. Within disciplinable fields of activity, the respects in which diffuse disciplines differ from compact ones, and would-be disciplines from both, provide an index not only of the theoretical character of diffuse and would-be disciplines, as compared with our idealized model, but also of the practical distance they still have to go before achieving the disciplinary 'maturity' at which they properly aim.

It is no criticism of carbon dioxide (say) to point out that it departs further from the physicist's conception of an 'ideal gas' than hydrogen does. Carbon dioxide is not trying to be an ideal gas, so no question arises of its failing to achieve that ideal; rather, it is physicists who are trying to explain the properties of different gases, and the gap between the actual and the 'ideal' properties of any actual gas is a challenge to their theories. By contrast, psychologists and sociologists are presumably aiming at the same disciplinary maturity that most of the physical sciences have already achieved. To that extent, the gap between actual diffuse and would-be disciplines and our present 'ideal' of compactness is a measure of adequacy or inadequacy in the structure and organization of the sciences concerned. At present, their intellectual goals may not be well defined or generally agreed, but in this respect they pursue intellectual ambitions which they have—as yet—failed to achieve. Like physics before A.D. 1700, and biology before 1860, the 'would-be disciplines' of today are on the road to becoming fully-fledged 'compact' disciplines, but have not yet found out how to do so.

In general, however, to say that any particular enterprise is 'disciplined', and that its historical development can be analysed in terms of our evolutionary pattern, is not to congratulate it, but is merely to understand its particular rational structure and aim. The ways in which the procedures of such an enterprise are directed towards its collective ideals, and the manner in which the resulting patterns of 'reasoning' reflect those ideals, may not be of universal application or validity; we cannot automatically look for the same kinds of 'reason' and historical development in other, non-disciplinary activities that we have found in the case of 'disciplinable' enterprises. Still, the kinds of ecological relationship we have discovered, between the methods of a discipline and its goals, may in due course teach us something of more general significance. For they can encourage to see how far, in other cases also, the intellectual organization of each field—the definition of its problems, its procedures of judgement, its selection-criteria and so on—tends to become, in the course of history, more functionally well-adapted to the intellectual or practical requirements of the enterprise in question.

With appropriate substitutions, our conclusions about 'diffuse' and 'would-be' sciences or technologies can also be applied in the sphere of law. If the effective development of scientific disciplines and professions requires certain intellectual and social conditions, so too there are necessary intellectual and social conditions for the effective development of a humane and well-adapted judicial system. Within a well-structured judicial system, the varied purposes of judicial procedure must be clearly delimited, both in the theoretical language of the law, and also in its practical conduct. It is a poor judicial process that fails to provide, and apply, effective rules of 'due process'; but it is, likewise, a poor judicial process that lacks a clear distinction between the different functions of criminal indictment and civil arbitration. In textbooks of jurisprudence and in judicial practice alike, our historical experience in differentiating legal issues of different kinds for separate treatment shows itself in two ways: in a conceptual differentiation between civil and military, criminal and international law (or between the law of personal status, corporation law, and the law of negligence), and in a separation of jurisdictions between the various courts and procedures through which these conceptual distinctions find institutional expression.

Correspondingly, the failings of a defective legal system are similar to the failings of 'diffuse' and 'would-be' sciences. They may be primarily intellectual: showing themselves in an inadequate grasp of the differences between the various judicial functions and their consequent 'demands'. In the early stages of social life, even the differences between legal, moral, and political issues (say) were far less clear-cut than they have since become, and no real provision existed for giving practical effect to these distinctions. It was not that early societies considered and rejected our modern distinctions: rather, their historical experience had not yet led them to articulate such differences at all. The advantages of having distinct and separate procedures and sanctions for dealing with legal, moral, and political issues, was (we might say) a prior, categorial discovery, more fundamental to the workings of civilized society than the subsequent development of substantive concepts or doctrines within law, morals, or politics. But, equally, the shortcomings of a legal system may be primarily institutional, rather than intellectual. It is not enough for professors of jurisprudence to formulate the doctrines and principles proper to each distinct type of legal function with ever-increasing refinements. A well-structured judicial system requires a correspondingly complex organization, if these basic distinctions are to be given a practical expression. In this respect, likewise, the differentiations between police and magistrate, judge and prosecutor, judiciary and executive, represent a prior, functional 'separation of powers' more crucial to the civilized conduct of social life, more 'constitutive' of law as we know it, and more far-reaching in its consequences, than any subsequent judicial decisions about particular issues of law.

6.3: *The Spectrum of Non-Disciplinary Activities*
In order to complete this outline survey of the different varieties of rational activity and enterprise, we must look at one last class of cases: those to which our disciplinary model of conceptual change is applicable, at best, only in part. In activities which are directed by more or less specific and well-defined goals—whether intellectual or practical—and which are therefore clearly disciplinable, it is not hard to see how the incomplete fulfilment of these collective aims gives rise to recognizable problems, or how the nature of the goals themselves determines the 'demands' of the resulting problem-situation. In activities which are by their very nature non-disciplinable, or disciplinable only in certain limited respects, the situation is different.

In these non-disciplinary cases, the very questions at issue are liable to be more complex, changeable, or even personal, than in a normal discipline. As a result, both the ecological demands of the particular situation and the criteria for judging conceptual novelties will be that much the less well-defined, settled, or agreed; and it becomes that much the harder to demonstrate a direct link between our criteria of 'rationality' and the intellectual or practical demands of our situation. Where the nature of the case neither imposes an agreed body of disciplinary ideals, nor calls for a corresponding set of collective procedures, the task of arriving at 'rational' judgements is liable—for understandable reasons—to be subtle and debatable. Yet this is not, by itself, a reason for denying the relevance of our analysis in such cases. Rather, it is a reminder that, in these cases, no single recognizable set of specific and easily defined collective 'demands' is operative.

Accordingly, we may here take some samples from the whole spectrum of non-disciplinary and non-disciplinable activities and ask, in each case, why the particular types of problems to which they give rise have the kinds of additional complexity they do. These samples will be of several types. To begin with, we can consider those activities whose directing goals are not communal or collective, but personal; next, those types of enterprise whose goals are a matter, not for some professional class of specialists or experts, but for everybody; thirdly, those problems which do not lend themselves to analysis in abstract or general terms (like those of a scientific discipline) but call rather for concrete decisions; and, finally, those problems which are essentially multi-valued, rather than single-valued, and which therefore require us to weigh up and balance off human concerns and ideals of several kinds.

(1) Literature and the fine arts occupy an interesting position: halfway between the fully-disciplined enquiries of physical science, on the one hand, and such non-disciplinable fields as ethics and philosophy on the other. In certain respects, indeed, these activities have more in common than we sometimes remember with full-scale 'disciplines', and could even be labelled 'quasi-disciplines'. Thus, in the development of every fine art, there is something of the continuity-through-change

that is characteristic of a natural science or a technology. While the actual techniques of (say) musical composition, as practised by Boulez and Stockhausen, are no more identical with those of Verdi or Bach than the physics of Feynman is identical with that of Maxwell or Newton, the composer's enterprise has—as such—preserved a recognizable unity and coherence, so that the changing techniques of successive generations again represent successive temporal cross-sections of a 'historical entity'.

Furthermore, the fine arts still have a lot in common, both in their methods and in their historic origins, with the earlier crafts from which both they and the technologies sprang. This fact is particularly evident in such marginal cases as architecture, where strong links remain with engineering and the builder's craft. Indeed, despite our own inherited image of the artist as an unconstrained individualist, and of the fine arts as vehicles for personal self-expression, our earlier analysis of the technologies might itself have been expounded, until not long ago, using the word 'arts' and its cognates. Until the late nineteenth century, technological designs and manufacturing procedures were still referred to as industrial 'arts'; the craftsmen who put these arts to practical use were called 'artisans'; while organizations for encouraging technological innovation were, quite naturally, given names like that of the Royal Society of Arts. Only gradually did a sharp distinction develop between certain 'fine' arts, which called for exceptional skill, and the general run of everyday, not-so-fine crafts; so differentiating the 'artist' from the run-of-the-mill 'artisan'. And even this distinction served initially only to mark off (say) a Benvenuto Cellini from a workaday jeweller, separating the artist who turned out costlier, made-to-order ('bespoke') products for a luxury market—in which the individuality of every new item was an implicit part of the contract—from the artisan, with his cheaper, ready-made products, designed for a mass market which set more store on practical utility.

By confining our attention to the *techniques* of a fine art, therefore, we could demonstrate a quasi-disciplinary continuity and coherence in its historical development. Draftsmanship and drypoint engraving, counterpoint and orchestration, camerawork and film-editing, are 'arts' in the original sense of the term: repertories of practical skills embodied in the persons of the artists concerned and transmitted by apprenticeship within a particular 'school'. And a history of fugue (say) could apply, with little modification, the same pattern of change by 'variation and selective perpetuation' that applies in the history of a science or a technology. Such a history of artistic techniques would, of course, differ strikingly from the more familiar historical procession of Great Poets, Masters of Sculpture or Giants of Music. It would have, for instance, to give proportionally more space than usual to composers who appear in retrospect to be of second-rank ability, discussing Mozart (say) in relation to Stamitz, Richter, and Hoffmeister, and Beethoven in relation to Cherubini, Albrechtsberger, and Hummel. But it need be none the worse for that. For too long, most critics and aestheticians, and even some artists, have confused 'originality' in matters of technical innovation—i.e. the invention of novel techniques, and the refinement of previous ones—with 'originality' in the sphere of personal communication—i.e. the discovery of 'something new to say'. So it is worth recalling here how far the fine arts have in fact been organized around the developing sets of technical methods which form their collective transmits, and how far—in consequence—the historical evolution of these collective techniques has been one central theme in the history of every fine art.

We must, however, say *one* central theme, not *the* central theme; and at this point we can turn to the differences between the fine arts and other, more disciplinable activities. The collective methods of an art may develop in a quasi-disciplinary way; but the manner in which those methods are employed to serve the individual artist's own goals is very different from what it would be if—by our standards—the fine arts were fully-fledged disciplines. The unity and coherence of a scientific or technological discipline are maintained, as we saw, by the collective ideals and ambitions of its practitioners; and no corresponding collective ambition unites the professional practitioners of a fine art. There is no single task engaging every painter or poet, composer or film-maker, as such—no shared 'poetical' goal or point of view (say) by which a man's preoccupations *qua* poet are delimited. For the natural scientist or research technologist, devising better explanations or design procedures is not just *an* end of his work but *the* end. (Discovery is 'the name of the game'.) Just what use an artist puts his techniques to is, by contrast, a matter over which the artistic profession collectively can claim no authority. The creative autonomy and independence of the fine arts gives every artist the liberty to employ the collective techniques of his profession in the pursuit of essentially individual goals. The artist's 'medium' is, thus, the *means* through which he achieves some end quite other than the improvement of the medium itself. And, in the last resort, it is this essential multiplicity of artistic aims that rules out all possibility of regarding any fine art as a 'discipline'.

Notice that this essentially personal element in the activities of the poet or the artist can be characterized perfectly well, without resort to any psychologically oriented 'subjectivism'. The manner in which literature and the fine arts differ from other more disciplinable activities is described only misleadingly by appealing to the contrast between the 'affective' and the 'cognitive', or between 'feelings' and 'facts'. For the true differences lie, not in the realm of individual feelings, but rather in the constitutive goals of the collective activities concerned. The biochemist *qua* biochemist has one specific set of collective concerns; these form the current *raison d'être* of biochemistry as a discipline, and it is his task as a professional biochemist to help in solving the resulting problems. The poet has no such professional commitments. Certain techniques may be the shared possession of sonnet-writers, epic poets, or tragedians, but the way in which each man puts those techniques to use is a matter of his own choosing. (Whether this choosing does or does not reflect, also, the individual's feelings is quite another issue.) In 'criticizing' an individual poet's imagery or themes, accordingly, we do so always with an eye to his personal aims. Our judgement turns, that is to say, not on the question whether his poetic innovations were sufficiently well-adapted to 'the' aims of poetry—for poetry-as-such has no single set of aims—but rather on the question whether they were well-adapted to his particular poetic aims.

(2) The contrast between disciplines organized around a communal goal, and non-discipline directed at individual ends, defines the boundary between disciplines and non-disciplines

in one direction; but, in other cases, human activities may be 'non-disciplinable' for quite other reasons. A rational enterprise achieves disciplinary status, for instance, not merely by having *at least one* set of well-defined collective goals and selection-criteria, but by having *one and only one* set of well-defined goals at a time. If natural science is fragmented intellectually into independent sub-disciplines, and sociologically into specialized and autonomous professions, this fact expresses the deliberate 'division of labour' which has become possible, thanks to the differentiation of scientific problems into distinct sets, each organized in terms of some one set of fundamental goals. That intellectual Balkanization—that policy of creating new, self-governing specialisms wherever the occasion arises—is possible and justifiable, however, only because the basic issues involved are abstract, theoretical ones.

So long as the concerns of scientists are simply with better understanding—better explanation, better intellectual grasp—they retain the liberty, as pure scientists, to organize and sub-divide their disciplinary and professional activities in ways directed solely at the development of more powerful explanatory concepts. Once we broaden our concerns, however, we no longer have that liberty, and a policy of abstraction and fragmentation is no longer open to us. In practical life, we have to deal with problems as we find them, in all their specific concrete complexity; and we cannot solve them in a way that does justice to this complexity, if we start by sub-dividing the issues involved and deliberately ignoring those aspects for which we have no well-established type of procedure. In such practical situations, accordingly, the rational assessments demanded of us are non-disciplinary and unprofessionalizable, not because they turn on personal or individual choices, but rather because the issues involved are multiple, specific, and interdependent.

To some extent, of course, there is room for a certain 'quasi-disciplinary' division of labour and problems in practical affairs, as in the fine arts. Consider, for instance, the problems involved in an engineering project. There, questions of policy are closely bound up with questions of technique, and we can tackle—or even formulate—the relevant policy choices, only in the light of technical assessments made from several complementary points of view. A large-scale hydro-electric installation such as the Aswan High Dam may raise a score of technical sub-problems, each of which is the professional concern of some group of specialists: ranging across the board from civil, mechanical, and electrical engineers to freshwater biologists, epidemiologists, and sociologists. Yet the decisions it calls for, on the level of policy, differ essentially from each and all of these specialized technical problems. They are concerned, not with general theoretical issues abstracted from the entire situation, but with concrete, practical choices: whether to go ahead with the project or to abandon it, how to compromise between (say) the demands of initial economy and those of long-term efficiency, how to handle the ecological side-effects of the installation, and so on. Such policy decisions oblige us, not just to consider the particular aspects of the project which are relevant from some one specialized point of view, but to face all aspects of the decision which carry weight from any point of view. The civil engineer's task may be to assess those selected aspects that depend on (e.g.) the general properties of reinforced concrete, and may therefore entitle him to abstract out those particular aspects which are relevant from his professional point of view. A decision of policy, by contrast, always involves dealing with some specific and complex tangle of issues as it stands; and the essential character of the problem is falsified by any prior abstraction.[1]

(3) If we pursue this last point further, we can make explicit another characteristic difference between disciplinable and non-disciplinable enterprises. It is a striking feature of disciplinary enterprises—whether intellectual or practical, technical or judicial—that we can normally elect to engage in them, or abstain from them, in isolation from activities of other kinds. This 'isolability' of disciplinary enterprises marks them off sharply from most of our everyday activities, in which actions and choices are meshed together in a complex spider's-web of interacting sequences, with equally complex consequences.

A politician, for example, makes a speech; this speech immediately becomes an element in a dozen different historical sequences, any or all of which may prove to be significant for

[1] This is a point repeatedly underlined by Lewis Mumford in *The Myth of the Machine* (New York, 1967–70) and others of his later books.

any judgement of the speech, considered as a 'political fact'. By speaking where, when and how he does, for instance, he may—at one and the same time—honour an engagement to his constituents, give unintended offence to some of his hearers, make an implicit promise to others, draw public attention to his skill as an orator, stake a claim to future office, enhance or damage his party's electoral chances, accelerate a thrombosis, precipitate a personal crisis of confidence and/or break an assignation with his mistress . . . We can thus view the speech from any of a dozen points of view, in the light of as many different sets of 'considerations'; and it is a mistake to single out some one of these points of view, or chains of circumstances, as having any 'essential' significance, while dismissing all the other sequences and considerations as historically irrelevant or 'accidental'. Considered as a concrete historical fact, the speech has as many aspects as there are relevant points of view from which it can be appraised, and which of these *is in fact* relevant to its political effects depends simply on which of them *proves* to be relevant in the context concerned.

The play of temperament, choice, and character that men demonstrate, in feeling their ways through the web of changing historical problems and circumstances, also creates the stuff of fiction. The life of a physical scientist—or anyone else who is professionally committed to some strictly isolable discipline—becomes 'a real story', and provides matter for moral reflection, only when his disciplinary activities become entangled with issues of other kinds: with political ideals and convictions, academic manœuvring and professional ethics, class prejudices and matrimonial stress. (Recall the plot of C. P. Snow's novel, *The Affair*.) More typically, the scientist, or other 'professional man', pursues the goals of his discipline in isolation from extra-professional goals; while the explanatory or other procedures characteristic of his discipline change and develop, on a collective basis, independently of extra-disciplinary procedures.

On what conditions then, does the isolability of disciplines and professions itself depend? Looking back at their history, we can usefully ask how certain representative disciplines became identified and differentiated in the first place. This status was never established quickly or easily, nor could the proper limits of the eventual discipline be defined exactly, or *a priori*, in advance. In retrospect, no branch of science appears to have a more secure disciplinary status than dynamics; yet the proposal to 'isolate' it met strong initial objections, and no-one could demonstrate clearly beforehand just how the disciplinary standing could be established. Aristotle, for instance, was ready to see 'distance', 'volume', and 'time' treated as general mathematical terms; but he was far less happy about analysing qualitative attributes like 'weightiness' in the same numerical manner.[1] If the concept of 'mass' faced difficulties, so too did the concept of 'force'. Galileo succeeded in moving scarcely any distance beyond kinematics, into dynamics proper; and for a whole century after Newton the measure of 'quantity of motion', or *vis viva*, remained a subject of confused debate.[2] Even near the end of the nineteenth century, Heinrich Hertz was still strongly in favour of eliminating the term 'force' entirely from dynamical explanations.[3] So if, in dynamics itself, scientists had to find out as they went along how far, and for what purpose, the use of abstract concepts could be justified, in other narrower fields of scientific enquiry this was even more the case.

The isolability and disciplinability of any intellectual enterprise must accordingly be proved in the course of our experience. In this respect, the very possibility of creating dynamics as we know it—of abstracting out a coherent group of issues concerned with 'mass', 'momentum', 'inertia', and the like, and studying these without immediate reference to questions of (say) colour, age, health, nationality, or psychology—was an empirical discovery. Experience has shown us how far this can be done; every specific statement in dynamics presupposes that it can be done; and the continued relevance of dynamical concepts and calculations assumes, and in turn confirms, the legitimacy of that presupposition. Yet—speaking 'empirically' —things might conceivably have turned out otherwise. Before any logic-book generalizations of physical laws of nature could be formulated in dynamical language, men had to satisfy themselves that an isolated 'discipline' of dynamics was possible

[1] T. A. Heath, *Mathematics in Aristotle* (Oxford, 1949), Introduction.
[2] For the debate about *vis viva*, see L. Laudan, 'The *vis viva* Controversy, a Post Mortem', *Isis*, 59 (1968), 131–43.
[3] See H. Hertz, *The Principles of Mechanics* (above, p. 249, n. 1).

at all; and they subsequently had to find out, as they went along, just where the boundary runs between the 'dynamical' aspects of the natural world and those other aspects which must be handled in other terms and by other explanatory procedures.

(4) As well as being abstract and isolable compact disciplines are also comparatively single-valued; and this feature, too, differentiates them from other, non-disciplinable activities. This is not to say that disciplines are value-free—i.e. *neutral* with regard to all ethical and moral issues. When an apprentice scientist chooses to take up a particular discipline, he is not deciding to consider only 'facts' as opposed to 'values', or 'means' as opposed to 'ends'. On the contrary: his choice involves an explicit selection of one particular end or ideal. As we saw earlier, the scientific professions have frequently demanded the single-minded devotion of their practitioners, in a way that has been justly compared to the single-minded devotion of the monastic orders.[1] Disciplinary commitment and integrity are, thus, to modern science what sanctity and loyalty to the order were to monasticism. In each case, the resulting patterns of life are not morally neutral, so much as morally self-limited: they demand a readiness to set aside all doubts about the preliminary abstractions, or assumptions, and to focus single-minded attention on the consequential tasks and ideals.

In the natural sciences, at any rate, it has until recently been fairly easy to maintain this single-valuedness and single-mindedness. In technology, its limitations have been more obvious, for we can successfully 'isolate' our disciplines, in practice, only to a first order of accuracy and for the special purposes of the rational enterprise in question. Once we press on beyond this order of accuracy, or take up alternative points of view, it becomes necessary to pay as much attention to the 'second order' side-effects of a discipline as to its internal procedures. So, at a certain point—both in history, and in the growing complexity of industrial organization—the very activity of technological innovation began to create whole new ranges of intellectual possibilities, which faced men with brand-new social, political, and even moral choices. Beyond that

[1] John Ziman, *Public Knowledge* (Cambridge, England, 1968), pp. 138ff.

point, issues of different kinds—theoretical and practical, technical and social—became, in effect, inseparable. At that point, accordingly, technological decisions lost the abstract, isolable character of academic questions, and acquired the irreducibly concrete character of ethical issues; while the abstractions involved in treating the technology concerned as a totally isolable discipline lost their validity.

To cite an extreme case: one could, in theory, consider the development of chemical and bacteriological weapons as a legitimate, self-contained technology, entitled to its own disciplinary ambitions and professional organizations. In the abstract, then, we might have to concede the same professional status to an Institution of Toxicological Engineers as are granted to the existing Institutions of Civil, Mechanical, and Electrical Engineers. (By the same token, the Royal College of Nursing could—in point of abstract theory—be paralleled by an equally professional Royal College of Prostitution.) If our minds rebel at such a suggestion, this only confirms that, in conceding an intellectual or practical enterprise disciplinary standing, we take its positive value for granted. Mechanical engineering and nursing are accepted as legitimate concerns of a discipline or profession, only because the goals of better nursing, or better mechanical engineering are themselves accepted without dispute; an ethical repugnance to nerve gases and prostitution alone disqualifies them from the same status. So the existence of an autonomous 'discipline', and of its associated 'profession', depends both on a preliminary intellectual abstraction—viz. our discovery that the goals in question can be effectively pursued in isolation from other goals —and also on our approval of those particular goals.

We are now ready to take the next step in our argument. We have seen that the boundary between disciplinable and nondisciplinable activities runs where it does because, in the course of their practical experience, men have discovered that it is both functionally possible and humanly desirable to isolate certain classes of issues, and make them the concern of specialized bodies of enquiries; while with issues of other kinds this turns out to be either impossible, or undesirable, or both at once. We must keep that diagnostic difference in mind now, in considering the status of *ethics*: in particular, in considering what weight is to be attached, in the light of our analysis, to the Hegelian dictum that Ethics is not a Discipline.

The situation in the ethical field is, in fact, by no means simple, and the claim that ethics is 'non-disciplinable' is in danger of ambiguity as a result. We may be tempted, for instance, to follow Kierkegaard and the existentialists in regarding ethical issues as 'essentially personal'. And it is true that, in appropriate cases, the non-disciplinary character of ethical problems and decisions has something in common with (e.g.) the non-disciplinary character of poetry. At certain points, that is, ethical problems do indeed involve 'irreducibly personal' choices and considerations; and any proposal to treat the whole of ethics as a single compact discipline would be vain, on that account alone. But this is not—and never could be—the whole story. On the contrary: alongside these personal problems, there are other equally legitimate ethical 'points of view', within which the issues raised are unquestionably communal or collective. Alongside a man's individual problems—whether about his choice of career, spare-time activities, private life or individual death—ethics-as-a-whole embraces also collective problems, having to do with family relationships and contractual obligations, the integrity of social administration, the treatment of infants or animals, and the protection of the sick or the poor.

For that matter, again, even if we do speak of ethics-as-a-whole as 'non-disciplinary', we need not despair of treating particular classes of issues, at any rate, in a 'quasi-disciplinary' manner. For, while it may be impossible to abstract and isolate particular types of ethical issues as completely as we can do those of the natural sciences—so that there is no question of ethical issues being *in general* disciplinable—we can none the less differentiate between the various different points of view from which ethical problems can be approached, and the classes of consideration that are relevant to each such point of view; and certain of these will turn out to be more nearly 'disciplinable' than others. Suppose, for instance, that we set out a categorical 'taxonomy' of ethical problems, issues, and points of view, comparable to that by which we distinguish and separate different intellectual disciplines. Just as we distinguish the 'points of view' of (say) biochemistry, thermodynamics, and neurology, we can then characterize the ethical problems which arise from each alternative standpoint, by reference to its own characteristic rubric—e.g. 'as a question of etiquette' or 'professionally speaking', 'as a matter of sheer humanity', or 'from a career standpoint'—and in terms of its own corresponding repertory of concepts and justification procedures. And we shall then be able to replace such undifferentiated, multi-purpose ethical terms as 'good' and 'bad', 'right' and 'wrong', by terms which are much more specific, determinate, and revealing—such as 'conscientious' and 'unfair', 'in good taste' and 'out of order', 'saintly' and 'disgusting', 'well-judged', and 'unfilial'.[1]

Considering these parallel ethical standpoints one at a time, indeed, we shall find that, in certain of them at least, men have already gone a long way towards developing 'quasi-disciplines', analogous to those that exist in the fine arts. With the growth of human experience and understanding, that is to say, a few groups of ethical issues have crystallized out which can with advantage be treated as technical issues, and made the concern of professional 'experts' or 'authorities'. This is most obviously the case with those types of issue that raise questions of fact too complex to be mastered, with equal discrimination, by all moral agents alike: e.g. those having to do with the ethics of punishment. When divorced from the special insights of penology and penologists, indeed, discussions either of the theoretical morality of punishment, or of the practical administration of prisons, very soon become crude, uninformed, and irrelevant. Even within the general field of ethics, therefore, a few specialized areas of enquiry exist within which it is practicable to harness collective human experience effectively, only on a quasi-disciplinary basis; and the technicality of these specialized fields is reflected, equally, in the development of corresponding professions.[2]

[1] On this point, see my paper, 'Principles of Morality', *Philosophy* 31 (1956), 142–53.
[2] A most illuminating discussion of the relevant topics is that by Glanville Williams, *The Sanctity of Life* (London and New York, 1958): this book shows how much can be learned about ethical issues by studying the ways in which the legal systems of different peoples and jurisdictions have taken account of them, in one milieu or another.

Historically speaking, of course, the intellectual mastery and professionalization of complex ethical issues has been halting, uncertain, and only partly effective. (Penology, like economics, can claim at best only partial theoretical understanding and practical success.) Still, if we set aside the temptations of moral relativism, and re-examine the varied tasks and methods of ethics in the same evolutionary perspective that we are more ready to apply in the cases of science and technology, we can —surely—recognize a slow but genuine improvement in the understanding of ethical, as well as of intellectual issues. Most of us would hesitate to accept a one-way journey in a time-machine to fourteenth-century Europe or third-century Rome; not merely for technological reasons—because of the smells, the poverty, and the disease against which we should have no protection—but also, in large measure, because we should be unwilling to forgo the extra degree of moral discrimination, flexibility, and sensibility that men have developed, slowly and painfully, over the last 500 or 1500 years. Leaving aside for a moment the political horrors of the twentieth century, how many of us would willingly go back to a situation in which mischievous children were treated as having an evil spirit in them, or in which lonely old women with involutional melancholia were condemned as witches? How many of us could, any longer, comfortably live with institutionalized slavery or abandon the hard-won claims of the individual conscience? However thorough-going and hardheaded our cultural relativism may be in point of theory, we should still find it hard to live with its implications, if this entailed acting in an infinitely tolerant manner towards people who—unthinkingly, and for want of experience or understanding—poked fun at spastics, brutalized prisoners, or treated women as beasts of burden.

If ethics-as-a-whole is not a discipline, accordingly, this is for two separate groups of reasons: (i) the entire field of ethics embraces a complex aggregate of sub-issues and sub-judgements —some of them personal, others collective, some of them quasi-disciplinary, others matter for the individual conscience; and also (ii) any final ethical decision, as such, demands the integration or reconciliation of all relevant sub-issues and sub-judgements in terms that bear on some specific concrete situation. Regarded in the same terms as a fine art, the ethical life may perhaps have its own quasi-disciplinary techniques and procedures, which are no more matters for 'merely subjective' attitudes than (say) the techniques of music or painting; but, in a specific situation, the final ethical decision itself goes beyond the scope of all such restricted and quasi-technical matters.

Alternatively—to change the metaphor—we might say, 'Ethics is the politics of the individual life': all the varied sub-issues and considerations that bear on an ethical decision do so in the way in which the various distinguishable technical and social aspects of (say) an engineering project bear on the ulti-mate decision of policy. The 'existential' decision is not a judgement about the relevance or irrelevance of certain con-siderations, but a decision *what to do*—like the question whether or no to go ahead with the engineering project, and if so in what form. Any final ethical decision is thus taken 'in the light of' many different sub-issues, each of which involves general considerations of a correspondingly different kind. But the decision itself calls, in the last resort, for a specific, concrete choice between alternative policies, of a kind that can never be arrived at as a matter of 'expert professional testi-mony' alone. Like decisions of practical policy, our final ethical decisions are concrete and particular, not abstract and general. While they are, in part, dependent on abstract, general con-siderations, they are—in the nature of the case—concerned with these abstract general matters, only as they bear on this particular concrete situation.

So the maxim, 'Ethics is not a Discipline', reports not one single truth but many linked truths. At one extreme, our ethical choices do indeed involve personal—even 'subjective'—prefer-ences, relevant to no collective enterprise, about which no one but the agent himself can speak with any kind of authority: in such cases, the most that other people can legitimately say is, 'If I were you . . .' At the opposite end of the spectrum, any final ethical decision requires us to view a situation, not just from some one particular point of view, but from many relevant standpoints at the same time; and then to arbitrate between the alternative verdicts suggested from these different, coexist-ing viewpoints. If we fail to recognize the full complexity of the ethical field, and look for one single, entirely general reason why ethics is not a compact discipline, we shall then be exposed to the fallacy involved in the more naive forms of 'existentialist' philosophy: we shall be tempted, that is, to equate the *imperative* character of concrete ethical decisions with the *personal* character of individual human choices, and end by using the term 'existential' confusedly, to refer to both features at once.[1]

Perhaps the most crucial contrast between ethics and a typical disciplined enterprise is that between the single-valued pursuit of any particular discipline and the multi-valued character of concrete ethical issues: i.e. the absence—in the ethical case—of some one collectively presupposed goal, and the need to arbitrate between the different types of 'value' associated with coexisting ethical points of view. In any culture and generation, men acknowledge the authority of a dozen inherited approaches to ethical questions. Each of these approaches has its own rubric—'as a matter of self-respect/morality/loyalty/etiquette/integrity/equity/religious commitment/simple humanity . . .'—and each defines a particular set of issues, considerations, and modes of argument. In any chosen culture and generation, furthermore, men do not merely continue applying all these different considerations and arguments in exactly the same way as their forefathers; they also attempt to refine their application, and to re-order their relative priorities, in the light of changing needs and conditions of life. If a particular society has its own characteristic ethical 'tone' or 'quality', indeed, this will commonly be determined less by its individual preferences in (say) matters of etiquette or civic responsibility, than by the relative priorities it allots to alternative 'points of view', when these conflict in actual practice. Where considerations of personal integrity (say) run counter to humanitarian considerations, how do the people concerned resolve this conflict? How much in the way of physical cruelty, for instance, do they regard as excusable by considerations of manliness or honour? These questions of priority are the test-questions by which the ethical attitudes of different cultures are differentiated most profoundly from one another. And the

[1] There is, for instance, an inner tension within the views of J. P. Sartre, who is unwilling to follow Kierkegaard into the extreme individualism required by the original existentialist arguments, but who is able to reintroduce a 'collective' element into his own supposedly 'existentialist' account of moral obligation, only by artificial devices.

integrative character of 'ethics-as-a-whole' arises from our need for a practical order of priorities, capable of cutting across the boundaries between disciplines and quasi-disciplines, of a kind that cannot be developed in the methodical, systematic way appropriate to any single, well-defined—or 'disciplined' —class of problems.[1]

[1] Cf. A. MacIntyre, *A Short History of Ethics* (London, 1967).

CHAPTER 7

The Apparent
Invariants of Thought ,
and Language

Two closely related topics remain on our agenda. The first of
these is the supposedly universal structure of our 'everyday
conceptual framework'—of space and time, causality and
substance, etc.—as discussed by philosophers and psychologists
from the time of Kant to the present day. The other is the
status of the so-called 'cultural and linguistic universals' that
have figured so prominently in recent anthropology and
theoretical linguistics. These topics form a necessary sequel to
our discussion of conceptual change. For, if we are to end up
with an adequate theory of collective concept-use, we cannot
be content to explain the stability of our concepts in one set of
terms, their variability in quite another. Instead, we have set
out here to show—in one and the same set of terms, and within
the framework of a single general account—both why some of
our concepts are so evidently subject to historical change and
cultural diversity, and also why other collective concepts give
the appearance of being universal and invariant as between
different cultures and epochs. Having attempted, so far, to
throw light on the rationality of conceptual change by paying
attention to those aspects of collective concept-use which are
most clearly mutable and non-universal, it is now time for us
to go on and ask the further question: What relevance does
our account have at the other end of the spectrum, as applied
to those features of collective concept-use which are seemingly
most immutable? What can we now say, for instance, about
those features which are apparently represented equally in the
intellectual frameworks, practical procedures, and languages
employed by men in all milieus?

Taking the manifestly variable aspects of concept-use as a

model for understanding the seemingly invariant aspects, instead of vice versa, means of course standing the basic assumptions of earlier theories on their heads. This may be no bad thing. By reversing the burden of explanation—recognizing that phenomena and processes hitherto regarded as simple, natural or self-explanatory (e.g. rest, or solidity) may in fact be more complex, and more in need of explanation, than those which had earlier appeared mysterious (e.g. motion or fluidity) —men have often ended by reorganizing their ideas in a way that brings to light relations, interactions, and interdependences which had, until then, been overlooked or misunderstood. So here: the effect of generalizing our present analysis will be to reverse the implicit presuppositions about human understanding, also. In philosophical epistemology, especially since Kant, the existence of some fundamental and unchanging framework of concepts and principles, which forms the universal and compulsory skeleton for all more technical and empirical 'world-pictures', has widely been taken for granted. By contrast, the novel and transient concepts of the empirical sciences— 'gene' and 'field', 'atom' and 'nerve-impulse'—have been set aside as of little philosophical interest. Just because those scientific concepts are mutable, and there has appeared no 'sufficient reason' why they must be as they are rather than otherwise, they have seemed to lack any deeper philosophical foundation, and have been dismissed as 'merely empirical'. It has seemed reasonable enough to seek deeper, 'transcendental' reasons for the validity of any universal concepts, categories, or 'forms of intuition'; given scientific concepts of a manifestly non-universal and temporary kind, that enterprise has been clearly pointless. So a sharp line has been drawn between the fundamental framework of apparently 'universal' ideas, which is presumably of philosophical significance, and the transient superstructure of scientific theory, which is not.

Our own line of approach rejects this distinction, and reverses the order of explanation. Once we have called in question the philosophical necessity for grounding human understanding on fixed principles, there is no longer any *a priori* reason to presuppose the existence of a universal and compulsory framework of intellectual forms, having a non-empirical status totally unlike that of the empirical concepts of scientific theory. If psychological, linguistic, or conceptual universals are *in fact* found to exist in the thought or language of men in all cultures and epochs, that will now become a highly interesting and unforeseeable fact, of which some special explanation must in due course be found. For from what source could such universal forms and categories derive their universal authority? Are we to think of such 'universals' or 'invariants' as rooted, pragmatically, in the inescapable needs of human life and activity? Or are they simply a 'phenotypic' expression, in the field of intellectual behaviour, of the common 'genotypic' characters of the human species? If all men alike organize their activities and thoughts according to the same general patterns of geometry, grammar, and so on, does this propensity depend on built-in psychological and physiological features of the human frame, like our built-in capacity to discriminate between neighbouring shades of colour? Or should such general human propensities be thought of as the current products of a slow, long-term cultural development? These must be our next questions.

Having adopted an evolutionary approach to other aspects of collective human understanding, indeed, we are obliged to take seriously the possibility of giving a long-term historico-cultural account of all apparent conceptual and linguistic invariants. The comparative rapidity of conceptual change in scientific disciplines depends (as we saw earlier) on the existence of specialized and protected 'forums of professional competition', and these have no obvious counterpart in the case of our every-day conceptual framework: so there are good prior reasons for expecting the non-specialized concepts of everyday life to change much more slowly than the specialized concepts of the professionalized natural sciences. As a hypothesis for discussion, therefore, we can suggest that any apparently invariant everyday concepts may represent merely those intellectual forms which have been protected most completely against the effects of innovation and evolution as a result of their unrestricted currency and unspecialized functions.

In any field of enquiry, after all, an evolutionary theory must explain not just why some historical entities have changed very rapidly, but also why others exposed to similar influences have apparently failed to change at all. The Chinese conifer, *Metasequoia glyptostroboides*, and the deep-sea 'fossil' fish from the waters off Madagascar, have apparently survived as species throughout entire geological eras; yet this in no way entails that the evolutionary process has been suspended for the benefit of the coelacanth or the metasequoia. Presumably, these fishes and plants are subject to the same genetic variability as any other organic species, but their ecological conditions of life have evidently been so stable that an equilibrium has been established: all novel variants have been eliminated as comparatively 'disadvantageous', and the pre-existing forms have been able to retain a dominant position within their respective niches. In the case of intellectual evolution, a similar equilibrium is not only perfectly conceivable, but is something we should expect to find developing in suitable circumstances. Alongside the rapidly changing concepts of science and technology, we should naturally expect to find other less isolated patterns of thought whose rate of development was, by comparison, insensibly slow; without being forced to conclude that their use was absolutely universal and unchanging—still less, compulsory. So understood, the everyday framework of concepts, categories and intellectual forms—which provides the common fabric of our ordinary life and thought, as expressed in the familiar language of space and time, causes and effects, etc.—will simply represent a particularly stable and well-adapted plateau in the development of men's intellectual activities and conceptual equipment.

So, once again, the 'burden of explanation' is shifted. Less than 200 years ago, Immanuel Kant could still assume that our everyday ways of handling spatio-temporal ideas, subject-predicate logic, and even Newton's dynamics and gravitational concepts represented the one and only 'rationally coherent' treatment of experience, and so were mandatory for any rational thinker. This option is no longer available to us today. While certain basic intellectual structures may yet prove to have played a universal part in all phases of intellectual evolution—as the nucleic acids (say) have done in organic evolution—this must first of all be demonstrated as a fact; and, even if it does turn out to be the case, we shall still have to recognise that the unique role of those basic 'rational forms'—like the unique role of DNA—is, first and foremost, an ecological fact. In this case, too, we shall have to go on and ask the further questions:

How far and in what respects does the survival of such 'universal forms' reflect a common genetic endowment which all men bring to their intellectual development and experience? Alternatively: How far and in what respects does it reflect the universal character of the problems on which the human intellectual 'genotype' must be successively exercised, in order to achieve its standard 'phenotypic' expression?

If we can reinterpret the supposed 'invariants' of thought and language in this way, that will enable us to dismantle the traditional dichotomy between *a priori* and empirical concepts, having quite different philosophical and epistemological bases; and to bring concepts of both types within the scope of a single theory, as the extremes of a continuous spectrum or gradation, along the whole of which the same underlying principles apply. Instead of seeing the changeable and transitory concepts employed in historically developing disciplines as essentially contrasted with the universal, *a priori* concepts of mathematical and formal philosophy, these 'opposites' will then prove to differ from one another only in degree. The intellectual structures of Kant's 'necessary forms of reason', Levi-Strauss's 'cultural universals', and Chomsky's 'structures of deep grammar' will no longer be distinguished absolutely from the shorter-lived concepts and principles operative in the natural sciences. They will be understood, instead, as the current end-products of a long historical development: i.e. as the outcome-to-date of men's successive responses to the problems by which their arithmetical, logical, and linguistic procedures have been confronted over the millennia of cultural and linguistic development. And then we shall be able to look, in the case of these invariants too, for the same general relationships between the criteria of rationality and the demands of intellectual adaptation as hold for the more rapidly growing concepts of science.

Granted that the formal structures of logic and grammar, geometry and arithmetic have evolved far more slowly than the concepts and theories of physics, zoology, and electrical engineering, the only operative question will then be, why the concepts at each end of our spectrum have developed historically at such different rates. Yet this is, once again, a question about the manner in which a single set of principles applies in two extreme cases; it is not a reason for seeing these two cases as involving principles of totally different kinds. Perhaps it has been perfectly rational to go on using our current concepts of a 'plane' and a 'straight line', or of 'subject' and 'predicate', for a very long time with very little change; but this need not mean that logic and geometry possess their own special kind of rationality, distinct from that which has demanded much faster changes in the concepts of theoretical physics and molecular biology. It need only mean that the practical demands facing the formal procedures of logic and geometry have themselves changed, and differentiated, much less markedly than the intellectual demands governing the development of the natural sciences. To put the point in a word: where the outstanding conceptual problems are few and rare, there are equally few 'occasions' for making conceptual changes to serve as the solutions of those problems.

7.1: *The Rationality of our Everyday Conceptual Framework*
The 'structuralist' thesis—that all mature experience and knowledge possess a universal, necessary structure, and that this structure is derived, not from the empirical properties of the 'external' world of objects towards which experience is directed and about which knowledge is claimed, but from the manner in which human thinkers impose order on their own 'internal' world of perception or thought—has been familiar since the time of Kant. From the start, however, this thesis has been afflicted by certain ambiguities, which have not yet been entirely eliminated, even today. On the one hand, these short-comings have helped to keep alive the misleading geometrical model of the human mind—as an 'internal', mental world confronting an 'external', material world—together with all the confusions that flow from it. (As we shall see again in Part II of our enquiries, Kant himself was as anxious as anyone to escape from the Cartesian trap, but the very language in which he renounced Descartes finally led him back into the same old difficulties.) On the other hand, the 'transcendental' status of Kant's universal forms of experience has distracted attention from the continuities that link the concepts of ordinary life to the more technical, but 'merely empirical', concepts of science proper. It is of course these *continuities* that are our concern in the present section.

Furthermore, the Kantian thesis is capable of great flexibility; its intuitively attractive core survives a great range of peripheral variations as between one interpreter and another. Kant himself originally gave his complex and elaborate account of the structure which all genuinely 'rational' experience or knowledge must possess, in the hope of explaining how the formal systems of Euclid's geometry, Newton's dynamics, and his own inherited Puritan moral code could claim a binding and universal authority.[1] He thus attempted to prove, not just that certain spatio-temporal, dynamical, and ethical *concepts* are indispensable for coherent perception and practice, but also that certain fundamental spatio-temporal, dynamical, and ethical *propositions* must have the status of necessary truths—at any rate, for any 'rational' thought or action. And the ambiguity between 'necessary', in the pragmatic sense of 'indispensable', and 'necessary', in its logical or quasi-logical sense, has played a powerful part in confusing the subsequent debate. For it is one thing to argue that a certain concept is something that a rational thinker or agent cannot very well get along without; it is quite another thing to establish the necessary truth of certain substantive ethical or scientific principles.

Kant's successors have redistributed their emphasis. During the nineteenth century, Kant's unqualified commitment to Euclid's geometrical theorems and Newton's dynamical theories naturally became an embarrassment. So, by now, more weight is generally placed on the indispensability of our basic concepts for the organization of language and practice, than on the logical or quasi-logical necessity of any particular propositions. Some contemporary analytical philosophers—for instance, P. F. Strawson in his book, *The Bounds of Sense*—discuss a system of everyday concepts, e.g. 'material object' and 'person', about which we can seemingly claim some sort of 'non-empirical' or 'pre-empirical' knowledge; at any rate in this sense, that any substantial empirical assertion whatever involves language of a kind that presupposes the validity of those concepts.[2] Yet Strawson does not rely on these basic

---

[1] I. Kant, *Prolegomena* (Riga, 1783), para. 38; and also *Critique of Pure Reason* (Riga, 1781, 1787), esp. 'Transcendental Aesthetic' and 'Transcendental Analytic'.
[2] See P. F. Strawson, *Individuals* (London, 1959) and *The Bounds of Sense* (London, 1966).

concepts to demonstrate, by 'transcendental deduction' or otherwise, that any specific formal system of propositions has a unique intellectual authority over us. Meanwhile, the cognitive theories of some contemporary psychologists, notably Jean Piaget, still embody a substantial part of Kant's central thesis. Thus Piaget describes his research on the development of a child's intellectual and ethical ideas as an attempt to discover 'how the child progressively recognizes the necessity' of employing certain concepts and patterns of thought about space and time, causality, conservation, and morality.[1] Once again, however, Piaget sees the child, primarily, as recognizing the necessity of operating in accordance with certain very general intellectual forms, rather than the necessary truth of some complex and specific propositions in Euclidean geometry and Newtonian dynamics.

Despite variations in detail, an enduring core of the original Kantian position survives in all these later incarnations. Quite as much as Kant, Strawson and Piaget are committed to the idea that certain very general patterns of thought or action are indispensable elements in all genuine experience and knowledge of the world—or, at any rate, in any experience capable of being articulated in language. At the very least, Piaget suggests, these patterns are the universal long-term goals, or final destinations, of intellectual or moral development—towards which the collective thought and practice of mankind has been developing in history, and the individual thought and practice of every human child still develops today.

What are we ourselves to say about these different claims, in the light of our present analysis? In answering this question, we must consider both (i) what exactly is in each case said to be 'necessary', and also (ii) what each man sees as the source of that 'indispensability'. For the supposed necessity of our every-day conceptual framework might be presented, on the one hand, as a pragmatic matter: the implication might be that our lives would not be half as effective if we made a thoroughgoing

---

[1] This central preoccupation is implicit in all of Piaget's theoretical interpretations; it was made entirely explicit, e.g. in a lecture for the Loyola University Centennial Symposium on *Brain and Behavior*, in October 1969 (ed. A. Karczmar, forthcoming).

attempt to do without them. Or it might be presented, alternatively, as a linguistic or quasi-linguistic matter; the implication might be that all our actual modes of everyday speech take the authority of this framework for granted, and that we should lapse into unintelligibility if we did not. Or again, it might be presented as a physiological or psychological generalization about the human species: the implication being that, try as we might, our psychophysiological nature prevents us from disregarding these forms, except in cases of extreme immaturity or abnormality. Fourthly, the necessity might be elevated to a logical or quasi-logical level: the implication being that any abandonment of the framework would land us in outright logical inconsistency or self-contradiction. Or, finally, it might be claimed that all these different considerations—pragmatic indispensability and genetical inheritance, linguistic adequacy and logical consistency—somehow stand or fall together. The joint implication would then be that, however many rival conceptual frameworks might be conceived in theory, the physiological unnaturalness and pragmatic inefficacy of those alternatives would, in practice, manifest itself in unintelligible inconsistencies, the moment we abandoned our actual basic concepts.[1]

We can highlight the differences between Kant, Piaget, and Strawson, by contrasting their attitudes to our own central problem here, viz., the problem of *conceptual diversity*. Suppose, then, we take into account the full range of concepts, categories, and forms to which—on Kant's original, fully-fledged account—every 'truly rational' thinker or agent is committed: this embraces not only common arithmetic and the categorical imperative of morality, but also Euclidean geometry, Newton's laws of motion, and even the inverse-square law of universal gravitation. One thing is then at once apparent. As a matter of historical and anthropological fact, it is just not the case that all of these particular forms of thought and practice have been current in all cultures and nations, at every stage in intellectual history. In Kant's own time, as he himself well knew, the basic concepts of Newton's dynamics had been current among

---

[1] In this connection, recall Aristotle's argument in the *Metaphysics*, Book IV, iv. 1–6, 1006a, to the effect that the Principle of Non-Contradiction is 'necessary' because no denial of it can be stated *significantly*.

physicists for only a few decades; while even the concepts and theorems of Euclidean geometry were, by historical standards, a comparatively recent achievement. Far from the detailed concepts, propositions, and principles of our 'everyday conceptual framework' being the universal possession of mankind in all milieus (that is to say) they have in fact achieved currency only for a very limited period of history, in certain sub-groups of the whole human race. What, then, do Kant, Piaget, and Strawson respectively make of this fact?

(1) To begin at the beginning: Kant himself—like Plato, Descartes, and Frege—was ready to sweep the facts of historico-cultural variety aside. Let it be the case that the ancient Pomeranians, or the contemporary Society Islanders, were found attempting to operate with ideas of 'causality' or 'morality' markedly unlike our own: what relevance could such discoveries have for the critical philosopher? Kant's thesis had not rested on empirical discoveries about the manner in which particular groups of human beings were in fact *found to* organize their knowledge and experience; such empirical reports were (as Kant put it) matters of 'mere anthropology'. For Kant as much as for Frege, the philosopher's business had nothing directly to do with the contingent facts of cultural history and ethnology; rather, he had to go behind these observations and demonstrate, in a purely theoretical manner, the existence of one unique set of forms having a rational authority denied to all seeming alternatives. In the resulting *a priori* enquiries, empirical observations about men's actual thinking habits could be only a distraction. The philosopher's concern was entirely with the necessary character of all 'rational' thought and action, properly so-called.

In one significant respect, therefore, the fully-fledged Kantian position differed from its present-day successors. In Kant's view, there really was one and only one genuinely coherent way of thinking about any particular subject-matter, whether in scientific theory or everyday life. In point of historical fact, men evidently did not arrive at this one and only coherent way of thinking in all possible subjects straight away; otherwise they would have had no need for an Aristotle in formal logic, a Euclid in geometry, or a Newton in theoretical physics. Yet, in a sense, the work of all these men could be understood as consolidation, not innovation. Their achievement was to articulate explicitly patterns of thought and procedures already implicit in their predecessors' practice. And only when they had been formulated explicitly in this way could those forms of thought be recognized as entitled to a unique status and authority. Men had been arguing, calculating, geometrizing, moralizing for millennia in a more or less confused and unexplicit way. As presented by Aristotle, however, the notions of 'subject' and 'predicate' clearly became the one and only concern of a coherent formal logic; as handled by Euclid the concepts and theorems of geometry became the one and only proper system for analysing the spatial 'order of coexistence'; while Newton's more recent analysis of 'force', 'mass', and 'quantity of motion'—to say nothing of the 'inverse-square attraction' which imposed order on our picture of the astronomical cosmos[1]—had been equally final and definitive. If Kant had lived long enough, he would probably have applauded Maxwell's electromagnetic theory, too, as a further step along the same road, which had made explicit the one and only genuinely coherent way of thinking about 'charges', 'fields', 'currents', and 'radiation'.[2]

To sum up: any field of knowledge achieved the rational status of a 'science', in Kant's eyes, only when finally organized into a system of concepts, categories and forms whose intellectual authority could be confirmed by *a priori* demonstration. (The central mission of Kant's Critiques was, in fact, to show how such *a priori* demonstrations might be given.) These *a priori* concepts then determined the scope and bounds of the reason in a non-empirical manner, regardless of all empirical questions about who and what, where and when. Indeed, the same 'necessary structure' of rational thought had as much authority, in Kant's view, over the thinking of angels, or the inhabitants of other planets, as they did over human thought.[3] His transcendental deductions, proving the unique validity of (say) our everyday concepts of 'substance' or 'causality' were thus intended to demonstrate that these concepts were indispensable

---

[1] *Prolegomena*, para. 38.

[2] See the essay by W. Heisenberg on 'Recent Changes in the Foundation of Exact Science', *Naturwissenschiften*, 22 (1934): English trans. in his *Philosophical Problems of Nuclear Science* (London, 1952), Chapter 1, esp. pp. 21ff.

[3] Kant, *Allgemeine Naturgeschichte* (1755): cf. p. 214, n. 2, above.

for any rational thought whatever; whether in eighteenth-century Königsberg, on the planet Jupiter a million years ago, or in the eternal and infinite reaches of Heaven. We must be careful not to take Kant's claims for the uniqueness of 'pure reason' too lightly. Rational coherence was for him an all-or-nothing affair. Just as one and only one setting of the synchronizing controls on a television set transforms the incoming signals into a recognizable picture, rather than an unintelligible blur, so one and only one system of 'rational forms' could yield a genuinely coherent interpretation of any field of knowledge. Men like Aristotle, Euclid, and Newton simply articulated explicitly the forms implicit in the organization of their respective fields; if the eighteenth-century Tahitians or the primitive Pomeranians failed to recognize those forms, so much the worse for them. Nothing followed from this of interest to philosophy: only that their ideas of 'predication,' 'prime number' and 'parallelism', 'mass', 'causality', and 'obligation' were still in a state of confusion.

(2) A century and a half later, Jean Piaget might start with a basically Kantian training, but he could not ignore the results achieved by history and anthropology during the intervening years. So we find him adopting in psychology a historicized version of Kantianism reminiscent of that advocated by Max Planck in theoretical physics.[1] For both Planck and Piaget, complete rational coherence was no longer an all-or-nothing affair, but rather an ideal to which men approached by successive approximations. It is not just that the ideas of children and primitive peoples are philosophically confused; they really do begin by ordering their thought and action in different ways from adults and more advanced nations. The only uniquely authoritative system of concepts, categories, and forms will therefore be the one towards which rational thinkers and agents are progressively moving, both in cultural history and in their individual lives. In Piaget's hands, the Kantian scheme thus becomes an ideal or final system towards which all rational thought develops as its inescapable and unique destination.[2]

[1] Cf. 'The Unity of the Scientific World-Picture', in *Physical Reality*, ed. S. Toulmin (New York, 1970), pp. 1ff.
[2] It is easy to overlook the fact that Piaget's use of the phrase *épistemologie genétique*,

If we ask Piaget about the thoughts and activities of young children or primitive societies, accordingly, we find him displaying none of Kant's indifference. On the contrary, he sees children and primitive cultures as organizing perception and experience according to their own alternative, more or less systematic procedures; and these procedures have a 'grammar', and even a 'logic' of their own, neither identical with, nor totally incomparable with, the grammar and logic of mature speech and thought. So the thought of children and savages differs from that of adults in advanced societies, not in being entirely formless and confused, where the other is entirely rational and coherent. The difference is merely one of degree. We are not justified in regarding the provisional procedures of juvenile or unsophisticated thought as incoherent, or un-intelligible, or non-rational; the most we can say is that they are less coherent, less intelligible, or less rational than the mature forms towards which they are (or are presumably) developing.

This last qualification is important. Although Piaget abandons Kant's absolute distinction between 'rational' and 'non-rational' thought-procedures, he nevertheless remains a historicized Kantian. Whether writing about psychological development in human individuals, or about the historical development of human cultures, in either case he interprets this process as having a single, definite long-term goal—and this goal is largely the same as Kant's own. There may not be one and only one coherent recipe for thinking about any par-ticular subject *at all*; still, even for Piaget, there does remain one and only one finally coherent way in which we can *end* by thinking about a particular subject-matter, and this final set of forms or operations represents the ultimate goal of rational development in cultures and individuals alike. That the result-ing 'rational forms' will present themselves as in some sense 'necessary' once they have been recognized, Piaget never questions. All he sets out to demonstrate is the sequence of developmental stages, whether in individual maturation or in

refers to the 'intellectual phylogeny' of human cultures, on a collective level, not to the 'intellectual ontogeny'—or developmental psychology—of individual human beings; I am grateful to Bernard Kaplan for this clarification.

cultural discovery, by which children and communities progressively come to recognize that necessity.

From the philosophical standpoint, Piaget's amendments need not have troubled Kant much. On occasion, Kant was quite happy to lecture on history or anthropology, instead of on metaphysics or ethics; and, in doing so, he might well have granted that adolescents and semi-civilized peoples have their own primitive, half-formed grammars and logics. He was not committed to the thesis that, in actual fact, men succeed in grasping the 'pure forms of rationality' at a single stroke, rather than bit by bit; that was one more empirical question for historians, anthropologists, or psychologists. All he had set out to do was to establish the rational necessity and authority of those 'pure forms', regardless of whether any particular groups of men acknowledge it or not. For both Piaget and Kant, therefore, the final 'necessary' forms and operations map the bounds of any thought which can claim to be fully 'rational'. The two men differ only in this: that, after 150 years of historical and psychological enquiry, Piaget prefers to speak of these bounds as defining, not the inescapable structure of any properly 'rational' thought, but rather the common destination of rational development in human individuals and communities alike.

(3) In quasi-Kantian analytical philosophers such as Strawson, we find a further significant variation. Like Kant, Strawson regards a certain set of basic terms and concepts as constitutive of all coherent or intelligible speech and thought. Just as the intelligible discussion of dynamics involves certain inescapable theoretical concepts (e.g. 'mass' and 'inertia') so one can talk or think intelligibly about anything whatever, only in terms of the corresponding everyday concepts: e.g. 'here' and 'there', 'before' and 'after', 'material object' and 'person'. But, in one respect, Strawson's view of this everyday conceptual framework is less grandiose than Kant's: the basic framework of everyday concepts maps and delimits, primarily, not so much the eternal, atemporal bounds of coherent rationality as the actual, current bounds of intelligible sense. Instead of some mysteriously unique 'rational coherence', what is now at issue is the more straightforward and down-to-earth matter of sheer linguistic intelligibility. And Strawson argues that human language, as we know it, does rely for its intelligibility

and effectiveness on its users presupposing a sufficient common framework of basic everyday concepts.[1]

Unlike Piaget's, Strawson's argument is philosophical not psychological; and his amendment to Kant's original thesis—again unlike Piaget's—does make a crucial difference to its meaning. Once a philosopher considers seriously the intelligibility of actual human languages, rather than the *a priori* character of any rational thought, he does what Kant was most anxious to avoid doing: viz., he exposes his flank to the historians and anthropologists. For suppose that we ask Strawson the same questions as before, about the implications of historical and cultural diversity; his position leaves open one possibility which neither Kant nor Piaget admits. If language as we know it relies for its intelligibility on its users presupposing a sufficient common framework of basic everyday concepts, it is still open to us to enquire: 'What if on some other planet, or in the remote future or past, there existed communities of thinkers and agents having their own shared conceptual frameworks—but ones which did not overlap with ours enough to allow intelligible communication between them and us?' For anyone who sets out to delimit the proper philosophical boundaries of any *rational* thought or action, this hypothetical supposition is no more relevant than the actual facts of history and anthropology. But a philosopher who claims only to be demonstrating the bounds of *sense* cannot so well share Kant's cavalier attitude to thinkers and agents in other cultures or planets.

In the long run, the implications of Strawson's argument have more in common with Vico's, or even Collingwood's position, than they do with the original fully-fledged Kantian system. If we can understand the historical events and attitudes of other nations and ages, Vico argued, that is because human beings at all places and times bring similar temperaments and inclinations to bear or similar problems; our historical understanding thus reflects the problems and personalities we share with the men whose actions we are seeking to understand.[2] With appropriate modifications, however,

---

[1] See *The Bounds of Sense*, pp. 150–2: also *Individuals*, p. 247.

[2] Cf., the discussion of Vico's views in R. G. Collingwood, *The Idea of History* (Oxford, 1946), and also the essay on Vico by Isaiah Berlin quoted above, p. 23, n. 1.

exactly the same can be said about linguistic understanding. We understand the writings and utterances of men from other milieus in the same general way, and on the same general conditions, that we understand their other actions; and if the general behaviour of men whose modes of life are sufficiently far removed from our own can be only partly intelligible to us, the same is surely true of their linguistic behaviour also.

Strawson's final conclusions are, therefore, much more modest than those of Kant's position. The bounds of sense reflect the conditions of linguistic intelligibility, given human life and language as we know them. Conversely, given a non-human life and communication-system (or 'language') sufficiently unlike those we know, it becomes quite conceivable—even if only hypothetically, as a matter of science fiction—that there might be other, independent and non-overlapping 'realms of sense'. Within such an alternative realm, thinkers might understand each other perfectly well, even though our own attempts to make contact with them were all doomed to frustration. There would, accordingly, be nothing intrinsically 'non-rational' about their thought and language; their modes of talking and thinking, reasoning and acting would merely be 'rational' in ways different from, and incommensurable with, our own.

Kant himself would have found this position obnoxious. By locating the philosophical foundations of our everyday conceptual framework in the 'bounds of sense', we rule out all possibility of a timeless, *a priori* demonstration that there is one and only one uniquely authoritative and coherent way of thinking. All we are now in a position to demonstrate is that one and only one way of thinking about such everyday matters as places and times, material objects and persons, is compatible with our actual language, and with human life as we in fact know it. Maybe this is all that Kant was ever really justified in claiming, yet—justifiably or no—he certainly did claim much more than this; and he would have dismissed the Strawsonian amendment as involving a fatal surrender to empirical anthropology. Language (for Kant) was merely the instrument of the reason. Linguistic intelligibility was bounded, not merely by the limits to our human sympathies, but by the more fundamental, universal, and inescapable limits of that 'rationality' which language exists to serve. As a first step in understanding why our everyday conceptual framework takes the form it does, we can perhaps invoke the demands of linguistic intelligibility; but this will never be more than a preliminary to the more fundamental philosophical step, of showing how those demands of intelligibility reflect, in turn, the necessary structure of any rational thought.

The slowly changing framework of everyday human thought and language accordingly raises exactly the same problems for us as the more evidently and rapidly changing concepts of science, technology and law. Kant takes up the same absolutist position towards the 'pure forms' of rational thought that Frege did over the 'pure concepts' of arithmetic; and his scorn for mere anthropology plays the same philosophical role as Frege's dismissive attitudes towards history and psychology.[1] At the opposite extreme, a philosophical account which links the structure of everyday human concepts to the character of actual human languages leads us in the last resort to a relativism which differs only in its scope from that of Collingwood and Kuhn. Human life and language being what they are, maybe accepting this everyday conceptual framework is the subscription we pay in order to join the linguistic 'club', and so communicate with one another in an intelligible manner. Yet are terms like 'intelligibility' and 'sense' really the final links in the chain of argument? Does not Kant's emphasis on the rational functions of language quite properly oblige us to go beyond any purely linguistic account? In extending our analysis of collective concept-use from scientific disciplines to the 'conceptual framework' of everyday life, we are—evidently—still faced with the task from which we began: of providing, in the case of these everyday concepts too, a 'middle way' between the *a priori* absolutism of Kant and a relativism which cannot go behind the actual structures of natural languages.

Before tackling this problem directly, one preliminary digression is in order. For how (we may ask) did Kant ever become so convinced that the authority of his one and only one system of pure forms of rational thought could be *demonstrated*? The

---

[1] Recall the quotations from Frege on pp. 55–6, above; and see also T. Mischel, Introduction to *Human Action* (New York, 1969).

answer to this question is connected with the ambiguity we discussed earlier, in the Kantian notion of 'representation'. In certain respects, Kant was the last in the chain of eighteenth-century epistemologists, who debated the role of sense-perception in knowledge as involving the relations between 'impressions' and 'ideas'; in this capacity, we find Kant still toying with the same geometrical picture that had trapped his predecessors, of the Mind as an essentially 'inner' world. Yet it was certainly his ambition to move beyond these eighteenth-century conundrums; and, to this extent, his position looked forward to that of certain nineteenth and twentieth-century pragmatists. This Janus-faced position shows up particularly in Kant's attitude to the *a priori* concepts, categories, and forms. For we appear compelled to accept these as absolutely unique, only if we view Kant's position from an eighteenth-century standpoint. On a more pragmatic interpretation, all his other essential insights can be preserved, without entailing any claims to such a downright uniqueness.

The key idea of 'representation' once again encapsulates the ambiguity that concerns us here. Kant connects the 'necessity' of the *a priori* forms with their role in providing the indispensable structure for any coherent representation; and one of the best ways of understanding that 'necessity' is to ask what kinds of conditions such a coherent 'representation' indeed requires. At this point, we find Kant advancing arguments of two kinds. Some of these deal with the question, 'How are sensory percepts given a cognitive structure?'; and these arguments carry the eighteenth-century debate further, without totally abandoning its terms of reference.[1] Others deal, rather, with the question, 'How are empirical observations explained in terms of logically structured conceptual systems?'; and these arguments anticipate later debates in philosophy of language and philosophy of science.[2] The contrast between these two aspects of Kant's position can, once again, be made clear by emphasizing the very different idiomatic associations of the two alternative German words for 'representation', viz., *Vorstellung* and *Darstellung*.

(1) We may concentrate first on the sensory interpretation.

---

[1] Kant, *Critique of Pure Reason*, 'Transcendental Dialectic', Book I, Sec. 1. 'The Ideas in General' (esp. A320, B376–7).   [2] Ibid.

It had been a commonplace of eighteenth-century philosophy that all intellectual claims to knowledge—whether of concepts or of propositions—had their basis in earlier sense-experience; so the discovery of *a priori* elements in our knowledge naturally suggested that, by some organizing activity, our minds confer a 'cognitive structure' on sensory experience itself. Our experience of the world appears to us through our senses in the form of 'sensory representations' (*Vorstellungen*); all our cognitive achievements have their basis in elements of our various sensory fields; so the *a priori* structure of our knowledge must, somehow or other, be manifested in the organization of those sense-fields. So understood, Kantian representations come very close to Locke's 'ideas of sense', except that an element of 'cognitive structure'—or, as Wittgenstein would have put it, of 'seeing-as'—is built into our sense-fields from the very start. Understood in these terms, Kant's task is then to demonstrate some plausible connection between, on the one hand, the necessary forms of our knowledge and, on the other hand, the structural organization of our sensory representations.[1]

Kant claims to establish the required connection by invoking an attractive series of associations: this apparently links the various formal systems around which our knowledge is intellectually organized with the different sensory modalities through which it is acquired. Where can we find the basis for our *a priori* commitment to the Euclidean system of geometry? The answer takes three steps. Leibniz had described spatial magnitudes as expressing the 'relations of coexistence' in terms of which our understanding of 'external objects' is organized; the sensory modality by which we form our conception of an 'external' world, as comprising objects having spatial extension and linked by relations of coexistence, is that of vision; so the *a priori* forms of spatial knowledge which Euclid made explicit were presumably implicit, even beforehand, in the procedures or processes by which our visual inputs are cognitively organized into a spatial array. In a similar way, Kant associated formal arithmetic with Leibniz's 'relations of succession', and so with our awareness of time; found a sensory basis for these relations

---

[1] See above, pp. 192–9: the term *Vorstellung* is the word regularly used in German versions of Locke, and the other British empiricists, to translate the term 'idea'.

in the modality of hearing; and so explained the *a priori* structure of arithmetic as making explicit the mental structures which are implicit in the cognitive organization of our auditory inputs. Euclidean geometry thus became the formal structure of our 'outer-directed' sensibility, while arithmetic specified the formal structure of our 'inner-directed' sensibility. So Kant was apparently able to do what the eighteenth-century philosophers had demanded—viz., give the *a priori* concepts, categories, and forms a foundation in the structure of sensory experience—without reducing them to the level of empirical generalizations from sensory experience. The formal systems of arithmetic and geometry thus acquired an essential link with the sensory basis of knowledge, while still preserving their logical status as networks of 'necessary truths'.

In this way also—as it seemed to Kant—their claim to *uniqueness* could be underwritten. If the formal calculi of arithmetic and geometry were directly associated with the cognitive structures of the respective sensory modalities, then surely only one system of arithmetic or geometry could have any such application. On this sensory interpretation, our earlier analogy between Kant's necessary forms and the synchronizing circuits of a television set becomes something more than a misleading metaphor. For Kant is now implying that our sensory awareness can yield a consistent 'representation', and so a coherent body of knowledge, only in so far as, within 'experience', our sensory inputs acquire the predetermined cognitive structure which enables them to serve as 'evidence', or convey intelligible 'messages'. And, on this first interpretation it would be highly implausible to suggest that our sense perceptions could be organized into a coherent and intelligible pattern in any but the one and only one system of *a priori* forms.

(2) Let us now turn to the alternative, intellectual interpretation of Kant's position, as concerned with the nature of 'explanatory representations', in the sense of *Darstellungen*. Now, the question at issue no longer has anything to do with our 'sense-fields'; rather, it asks how, and on what conditions, theoretical systems having invariant logico-conceptual structures can be used to explain contingent, empirical observations about transient facts of nature. So let us leave aside, for the moment, all matters of sensory physiology and psychology, and

consider instead how the formal structures of arithmetic, geometry, dynamics, and other such sciences come to have any application at all—not to say any 'necessary' application—to the empirical and contingent facts of nature.

Taken in this second sense, Kant's approach to philosophy broke new ground, which was cultivated fruitfully in the late nineteenth century, when Hertz, Hilbert, and others focused attention on this question again. As a result, it became clear that formal systems of 'necessary propositions' can be relevant to empirical facts about the world, only on certain very definite conditions: in particular, only on condition that suitable 'application procedures' exist for identifying in Nature empirical objects, figures, or systems answering to the theoretical specifications of those formalisms. From this point of view, then, the operative questions have to do, not with the 'organization' of 'sensory inputs' within an 'inner', passively receiving mind, but with the active procedures by which human beings handle and deal with the objects in their environment. Instead of attempting to descry some uniquely Euclidean character in (say) the 'perceived organization' of the visual field,[1] the post-Kantian task will now be to discover—in (say) the practical procedures of carpenters, surveyors, and natural scientists—a pragmatic basis for giving Euclidean geometry a preferred status over its mathematical rivals.

By now, we can begin to see how such a pragmatic basis might be found. In a striking argument, for instance, Hugo Dingler has explained our everyday commitment to Euclidean spatial relations as a consequence of the practical application procedures by which we identify actual empirical shapes, surfaces or bodies as 'right-angled', 'plane', 'spherical', and so on.[2] How could we set about manufacturing (e.g.) an actual physical surface conforming to our mathematical ideal of a 'plane' surface? Suppose that we begin by taking three pieces of (e.g.) stone with flattish surfaces (A, B, & C), and continue rubbing these surfaces together in pairs (A & B, B & C, and then C & A), until they are worn down to such an extent that all three pairs of surfaces touch at every point. Any single

---

[1] Cf. F. Waismann, *Ludwig Wittgenstein und der Wiener Kreis*, ed. B. F. McGuinness (Oxford, 1967), pp. 100, 162.
[2] Cf. H. Dingler, *Das Experiment*.

pair, if rubbed together in this way, will end up more or less spherical, with the one surface convex and the other concave; by insisting that all three pairs touch, we eliminate this sphericity. Then, Dingler argues, given what we mean by the term 'plane', the three resulting surfaces will end up by being—as nearly as we can make them—perfect planes. Furthermore, he claims, the resulting surfaces will be ones to which the Axiom of Parallels, which distinguishes Euclid's geometry from its mathematical rivals, can be applied with complete confidence: this is so, simply because our application procedure has the effect of guaranteeing its relevance in this case. And, so long as we identify physical exemplars of geometrical figures, magnitudes, and relations using this particular kind of recipe, Euclidean geometry will automatically hold good for the figures, magnitudes, and relations so identified. Conversely, to the extent that we use application procedures which depart systematically from that kind of procedure—as is done in astrophysics and physical cosmology—the Euclidean system will be correspondingly inapplicable.

Dingler's argument shows how the Kantian position can serve as a forerunner of later pragmatist views; and this interpretation is by no means foreign to many of Kant's own ideas. After all, Kant was one of the first to insist that all the perceptual and epistemic functions of the human mind are active rather than passive. If we set aside the Cartesian picture of the Mind as an 'inner world', this implies that perception and cognition are the products of operations or activities, rather than mere effects of the passive reception of input stimuli. We can accordingly restate the intellectual core of the Kantian thesis as follows. A coherent 'intellectual representation' (*Darstellung*) of the empirical world, conforming to the formal structures of (say) arithmetic, geometry, and dynamics, requires suitable application procedures for identifying (say) 'material objects', 'plane surfaces', and 'momenta' in the empirical world; but, once these procedures have been devised and employed, the resulting theories will comprise formal ('necessary') systems of general propositions whose empirical ('synthetic') relevance is nevertheless guaranteed.[1]

[1] H. Poincaré, *Science and Hypothesis* (English trans., London and New York, 1905), Chapters 3–5.

This second interpretation, of course, gives nothing like the same support to Kant's claim that there is one and only one coherent way of thinking about spatial relations, material objects, or morality. However much the supposed 'inner organization' of our sense-fields may seemingly possess one and only one form of 'cognitive structure', it could not plausibly be claimed that a single, uniquely authoritative set of application procedures must exist for relating the formal systems of mathematics to the actual behaviour of their physical exemplars in the natural world. On the contrary, any particular set of application procedures will have pragmatic advantages only for certain purposes, and on certain conditions, and the formal system in question will be applicable or inapplicable accordingly. Indeed, quite different sets of application procedures may have compelling advantages for different purposes; so that, in one situation or another, alternative mathematical systems of (say) geometry or dynamics may have to be employed. Though, for the practical purposes of carpentry and similar everyday activities, we may adopt procedures which commit us to using the concepts and theorems of Euclidean rather than Riemannian or Lobachevskian geometry, it does not follow from this that we are committed to Euclidean concepts and theorems unconditionally—that is, in all contexts and for all purposes. When it comes to triangulating the nebulae and surveying the cosmos on the largest possible scale, Dingler's recipes are clearly irrelevant; so the questions how we may identify 'planes' and 'straight lines' in outer space, and what formal system of geometry is most appropriate for astrophysical theory, can be reopened without inconsistency.[1]

Let me summarize these two interpretations of the Kantian problem. Firstly, if we understand the word 'representation' as a sensory term (= *Vorstellung*), we may claim that all sensory experience has a spatio-temporal character, with the objects of vision presenting themselves to us in a spatial array and those of hearing in a temporal sequence; and it will then be legitimate to ask how the visual and auditory sense-fields come to acquire their peculiar spatial and temporal characters. Alternatively, if we take the word 'representation' as an intellectual term

---

[1] See H. P. Robertson, 'Geometry as a Branch of Physics', in *Albert Einstein Philosopher-Scientist*, ed. P. A. Schilpp (Evanston, Ill., 1949).

(=*Darstellung*), we may claim that the axioms and theorems of the formal sciences analyse in an explicit form relations which are implicit in our current operations for identifying and handling physical objects or magnitudes; and it will then be legitimate to enquire, like Hertz and Dingler, just how, and on what conditions, each of our current mathematical theories came to acquire its peculiar empirical relevance and application.

The one thing which we can no longer do is to equate these two problems, and attempt—as Mach did—to solve them both in the same breath. Whether or no it makes any sense to say that the spatial relations apparent within our visual *Vorstellungen* have something specifically 'Euclidean' about them, that issue is clearly independent of Dingler's pragmatic assertions about the ways in which mathematical geometry is applied to physical exemplars. Likewise with the formal calculus of arithmetic: whether or no we can be said to use numbers or number-groups (. . . . . . .) to impose a 'mental structure' on drum-beats and other auditory inputs, that psychological issue is again independent of all pragmatic questions about the ways in which arithmetic is given an application in the everyday tasks of counting and ordering physical objects. Certainly, it will no longer be plausible to suggest that the pragmatic application procedures for allotting numbers to physical objects derive their 'validity' or 'necessity' from 'cognitive structures' inherent in the temporal character of all auditory experience!

The fact of the matter is that Kant succeeded in first raising two independent groups of questions, and subsequently confusing them. Both groups of questions were original and intriguing, and had great long-term importance for both philosophy and cognitive psychology; and, if Kant himself failed to keep them clearly differentiated in his own mind, that fact simply reflects his residual commitment to Cartesian epistemology. For our purposes, however, it is absolutely essential to keep these differences in mind. Some of the crucial questions about human understanding—as we shall see in due course—have, indeed, to do precisely with the relations between 'intellectual representations', such as we employ in geometry or physics (i.e. collective *Darstellungen* or demonstrations), and 'sensory representations', such as we study in sensory physiology and cognitive psychology (i.e. individual *Vorstellungen* or perceptions). If we fail to differentiate between these two types of representations from the outset, we shall not only risk confusing the issues that arise about each type. Worse, we may even end by depriving ourselves of the very terms we need in order to ask questions about the relations between them.

With this distinction in mind, we can return to the problem of ordinary language and the 'everyday conceptual frame-work'. If we were compelled to accept the sensory interpret-ation—according to which the formal structures of arithmetic and geometry are already implicit in the organization of our auditory and visual experience—the consequential questions would be perplexing to a degree: in just what sense, for example, can it be said that the elaborate formal relations analysed by Euclid and Peano were already 'programmed into' the mental operations and/or cerebral structures of primitive man? And, if they were 'programmed into' man's mind or brain, in this way, how then could that same mind or brain ever succeed in thinking up and applying any non-Euclidean geometry? This interpretation—as we shall see in the next section—has some even more unfortunate implications, when we turn and con-sider the historical antecedents of our current concepts of space, time, and number. For it entails, firstly, that the entire implicit system of Euclidean relations must have sprung into existence during human evolution all at once, as a complete 'cognitive structure'; and it entails, further, that the men of pre-Euclidean times, who applied cruder, practical rules of thumb in their measuring operations, were, in some sense, misunderstanding throughout the formal patterns already implicit in their own mental activities!

The pragmatic interpretation, by contrast, allows us to take a more historical view of everyday concepts. Dingler's argument certainly indicates that Euclid's idealized geo-metrical system was consistent with earlier practical procedures of surveying and mensuration, and even that it was a natural extension of them; but Euclidean geometry remains, on this account, a specialization from those more workaday procedures, rather than their retrospective articulation and justification. Formal geometry then appears to be merely a late and sophisticated product of a long historical development, which was preceded by other, less formalized sets of spatial concepts, with independent and more practical functions of their own. The stronger Kantian claim, that one particular formal system of geometrical concepts has always implicitly had an absolute claim on our intellectual allegiance, and so a timeless 'necessity' will then involve us in defending either or both of two additional hypotheses. Either, we shall have to suppose that our genetic endowment includes an innate propensity to organize our experience in a specifically 'Euclidean' manner; or, alternatively, we shall have to conclude that our current framework of everyday concepts constitutes in every respect the formalism best adapted both to all conceivable goals of 'rational' thought and action and to all plausible human situations, however remote.

We shall be discussing the first of these hypotheses—the 'nativist' doctrine, according to which our basic conceptual repertory reflects inborn patterns in the human mind or brain—at greater length in the following section. If that were the entire story, of course, the truth of the second hypothesis also would follow from it as a mere triviality; for how could it ever be 'well-adapted' to reason in ways contrary to the 'inborn patterns' of our natures? Taken separately from the nativist doctrine, on the other hand, the second hypothesis—according to which our current 'everyday conceptual framework' is uniquely well-adapted to all conceivable human goals and situations—appears to be untrue, simply in point of fact. The evidence of intellectual history and cultural anthropology, especially about the so-called 'ethnosciences', may not yet be entirely final or conclusive. Still, its general direction is clear. Whichever of Kant's *a priori* 'forms of thought' we choose to consider, cultures can be imagined—and even discovered—in which the accepted and 'well-adapted' conceptual frameworks deviate significantly from Kant's 'one and only coherent' scheme.

We may touch in passing on the concept of 'colour', which helps to show the range and complexity of the questions involved in all such cases. To begin with, there is undoubtedly a genetically controlled, physiological component in our capacity to perceive and discriminate colours, and the consequences of this component show themselves—as one should expect—in a 'culture-invariant' manner.[1] Thus, if human beings from different cultures are faced with coloured surfaces placed side by side, no significant cultural difference is found between their capacities to discriminate between neighbouring shades: they all succeed—or fail—equally in pointing out the boundary between the two colours. (Even the different types of colour-blindness take similar forms, and appear in roughly the same proportions, in all human populations.) For all this 'cultural invariance' on the perceptual level, however, the conceptual treatment of colours and colour-terminology differs markedly between human cultures and languages. Homer's colour-words and colour-descriptions, as has long been known, are extremely hard to reconcile with our own, suggesting that the categorization of colours in ancient Greek was at variance with that which is embodied in most modern European languages; and this conceptual variability in men's treatment of colours is confirmed by anthropological and linguistic studies of contemporary cultures in more remote parts of the world.

While sharing man's common physiological inheritance, accordingly, people living in different cultures commonly 'segment' the colour-continuum for linguistic and conceptual purposes in strikingly different ways, which are in turn transmitted culturally. Perceptually, a Mexican Indian and an Egyptian peasant, a mountaineer from the Philippines and a contemporary Londoner may discriminate exactly the same fine shades of colour; yet, conceptually, they may seemingly disagree about whether two such shades do, or do not, belong in the same broader colour-category. And it is not merely that men in different cultures subdivide the colour-continuum more or less finely—with one people giving special names to a great variety of browns, another to a great variety of greens—while all their larger units cut along similar lines. In some cases, these classifications too are plainly irreconcilable. For instance, in a country where plants are green only briefly in springtime, one single colour-word ('vegetation colour') may quite naturally be used for both green and brown. The different shades of green will then be referred to as—and even 'seen' as—alter-

[1] Cf. E. H. Lenneberg, *Biological Foundations of Language* (New York, 1967), Chapter 8, Section IV.

native shades of a single general colour shared along with summer wheat, sand, and camels.

In the case of colour-concepts, at any rate, an ethnoscientific comparison between different peoples and epochs can take us beyond the mere collecting stage, and suggest how we might set about explaining the existing variety in colour-classifications and terminologies. Eventually, we might even repeat for everyday colour-classifications the same kind of 'diachronic' analysis of conceptual change that we developed earlier for scientific concepts. Historically speaking, for instance, how did the accepted classification of colours develop down the centuries, from Sumeria to Homeric Greece, and so on to Hellenistic Alexandria, medieval Italy, and modern Europe? Is there (as Berlin and Kay have suggested)[1] a more or less regular sequence in which terms for primary colours are added to the linguistic repertory of different cultures? And are there any demonstrable correlations between the historically recorded changes in colour-classification and other, contemporaneous differences in the modes of life of the peoples concerned? Again, does anthropological evidence about differences in colour-classification between present-day cultures bring to light any synchronic correlations with other contemporary cultural differences? In this way the development of 'everyday' concepts can be brought within the scope of our evolutionary analysis.

With this example in mind, we can look next at some 'everyday' concepts and categories which do fall within Kant's *a priori* scheme: for instance, notions like 'substance' and 'causality'. In these cases, too, there may well prove to be some invariant core, as in the case of colour-perception. Up to a point, for instance, most living creatures are physiologically equipped to discriminate between separate physical objects, or between physical objects of different kinds; and human beings presumably share this same general inheritance. Yet, alongside any such universal, genetically-transmitted and culturally-invariant human capacity, we must also expect to find, on a finer conceptual level, detailed historical and cultural

[1] B. Berlin and P. Kay, *Basic Color Terms: their Universality and Evolution* (Berkeley, 1969); also H. C. Conklin, 'Hanunóo Color Categories', *Southwestern Journal of Anthropology*. 2 (1955), 339–44, reprinted in *Language in Culture and Society*, ed. D. Hymes (New York, 1964), pp. 189ff.

variations between the notions of 'causality' and 'substance' current in different milieus; and we can usefully ask how these variations are connected with other historico-cultural factors. In animistic cultures, for instance, the 'causal' ideas in terms of which everyday experience is commonly interpreted differ significantly from those generally current in most industrial societies. Even at the present time, again, the needs of specialized groups and milieus call for somewhat different conceptions of 'causality'; in the courts, the idea of causality figures in the law of negligence with implications quite other than those familiar in (say) clinical medicine or engineering.[1] In judicial arguments, for example, it is accepted that the 'chain of causes' can be 'broken' by the intervention of a fresh agent, in a way that would be unheard of in a scientific or technological discussion and that also runs directly counter to Kant's own belief in the essential 'continuity' of causal connections.

Changes of circumstance thus create occasions for modifying our concepts of causality, even at levels which Kant regarded as immutable and apodeictic. So any attempt to argue that our 'everyday framework' of ideas about causation represents a natural functional equilibrum, which will not be upset by any foreseeable changes in human life and practical activity, faces an almost intolerable burden of proof. For we cannot demonstrate that our 'everyday conceptual framework' is uniquely adapted to the demands of all conceivable human life, without first performing a preliminary task: that of showing how, as a matter of history, our current conceptions took their present form, and what 'ecological demands' their development has satisfied hitherto. And, in carrying out that preliminary task, philosophers must begin taking the evidence of cultural anthropologists and intellectual historians more seriously than they have done up to now.

We can drive home a similar point about the concept of 'substance'; this time against the theories of such quasi-Kantian psychologists as Piaget and Bruner. To claim, for instance, that children in all cultures come to recognize for themselves that 'conservation', in some sense or other, is an essential characteristic of 'material substance' is easy enough. Stated in so general a form, however, this claim is too vague to

[1] See H. L. A. Hart and A. M. Honoré, *Causation in the Law* (Oxford, 1959).

lend support to a fully-fledged Kantian position.[1] For the notion of conservation has itself changed substantially in the course of intellectual history, and even today people from different backgrounds will understand it in very different ways. So just what conception of 'substance' or 'conservation' are we talking about? Do all children recognize for themselves (e.g.) the 'substantial' character of gases and vapours? And, if so, why did not European adults themselves acknowledge this character explicitly until the eighteenth century A.D.? Are children who have no occasion to use sophisticated procedures of measurement and calculation compelled, nevertheless, to regard the liquid in a tall, thin glass as 'equal' in amount to that in another squat glass, in just the same circumstances as we do ourselves? Why should they not, in their own terms, speak of the liquid as 'expanding' when poured into a narrower tube, just as we speak of a solid, in our own terms, as 'expanding' when heated?

In such problematic situations, it may be simply misleading to describe the experience of infants or men from other milieus entirely in our own language. The question, whether or no they have exactly the same 'everyday conceptual framework' as ours, is then secondary to questions about the full variety of ways in which they *speak about and handle* material substances; and it is no longer so clear that the concepts of 'matter', 'substance', and 'conservation' are truly cultural universals, in any but an excessively vague sense. On the contrary, it is far more plausible to argue that the concept of 'matter'—together with the associated notions, 'material object' and 'material substance'—has been transformed several times, in the course of the historical transitions from pre-scientific craft-technology to Aristotle and Descartes, and from Newton on to Rutherford and Chew; and to admit that the twentieth-century concept of 'conservation', on which Piaget and Bruner place such weight, is itself a sophisticated product of an elaborate conceptual history.

To return finally to Kant's central examples, viz., geometry and arithmetic: in other milieus, men have certainly employed numbering and measuring systems of much cruder kinds than

[1] See, e.g., the experiments by J. Bruner reported in J. S. Bruner, J. J. Goodnow, and G. A. Austin, *A Study of Thinking* (New York, 1956); also the very extensive literature on similar experiments by Piaget and others.

those that are generally current nowadays in the industrial countries. In doing so, furthermore, they have shown no sign of recognizing the 'necessity' of our modern arithmetical and geometrical formalisms. Given the modes of life of other cultures and environments, indeed, it may be perfectly proper to treat a distance of one mile (in our terms) across marshy ground as 'equivalent to' two of our miles through forest, or five of our miles across a dry plain—regardless of the consequences that such a criterion of 'equivalence' would have for any would-be Euclid. For, in a non-industrial culture, it is natural to link the concepts of 'space' and 'time' directly to those of 'effort' and 'work', rather than to the readings of yardsticks and clocks. And there are good historical grounds, also, for regarding the intellectual authority of clocks and yardsticks as a recent achievement of human culture—even perhaps, as Lewis Mumford has argued, as a by-product of the liturgical exactitude of mediaeval monasticism.[1] It may yet turn out, of course, that some residual minimum of arithmetic—say, the first four positive integers—forms a culturally invariant feature of every culture and language. But, if mathematical universality goes no further than this, it does little enough for Kant; for it throws no more light on the 'necessity' of our own fully-fledged arithmetical and geometrical systems than the universal perceptual capacity for colour-discriminations throws on the conceptual variety and development of colour-terminologies.

At this point, we have a choice to make. On the one hand, we can continue to claim that our own concepts of 'space', 'time', and 'number' have a uniquely valid status, and dismiss all the alternative frameworks of everyday concepts revealed by history and anthropology as rationally defective. (By now, however, such a reaction would be intolerably parochial; at any rate, until it is proved that our own spatio-temporal and numerical concepts do—after all—perform intellectual functions indispensable for all conceivable human lives and cultures.) Alternatively, we can accept those alternative frameworks as having a provisional validity and authenticity of their own, and then raise comparative, evolutionary questions about them. How, for instance, do the differences between the numbering

[1] L. Mumford, 'The Monastery and the Clock', in *Technics and Civilization* (New York, 1934).

and measuring procedures of different cultures relate to their other conditions of life? Do maritime cultures differ systematically, in this respect, from mountain or jungle cultures: do hunters speak of distances in different terms from cultivators; and do tropical peoples divide up time differently from arctic peoples? Again: has the development of modern industrial society affected our everyday concepts of number, distance, and time in any characteristic way?

If stated so generally, these questions almost answer themselves. For, of course, the transition from classical antiquity, by way of mediaeval Europe, to modern industrial society, has had *some* effect on the spatial, temporal, and numerical concepts characteristic of our 'everyday framework'. During the actual development of modern industrial society (as Mumford has argued) there were indeed pragmatic grounds to adopt our current system of spatial, temporal, and numerical procedures; even though, as with all pragmatic considerations, these did not necessarily hold good in exactly the same way in all cultural situations whatever.[1] So the question we should be asking is not, 'Did the transition to modern life make any difference to these procedures?', but rather, 'What exact difference did it make?' The cultural adaptedness of concepts and procedures, like the organic adaptedness of species, is a comparative matter; among an available pool of alternatives, one or another proves 'better-adapted' to the specific demands of some particular 'niche' or application. So considered, the general-purpose merits of one alternative may not rule out the special-purpose merits of another alternative in some special application or 'niche'. So the Euclidean patterns by which we analyse spatial relations, for the normal purposes of carpentry and surveying, are compatible, both with the pre-Euclidean conceptions of the hunter or cross-country runner, and with the post-Euclidean patterns of theoretical astrophysics.

The conclusion of this argument is easily stated. Men's everyday spatio-temporal, numerical, and causal concepts have less technical uses and a wider currency than the specialized concepts of natural science; as a result, they change more slowly. Yet these everyday concepts, too, are open to change,

[1] Ibid. cf.; also Mumford's more recent books on *The Myth of the Machine*, Vol. I (New York, 1967), Vol. II (New York, 1970).

and therefore have conceptual genealogies. And, if we were to reconstruct those genealogies from the evidence of history and anthropology, we might finally succeed in answering questions that we can, at present, do no more than state; e.g. the 'every-day' counterparts of Collingwood's two crucial questions about conceptual change. In this way (that is to say) we might succeed in identifying 'the occasions on which' one set of every-day spatio-temporal, numerical or causal concepts has been displaced by another; and bring to light also 'the historical processes by which' such changes have been brought about. If that could be done, it would confirm that our everyday concepts—like those of the sciences—are involved in a genuine conceptual evolution, as a result of which our familiar language and practice are progressively 'adapted' to changes in the detailed goals and circumstances of human life and activity.

Whether taken in its original *a priori* form, or with Piaget's 'genetic' amendments, or as a thesis about everyday language: in each case, the Kantian thesis evidently leaves the facts of historical change and cultural diversity mysterious. Maybe there are 'linguistic universals', of a kind that require to be explicated in a 'non-empirical' manner; but, then again, maybe men from different milieus really have less of an 'everyday conceptual framework' in common than philosophers have assumed. We can argue, if we please, that any human culture must operate with *some* notions of 'space', 'time', 'substance', 'causality', 'morality', and so on; but this, by itself, gives us little more than the general form of a claim, lacking any specific substance. What Kant himself believed was that all truly rational thought was built around just those *specific* notions of 'space', 'time' etc., at which Euclidean geometers, Newtonian physicists, and Puritan moralists had arrived in the 1780s. And, despite the other merits in Kant's account, his claims for the uniqueness of his one and only one coherent system of 'rational forms' appear, in retrospect, to have been highly arbitrary. Given the variety of numbering, measuring, and ordering procedures we now know about, it would be implausible to claim that every one of these—after all—'implicitly' embodies the formal characteristics of Euclid's geometry and Peano's arithmetic; and it would be still more fanciful to argue any longer that those formal structures have been shaping men's visual and auditory sense-fields throughout the centuries, without their being aware of the fact!

The general form of the Kantian claim—viz., that the everyday conceptual frameworks of all milieus must include spatial, temporal, material, causal and moral notions of some kind, whether or no these coincide in detail with the particular forms we ourselves employ—may yet be worth making explicit, as a step in explaining what kind of creature an 'everyday conceptual framework' really is. For, while we can set about imagining a language which provided only for aspatial, atemporal propositions (such as 'Jack is the uncle of Jim' and 'Twice three is six') and which therefore lacked any ways of referring to specific times and locations (as in 'Jack is now in the dining-room' and 'Six men were here yesterday'), the moment we ask what kind of a life such a restricted language would permit, we can see how poverty-stricken and counter-functional it would be. What the Kantian thesis then does for us—if interpreted liberally, as a purely formal claim—is to establish one further connection between the philosophical analysis of conceptual frameworks or conceptual change, on the one hand, and the psychological study of behavioural development and perceptual systems, on the other.

Despite the work of recent phenomenologists, such as Merleau-Ponty,[1] much still needs to be done before we can hope to see clearly all the possible relations between the properties of our behavioural and perceptual systems, as analysed by psychologists like J. J. Gibson,[2] and the conceptual expression which these systems have acquired in the course of cultural development. There can be little doubt that all human beings, except those with certain types of congenital brain-damage, start life with the native equipment for developing (e.g.) effective 'orienting' and 'postural' systems, and for bringing their perceptual and cognitive systems into play, in a vast variety of subsequent transactions and modes of behaviour. And it seems equally certain that we also learn to employ the conceptual procedures current in our particular milieu—both

[1] See M. Merleau-Ponty, *The Phenomenology of Perception* (Paris, 1945; English trans., London, 1962).

[2] The most recent and authoritative statement of Gibson's position is given in J. J. Gibson, *The Senses considered as Perceptual Systems* (Boston, 1966).

intellectual and practical—as specialized extensions of the simpler modes of behaviour through which our native behaviour and perceptual systems find primary expression. (It is interesting, for instance, to trace the origin of our geometrical concepts back to such primary spatial terms as 'left' and 'right', 'up' and 'down', 'forward' and 'back'.) When we consider the form of our current conceptual framework, accordingly, we once again have to take into account neither genetical factors alone nor cultural factors alone, but rather the whole historical sequence of forms through which our native intellectual and practical capacities progressively find—and have historically found—better-adapted functional expressions.

7.2: *Nativism, Functionalism, and the Language Capacity*

This last conclusion confronts us head-on with a problem that we encountered earlier in passing: that of distinguishing the respective contributions of 'nature' and 'nurture' (i.e. of physiological and cultural factors) to our conceptual and cognitive skills. Suppose we consider the patterns of collective concept-use in different communities—in particular, any widespread or universal features that may be discovered in our actual conceptual or linguistic capacities; how far should we explain these as products of genetically controlled ('innate') processes and mechanisms, on the one hand, and how far, on the other hand, as common effects of similarities in the cultural environments into which we all grow up? In what respects and to what degree of detail, can any 'cultural universals' accordingly be attributed to corresponding 'innate capacities' in the common genetic inheritance of all human beings, how far to the universal practical tasks facing men in all cultures?

This question needs to be stated and handled with great care. The entirely general question—whether it is reasonable to look for an innate basis to our linguistic and cognitive capacities at all—is scarcely worth asking. To some extent and in some respects, one can reply, all the end-products of evolution and development of course require a material basis of some kind; whether genetic—e.g., a particular arrangement of nucleic acids in the chromosomes of the gametes—or neurological— e.g., a particular structure and organization of brain-processes and mechanisms. In this, our cognitive capacities are, presumably, no different from any other. Until the last 100 years, that fact may not have been acceptable to everybody, and it still encounters pockets of reluctant scepticism even today; but— stated modestly enough—it simply records such well-proven observations as (e.g.) that cognitive, conceptual, or linguistic behaviour is associated with the integrity of certain elements in the human central nervous system, and is disrupted by diseases or injuries that interrupt, or frustrate the development of, the corresponding brain-processes.[1] In some weak or general enough sense, therefore, we may take for granted the existence of a psychological endowment, genetic inheritance and/or other 'native' basis for our collective conceptual skills, and turn instead to a more specific and detailed question: viz., how exactly that generalized 'native basis' finds behavioural expression, and to what extent the complex patterns to be found in the resulting behavioural end-products were implicitly present, also, in our original 'innate equipment'.

The reasons for concentrating on this more specific question will be familiar to anyone who has studied the relationship between genetics and evolution in cases of other kinds. Whatever the particular subject-matter under investigation, it is always essential to distinguish the specific 'genotype'—which corresponds here to the 'material basis' directly transmitted by genetic mechanisms—from the actual form of the adult organisms, or 'phenotype', into which particular individual gametes, with their characteristic 'genotypes', typically develop. There are, of course, close and significant relations between the eventual character of any actual phenotype and the nature of the genotype which serves as its material basis, and of which it is the end-product or 'expression'. But they are by no means the same thing. The representative 'phenotype' is the end-product to which a particular 'genotype' gives rise, as the outcome of a typical development in a standard environment.[2] And, however much we are tempted to characterize an organic species or population by its directly-transmitted 'gene pool', it is necessary to bear in mind that natural selection can

---

[1] N. Geschwind, 'Disconnexion Syndromes in Animals and Man', *Brain* 88 (1965), 237–94, 585–644.

[2] The distinction between 'genotype' and 'phenotype' is dealt with clearly in the authoritative works of E. Mayr, cited above.

influence only the statistical distribution of phenotypic alternatives found in an actual breeding population. From an evolutionary point of view, that is to say, the significant features of organisms are 'expressions' of genes, not direct 'properties' of them. In this respect the case of our cognitive, conceptual, or linguistic capacities can be no exception. Granted that these, too, have some 'innate basis' in the new-born infant, however generalized, these capacities are themselves represented directly by the intelligent adult behaviour which is their eventual 'expression'. And that adult behaviour is, once again, the end-product to which this 'innate basis' gives rise as the outcome of typical developmental processes—physiological, behavioural and cultural—taking place in a largely standardized sequence of environments or situations.

On the face of it, therefore, any 'universal' features actually found in the expression of these cognitive capacities in different milieus might be explained in one of three ways: by direct reference to invariant elements in the original 'innate basis', or as effects of the standard sequences of situations within which human beings develop, or else in terms of the reciprocal interactions between the innate basis and those cultural situations. We shall be having to look at the problems of cognitive development, or 'conceptual ontogeny', in more detail in part II of these enquiries. For the moment, we can concentrate on asking in what respects, and with what degree of exactness, one might reasonably expect any universal or typical 'phenotypic' patterns of cognitive behaviour to have exact formal counterparts in the native psychological endowment or physiological inheritance of which they are an expression; and how far one should look for the origins of such patterns, rather, in the universal exigencies of human life, and so in the standard sequence of interactions through which our 'genotypic' endowment finds its behavioural expression.

This question can best be attacked as it arises in theoretical linguistics and the philosophy of language. In recent years, the writings of Noam Chomsky and his colleagues have made some dramatic claims about 'universal grammar', and about its supposed basis on the shared 'native capacities' of all human beings. Chomsky himself has several times argued, with some vehemence, that the existence of the same grammatical patterns in all human language is sufficient evidence that the infants born into all language-using populations possess identical innate endowments. This 'nativist' thesis is, at the same time, so intriguing and so ambiguous that we can afford to look at it quite closely.[1] As it has been debated up to now, indeed, one can show that 'linguistic nativism' has telescoped two distinct doctrines. The stronger and more extreme thesis lands us in severe theoretical difficulties, but the weaker form is entirely acceptable, if not a truism. By distinguishing these two forms of nativism with sufficient care, we can then re-state, with greater precision, the central questions with which we shall be having to deal later on: about the relationship between the cognitive capacities embodied in the human genotype and the behaviour which is its phenotypic expression.

The task of tracing our linguistic capacities back to their innate basis in the physiological structure of our brain, and to the psychological endowment with which we are born, might appear at first sight a matter strictly for empirical science. The questions involved might appear (that is) to be answerable simply by bringing together, and setting alongside one another,

[1] Chomsky's views on the 'native' or 'innate' character of the human language capacity, and his arguments for believing that its historical origin is not explicable in terms drawn either from cultural history or from evolutionary zoology, are scattered throughout his writings and lectures. His initial attack on empiricist theories of language learning is best represented by his critical review of Skinner's *Verbal Behavior* (*Language*, 35 (1959), 26–58); but see also the symposium at the Boston Colloquium for Philosophy of Science in December 1966, reprinted in *Boston Studies in the Philosophy of Science*, Vol. 3, ed. R. S. Cohen and M. W. Wartofsky (1968), pp. 81–107. The most concise presentation of Chomsky's central philosophical thesis was that given in his John Locke Lectures at Oxford University in the Summer of 1969 (forthcoming); though important glosses to his general position are to be found in *Aspects of the Theory of Syntax* (Cambridge, Mass., 1965), especially Chapter 1, in *Language and Mind* (New York, 1968), and in *Cartesian Linguistics* (New York, 1966). To some extent, Chomsky has changed his views over the years and, by now, the 'strong' nativist thesis which I am criticizing in this essay may well go further than he himself would want to do, at the present time. So let me just add, that the *central* purpose of this section is to distinguish the natures—and the respective implications—of the two very different doctrines which can go by the name of 'nativism'; and to show that Chomsky's legitimate objections to the hard-line behaviourists can be defended perfectly well by a form of nativism which is, at one and the same time, *biologically* more intelligible than the view he advanced in the John Locke Lectures in 1969, and also *linguistically* more easily reconciled with (e.g.) the 'functionalism' of men like Bühler and Wittgenstein.

the separate results arrived at in linguistics and cognitive psychology, biochemistry and physiology, clinical neurology, and other related sub-disciplines. If that were the whole story, of course, philosophers could only wait for the different groups of scientists concerned to piece together an agreed account of the matter; meanwhile, these questions would raise no specifically philosophical issues.

In fact, it is highly doubtful whether an empirical attack alone can give us—as it were, by aggregation—the coherent understanding we here need of the ways in which adult language-use is related to its 'innate basis'. Before such empirical discussions can be theoretically fruitful, certain preliminary analytical problems must be sorted out. The reasons for this hesitation are of two kinds. In the first place, the theory of language cannot be disentangled entirely from the theory of mind. As Chomsky himself emphasizes—like Kant before him—psychological questions about how we do our thinking are inextricably tied up with linguistic questions about the 'forms'—propositional or grammatical—in which those thoughts are cast.[1] Indeed, if we look again at the ways in which the problems of 'brain and language' is commonly discussed today, we can see in it the traditional problem of Body and Mind writ small: the older and more grandiose conundrums about Mind and Matter being transcribed, in accordance with more astringent twentieth-century tastes, into modest-looking questions about 'language' and its 'neural correlates'. So, unless we incautiously assume that the whole of the Mind–Body problem can be disposed of by half a dozen industrious scientific groups working independently, the task of relating our linguistic capacities to their neurophysiological correlates still involves an inescapable analytical or philosophical element.

In the second place: if we consider the current developments in the special sciences, we shall find that, as normally stated, the terms in which their questions and results are posed are incommensurable, so that their mutual bearing remains quite unclear. As a result, a simple aggregation of their results cannot give us all that we require. In addition, one must analyse the questions arising at the interdisciplinary boundaries between the sciences, and so recognize what bearing the results

[1] See *Language and Mind*.

of each science, taken separately, could—in principle—have on those of the others. This demand may, at first glance, appear over-ambitious; and certainly, in the present state of knowledge, it would be premature to insist on a detailed and specific account of the actual relations to be expected between different neurolinguistic and psycholinguistic sciences. Still, it is not too soon to pose this question as a matter of principle, nor to discuss what general *kinds* of mutual relevance these different sciences can, and cannot, have to one another. On the contrary, if we pursue specialized enquiries in these different fields in total autonomy, and shut our eyes to all questions about their mutual relevance, we may be making sticks to beat our own backs.

To insist on analysing the human 'language capacity' in terms drawn from theoretical linguistics, for instance, while paying only lip-service to clinical neurology and neurophysiology—to concede merely, as Chomsky does, that there will 'definitely some day be a physiological explanation' for the mental processes revealed by linguistics, while refusing to allow that physiological or biological considerations may already be capable of reflecting back on current linguistics—is to disregard one important source of evidence about the most fundamental issues.[1] For, clearly, language demands some physiological prerequisites. If these are damaged, a man may no longer be able to exercise his linguistic competence, and for lack of them a child will never develop it in the first place; so our conclusions about the exact nature of the human language capacity cannot be wholly independent of discoveries about those prerequisites. And one must have a very low view of the current neurosciences to suppose that they have nothing useful to tell us about this subject, even of the broadest kind.

Similarly, to dismiss evolutionary zoology and cultural history as irrelevant to our understanding of language—to sweep aside the whole Darwinian account of organic evolution as 'one vast tautology', as Chomsky himself has done, while declining even to speculate about alternative historical origins for language—can again land us in a needlessly paradoxical

[1] In the John Locke Lectures, Chomsky went out of his way to dismiss existing physiological knowledge about the brain, as being far too crude to throw light on the language function in man.

position.[1] For, again, it is clear enough that the human species began to manifest the 'language capacity' at some stage in its organic evolution or cultural history; so this capacity is—at the very least—one whose appearance in the course of human evolution and history is, presumably, intelligible. The detailed origins of human language may well remain a mystery; all the same, general considerations about historiography and evolution may well restrict the range of hypothetical origins we should consider seriously at all. And if it turns out (as we shall claim here) that one of two alternative theories of language involves grave historico-evolutionary difficulties which the other easily avoids, that will certainly mean that the former theory faces —at least—a heavier burden of proof than its rival.

Here, then, we shall be content to ask in what general area the different lines of enquiry followed in the various neurolinguistic sciences—theoretical linguistics, cultural evolution, the neurology of aphasia, the physiology of perception, etc.—can eventually be expected to intersect; and so to identify the general types of problem that will arise within their area of intersection.

Let us begin by stating a question to which theoretical linguistics and clinical neurology at present imply definite but contrary answers. That is the question whether or no the 'language capacity' is to be thought of as a single, unitary capacity, which must be possessed either in its entirety or not at all. If we sharpen up the implications of this question, we shall see how necessary it is to keep in mind our longer-term goal, of an integrated theory of language, thought, and brain; and how important it can be, even for the most abstract purposes of theoretical linguistics, to make sure that our account of the 'language capacity' makes *biological* sense.

We may leave aside, here, the question whether or no a 'language capacity' of any kind is, in fact, unique to the human species. That further issue is, of course, the subject of an active and sometimes acrimonious debate. Are the things that domesticated chimpanzees have learnt to do, using deaf-and-dumb signals and/or coloured shapes, sufficiently complex in structure and similar in function to human language for us to regard them

[1] See the John Locke Lectures again.

as genuinely 'linguistic' modes of behaviour?[1] The question is fascinating, but marginal; and, for the moment, we can restrict ourselves to language of the fully-fledged kind at present known only in the case of human beings. About the fact of human linguistic uniqueness, we need not, at this stage, trouble ourselves; the crucial questions will arise, rather, when we begin looking for an explanation of that uniqueness. For, when we reach that point, we shall find ourselves led off in quite different directions, depending on whether we regard the 'human language capacity' as being unitary or not. That question divides ahistorical philosophers and linguists like Descartes and Kant, Frege and Chomsky, from evolutionary biologists, neurologists, and linguists of a more historical frame of mind; and it proves of particular importance when we set about embedding our theories of language in a larger biological setting.

(1) On the one hand, then, the unique linguistic competence of human beings might be interpreted as a single, all-or-nothing capacity, of an entirely unitary and specific sort, which humans alone happen to have. According to this view, the 'language capacity' can—in the nature of the case—be possessed not in part, but only in its entirety; and the question then arises whether it has, also, an equally unitary and specific physiological basis. Philosophically, this first view chimes in with a Cartesian position, since the language capacity can be cited as the unique characteristic of human 'mental' life; or, alternatively, with a fully-fledged Kantian position, by connecting up the unitary capacity for language directly with the one and only one set of authentically 'rational' patterns of thought; or, again, with a Fregean position, in which the underlying grammatical structure common to all languages is taken as representing the 'pure form' of concepts, with all historical and psychological 'accretions' stripped away.

(2) On the other hand, the unique human possession of language might be interpreted as expressing not a single,

---

[1] Both Chomsky himself, in the John Locke Lectures, and also E. H. Lenneberg, in his paper on 'Brain Correlates of Language', in *The Neurosciences: Second Study Program*, ed. F. O. Schmitt et. al., (New York, 1970), pp. 361–71, are highly sceptical about recent reports on the quasi-linguistic behaviour of the ape Washoe, which has been taught to use some of the gestures of American deaf-and-dumb finger-language.

unitary capacity, but rather a unique pattern or constellation of interrelated capacities—in the plural—all of whose constituent elements are present, in the required configuration, only in the case of human beings. This second position lends itself more easily than the first to a historical interpretation; in this respect, it permits one to break with ahistorical accounts of language, reason, and concept-use. Furthermore, it makes the unique character of human cognitive skills a matter of far less profound and long-term mysteriousness for theoretical biology and physiology. Indeed, it dovetails very neatly with what we have reason to believe in evolutionary zoology and neurology about the origins and functioning of the relevant brain mechanisms.

Both interpretations are compatible with the view that the 'language capacity' reflects—and is dependent on—some 'innate' genetic inheritance or psychological endowment possessed by all human infants; but they see this relationship in somewhat different lights. Supporters of the unitary thesis regard the innate equipment of the human infant as implicitly possessing the same formal complexity as the linguistic behaviour which is its eventual end-product. Supporters of the pattern view, by contrast, do not need to assume that this 'native basis' has any direct formal resemblance to that eventual behaviour—still less, that it possesses its entire formal structure and complexity. To put the point in a word: the unitary view requires a very *strong* interpretation of the nativist thesis, whereas the pattern view is compatible with a much weaker interpretation.

If we bring together Chomsky's statements about the historical and physiological aspects of language, it is reasonably clear that he has committed himself, more or less wittingly, to a unitary view of the human language capacity and to the stronger form of nativism. According to him, a man demonstrates an authentically 'linguistic' capacity only when his language-use conforms to the entire system of 'deep grammar'; since the elements of that system are complex and interrelated, there is no possibility of grasping or using it only partially. By contrast, if we consider the consensus of opinion among (e.g.) clinical neurologists, we shall be led to give a much less restricted account of the human language capacity. Their experience would encourage us to speak of brain-damaged patients losing some, but not all, the capacities required for the normal use and comprehension of language; this would imply that the notion of partial language-use does not—after all—involve an inconsistency or contradiction in terms.[1] Indeed, it is natural to classify different types of aphasia, precisely, by the particular combinations of those capacities preserved and lost in different syndromes. And a future biology of language will somehow have to reconcile these apparently contradictory insights: showing, for instance, how a genuinely 'linguistic' capacity, involving the complete, unitary grasp of deep grammar, might nevertheless display analogies to the perceptual capacity of animals—and even, possibly, be an evolutionary development out of it.

The choice between the unitary and the pattern view of the language capacity has biological implications of two quite different kinds: firstly, for the historical origins of language, and secondly for its neurological correlates. About the former issue, we can ask: 'Must human language have appeared—historically or evolutionarily speaking—suddenly and all of a piece? Or might the precursors of modern man have developed language progressively and bit by bit, beginning with cruder signalling systems and gradually refining them in the direction of true language?' About the latter issue, we can ask, similarly, 'Must the neurological prerequisites for language form a correspondingly unique and species-specific physiological system? Or might the capacity for normal language-use again be associated with, and dependent on, the existence in human beings of several distinct and co-operating physiological systems, alternative combinations of which are present in other species also, but with non-linguistic functions?'

In the case of both questions, outright commitment to the unitary view compels one to take a particularly hard line. Chomsky himself has more than once declared that the historical origins of language are a total mystery, about which he is not even prepared to speculate;[2] and, more particularly, that the search after evolutionary precursors of language is a delusion, since language—being unitary—is not the sort of thing that

[1] See the general survey by Geschwind referred to above, and also Eugene Green, 'Psycholinguistic approaches to aphasia', *Linguistics*, 53 (1969), 30–50, and 'On the contribution of studies in aphasia to psycholinguistics', *Cortex*, 6 (1970), 216–35. Cf. K. H. Pribram, *Languages of the Brain* (Englewood Cliffs, 1971), pp. 357ff.

[2] John Locke Lectures again.

could have any partial 'precursors'. The thesis that all genuinely linguistic behaviour requires a unitary grasp of 'deep grammar', as a single complex system, thus drives him to take a short way with Darwinism. Likewise for the neurophysiological mechanisms underlying language: Chomsky acknowledges that such mechanisms may eventually be brought to light, but in his view it would be quite premature to search for them at the present time. Physiology and linguistics can hope to come together only in the future, after physiology has developed theoretical categories as radically new as those which physics has acquired over the last 300 years. The discoveries of theoretical linguistics are —it seems—absolutely secure in their present form; if there is to be a reconciliation, it is physiology that has got to change! In complete contrast, the pattern view of human language is quite straightforwardly accommodated to the general principles of physiology and evolution theory, as we know them today. If all that is unique about human language is the peculiar constellation of capacities that it involves, it can then make perfectly good sense to talk about evolutionary 'precursors', both of man as a language-user and of human language itself. Some or all the anatomical structures called into play in the use and comprehension of language may then have been present, in one form or another, in Man's evolutionary ancestors;[1] and Man's behaviour may have developed into 'true' language as the historical outcome of a series of steps, rather than by a single drastic transition.

To see just how severe are the biological difficulties facing the unitary theory, let us examine these two issues—historical and physiological—in a little more detail. First, consider the evolutionary aspect of the matter. The historical origins of language may still, in point of fact, be largely mysterious; but, without insisting on premature guesswork about unknown events in the remote past, we are still entitled to ask what *possible* accounts are consistent with our broader ideas about evolutionary and cultural history. Taking the historical issue on this broader level, three reasonably clear forms of hypothesis are conceivable for explaining how, by intelligible natural processes,

---

[1] N. Geschwind, *The Development of the Brain and the Evolution of Language*, Monograph Series on Language and Linguistics, Vol. 17 (Washington, D.C., 1964), pp. 155–69.

non-language-using hominids might have been succeeded by language-using humans. Two of these forms can—at some intellectual cost—be reconciled with the unitary view of human language; the third is the only plausible one, from the general standpoint of contemporary biology, and also accommodates itself without strain to the pattern view.

(1) Starting with the hypothesis which is least plausible on general biological grounds: one might try to explain the transition from non-language-using hominids to language-using humans as the outcome of a one-shot genetic saltation, either a mutation or a recombination. Such a change might have conferred the 'language capacity' on individual hominids as a single, completely unitary character, and the advantages of that capacity might have been such that, very rapidly, the population in question achieved ecological dominance, with disastrous side-effects on all competing populations. This 'one-shot' hypothesis is certainly intelligible; at the same time, it is open to all the objections that biologists advance against any hypothetical 'saltations' having pre-adaptive consequences of an implausibly striking kind.[1] Quite simply, the change from non-language-use to full-language-use is too drastic to be the consequence of a single genetic change. Failing any alternative, supporters of the unitary view might feel bound to claim such a saltation as 'the' origin of human language, but the biological price of that view would then be a hard one to pay.

(2) An alternative hypothesis distinguishes the physiological prerequisites of language from language itself. On this hypothesis, one could argue that the physiological systems involved in human language-use developed gradually in the hominids over a long period, even though their behaviour displayed nothing we can recognize as 'true language'. On this hypothesis, no question need arise about behavioural precursors of language itself; instead, the physiological prerequisites for language will have been selectively perpetuated, during the transition from hominids to men, on account of other non-linguistic advantages. Presumably, then, natural selection gave rise to a population of proto-men whose bodily frames already incorporated those neurophysiological structures required for language, even though

[1] See, e.g., E. Mayr, *Animal Species and Evolution* (Cambridge, Mass., 1963), Chapter 17.

there had as yet been no opportunity for these structures to find 'expression' in linguistic behaviour. The physiological capacity for language (that is) pre-existed its behavioural expression; and the physiological evolution of the 'language capacity' has to be distinguished clearly from the subsequent discovery of language itself. The essence of this second view can be distilled into a word. If the physiological prerequisites of language evolved entirely for non-linguistic reasons, their linguistic expression was then a strictly 'cultural' discovery; and this discovery spread from language-using humans—who already had made it—to other, non-language-using humans, who had not yet made it, by 'cultural radiation'.

One might compare this human discovery with that of the Japanese monkeys which were given muddy potatoes and paper-wrapped toffees: the practice of washing the potatoes in the sea and unwrapping the paper from the toffees, before eating them, spread through the population progressively, beginning with the more enterprising and innovative younger male monkeys.[1] So too, on this second hypothesis, the possibility of using 'language' will have spread outwards by imitation, from an initial cultural focus or foci, to other human populations which were physiologically pre-adapted for language, but had not invented it for themselves. In each case, one can say—for what it is worth—that the relevant psychological 'capacity' existed in advance of the corresponding cultural 'discovery'. Evidently enough, the monkeys had the 'capacity' to wash potatoes or unwrap toffees, before the occasion for doing so presented itself; yet, just as evidently, neither activity played any significant part in the evolutionary development of the monkeys concerned. And, likewise, the 'capacity' for language might have existed in proto-human populations before its actual invention or discovery, even though 'proto-linguistic' activities were incapable—in principle—of playing any significant part in human evolution.

This second hypothesis accordingly makes the relationship between human language and its physiological basis an artificial one: as with the capacity of monkeys to unwrap toffees, or of sea-lions to balance large rubber balls on their noses. Once again, the hypothesis is perfectly intelligible; and supporters of

[1] The macaques on the island of Koshima also learned to swim and walk upright, when given problems and occasions of kinds their normal lives did not provide.

the unitary view of language might once again feel bound to pay the price for accepting it. Still, we can do so only if we allow that the physiological prerequisites of language developed, in proto-human populations, in a manner having nothing whatever to do with their subsequent 'linguistic' expression.

(3) The third and final hypothesis fits the general principles of current biology much more naturally and easily. This hypothesis interprets the evolutionary development of the language capacity as involving the gradual accumulation, in Man's precursors, of many distinct physiological and behavioural changes, some of which were presumably advantageous chiefly for 'proto-linguistic' reasons, others for non-linguistic reasons. The physiological changes that were progressively selected in this way were behaviourally associated, at first, with the emergence of a partial language function, later with the refinement of a fully-fledged language function; so the language capacity itself did not appear all at once—either through a drastic genetic saltation, or through an abrupt cultural discovery—but emerged slowly, and bit by bit. On this account, no sudden changes need have been involved, either on the genetic or on the cultural level. The necessary changes in both brain structure and behaviour will have taken place gradually, over a long period of time, approximating only gradually to 'language' as we know it today.

This hypothesis is compatible not only with our evolutionary and genetic ideas, but also with known neuro-anatomical differences between the brains of humans and of other higher animals.[1] The very terms in which it has been stated are, however, directly inconsistent with the unitary thesis. We can even put this third view in words at all, indeed, only if we accept a possibility which the unitary thesis denies—of a 'partial' language function. The transitional populations between the higher apes and the earlier men are then supposed to have possessed most of the neurophysiological prerequisites for true language, and to have manifested simpler modes of signalling behaviour which already shared, in a rudimentary form, some of the essential characteristics of fully-fledged human language.

The strict 'unitary' view is, accordingly, compatible only with the first and second of our three hypotheses: that which

[1] Geschwind, *The Development of the Brain*; see p. 456, n. 1, above.

associates the origin of language with a one-shot genetic saltation, and that which treats it as a single, equally drastic
cultural discovery. It is not yet clear which of these courses
Chomsky personally would take. He proclaims his complete
agnosticism about the actual historical origins of language;
when asked about its evolutionary precursors, he sweeps
Darwinism as a whole aside, as irrelevant and 'tautological';
yet, short of abandoning the basic thesis, that the human
language capacity comprises an entire unitary system of
transformational grammar, he has only three positions to
choose from. Either, the human language capacity was the
result of single genetic saltation; or language originated in a
single cultural invention, which exploited physiological mechanisms selected for totally different functions; or else the very
existence of language is—somehow or other—a standing
demonstration of basic conceptual inadequacies in our current
biological ideas. One way or another, the burden of proof to be
sustained by supporters of the unitary view is a substantial one.

Other, neuroscientific questions are closely linked to these
historico-evolutionary ones. These have to do with the different
kinds of physiological correlates, or prerequisites, implied by
different theoretical approaches in linguistics. So let us consider, next, what consequences follow from the choice between
the unitary and pattern view of language on a neurophysiological level: in particular, we must ask what degree of formal
resemblance is to be expected between the 'native mental
capacity' required for language and its underlying physiological
correlates. Unless we face these questions head-on, we risk
assuming that the native capacity for language has—and *must*
have—a far closer and more specific formal resemblance to its
physiological correlates than is plausible, on the basis of
current ideas in neuroscience.

For lack of explicit statements about the physiology of language, we must proceed by exploring the apparent neuroscientific
implications of the unitary view. Chomsky's own statements
alternate between two positions. Sometimes his support for
Cartesianism leads him to dismiss all physiological questions
about language as irrelevant. If Mind (or language) and Body
(or brain) are distinct 'categories' or 'substances', having no

intrinsic connection, he does not have to assume that the 'innate ideas' in his Theory of Mind have 'isomorphic' hereditary counterparts in the structure of the human body or brain. At other times, he is prepared to allow that the human language capacity should eventually have some physiological explanation, though this is not yet within our grasp. Still, many of Chomsky's readers certainly do assume that the existence of a unitary and specific innate capacity for language implies the existence of an equally unitary and specific physiological endowment, transmitted by genetic inheritance. So we must ourselves ask, here, just what kind of resemblance one can reasonably expect between this 'innate mental capacity' and its associated physiological mechanisms.

In order to sharpen up our own questions about the language capacity, we may use two other well-known 'native capacities' as objects of comparison. Consider, in the first place, the physiological basis of the human elbow-movement. It takes a new-born infant some months to learn to co-ordinate hand and eye with any exactitude; yet no corresponding learning-period is apparently required in order to co-ordinate the four separate muscles which operate at the elbow. The new-born infant performs the elbow-movement smoothly from birth, and apparently spends no time learning to avoid operating these four muscles out-of-phase, and so wasting energy. We can thus speak of the elbow-movement itself as involving a 'native' or 'innate' capacity, and contrast it with the 'learned' ability to co-ordinate hand and eye. In the second place, consider the capacity of honey-bees to manufacture combs with hexagonal cells, without any apparent need for instruction, still less any extended period of learning by trial and error. In our present sense, honey-bees too have the 'native capacity' to manufacture hexagonal cells: i.e. they already possess, at birth, whatever constellation of physiological prerequisites and psychological propensities eventually finds its behavioural expression in the production of combs with hexagonal cells. And in both cases, we may ask the same question: 'What resemblance, or degree of isomorphism, presumably exists between the native behavioural capacity manifested in each case, and the physiological endowment on which that capacity depends?' Must the physiological basis of each innate capacity have about it something correspondingly elbow-movement-like, or hexagonoid? Or might the same behavioural end-products be expressions of neurological structures, or systems, having much simpler or more generalized forms than these specific behavioural manifestations?

(1) To begin with the elbow-movement: here, considerations of medical ethics make our direct knowledge of the relevant human physiology necessarily incomplete. But studies on the neurology of aphasia and apraxia in humans, as well as on the physiology of other vertebrates—notably, salamanders—make it possible to infer something about it indirectly. In a classic series of experiments, for instance, Paul A. Weiss demonstrated that the electrical impulses activating the four muscles at the salamander elbow-joint are related in a stereotyped time-sequence. As a result, the signals passing down the motor nerve to these four muscles represent, not so many wholly independent impulses, but rather a single, co-ordinated bundle; and the smooth co-ordination in the movement of the elbow reflects this deeper temporal co-ordination in the motor impulses.[1] If you transplant one limb of a salamander to the opposite side of the body and re-establish the innervation to that limb—as is practicable with such amphibian vertebrates—you will observe the same stereotyped bundle of impulses causing exactly opposite (and functionally inappropriate) movements in the elbow of the transplanted limb; though, apart from its inappropriateness, the movement itself is as smoothly co-ordinated as it was before.

In this case, at any rate, we can directly compare the form of a 'native behavioural capacity' with that of its neurological correlates. Manifestly, there is very little in common between them. The behaviour in question is a smooth spatial motion; its neurological basis is a time-coded sequence of impulses. Given all the other relevant facts about the arm-structures and modes of life of vertebrates, we can—it is true—understand how such a time-sequence of nerve-impulses can serve as the physiological prerequisite for the production of a smooth spatial elbow-movement; any infant inheriting an appropriately time-coded nervous system will, physiologically speaking, 'have what it takes' to support the native capacity for a smooth

---

[1] Paul A. Weiss, *Self-Differentiation of the Basic Patterns of Coordination*, Comparative Psychology Monographs, Vol. 17, No. 4 (Baltimore, 1941), pp. 1–96.

elbow-movement. By looking at the time-sequence of motor impulses alone, however, no one could ever have told that it had anything 'smooth-elbow-movement-like' about it. The fact that this particular time-sequence leads to a smooth elbow-movement, as its behavioural product or expression, is intelligible only in one particular context: i.e. when it is related to all the other facts that bear on the developmental process by which the 'innate endowment' finds its behavioural expression in the typical life of the species concerned.

This first example, accordingly, suggests one question about the neural basis of the language capacity. Supposing that we conceded, for the moment, that the actual behaviour in which the innate human capacity for language is expressed universally manifests the same 'invariant' grammatical system: would it immediately follow that this same characteristic structure must also be 'projectively represented' in the central nervous system? Or might we look for relations between linguistic behaviour and its neural correlates of a much less direct and exact form? Is it enough, for instance, to suppose that the structures of 'deep grammar' are characteristic of language *only* as a behavioural product, and are no more directly represented in the central nervous system than (say) the smooth spatial pattern of the elbow-movement is? (Eric Lenneberg has suggested that the neural correlates of linguistic structure include not merely spatial neuro-anatomical structures, but also *time-coded* sequences and mechanisms; and this idea has some real attractions.[1] One neurophysiological element in language-learning might well consist in the establishment of regular, time-coded patterns in the neurological processes associated with writing and talking.) So the example of the elbow-movement refutes any general assumption that a 'native behavioural capacity' must necessarily share a common form with its physiological correlates.

(2) In a similar manner, the example of the honeycomb refutes any general assumption that the form of physiological endowment underlying a 'native capacity' need be anything like as specific and detailed as the forms of its typical behavioural expressions. Among all their other 'innate capacities', honeybees evidently have whatever it takes to manufacture combs with

---

[1] Lenneberg, 'Brain Correlates of Language'; see p. 453, n. 1.

cells of hexagonal cross-section. Yet this rather trite observation does not, by itself, imply any striking conclusions about the neurophysiology of honey-bees. In particular, we cannot, on this account alone, jump to the conclusion that 'the form of the hexagon' is itself somehow 'programmed into' the nervous system of honey-bees. This hypothesis is not inconceivable, and might in fact have proved to be the case; if bees of all species, working in all kinds of situation, secreted wax into cells of precisely this form, the hypothesis might even be quite appealing. In fact, however, social bees of other species are known, which produce cells having a wide variety of forms: notably, isolated spherical ones in the simplest cases, becoming hexagonal only when they are closely packed into a narrow space.[1]

One can easily show how behavioural propensities and physiological endowments of a much more generalized sort might find expression in the very same behavioural end-product. The hexagon is a particularly economical shape. Compress a raft of soap-bubbles, and the individual bubbles (or 'cells') will take up a hexagonal shape of themselves; they do so, not because the constituent molecules of liquid soap have anything 'hexagonoid' about them, but simply because the overall physical 'energetics' of the situation makes this the natural equilibrium. So let it merely be supposed that honey-bees have an innate propensity to construct the maximum number of cells of a given area and rigidity, from any given amount of wax, rather than hexagons-as-such. In most situations, this generalized propensity (and its physiological correlates) will be expressed in the production of cells having the familiar, hexagonal form; but in exceptional circumstances this will not happen. Where, for instance, a honey-comb is built in the open air—hanging freely from the branch of a tree, instead of being formed within an enclosed space—it will be physically uneconomical for the outermost cells to have hexagonal exterior walls, and one would then expect the comb to be rounded off, so as to produce a smooth exterior surface. In fact, there is a good deal of evidence that the actual capacities of honey-bees are of this second kind. Free-hanging combs do typically have smooth outer surfaces,

[1] M. Lindauer, *Communication in Social Bees* (Cambridge, Mass., 1961), Chapter 1. Cf. also the comparative work of C. Michener on other species of bees, especially those from Australia.

rather than preserving the same stereotyped hexagonal form throughout.

This second example, too, is highly suggestive for the theory of language. The striking form of the behavioural end-product in this case is unquestioned evidence of a highly specific native capacity; yet the resulting behaviour is more precise and detailed in form either than the inherited propensity of which it is an expression, or than the physiological structures which presumably underlie that propensity. The hexagonal end-product can be fully accounted for, in fact, only by relating it to the *objective external task* which is involved in making a comb. The hexagonal form has (that is) a biological significance in virtue of its 'functional adaptation', as the end-product through which a generalized psychological propensity is expressed in response to a specific type of task. In order to yield a hexagonal end-product, neither the nervous physiology of the honey-bee, nor its 'native behavioural capacities', need have any specifically 'hexagonoid' feature. And this prompts the question, just what sort of 'native capacities' we need presume—in the linguistic case, too—in order to explain the alleged occurrence of invariant formal characteristics in all human language.

If the grammatical structure of language were at all comparable to the spatial structure of the honeycomb, language might then turn out to be the behavioural end-product, not of a unitary and specific 'native capacity' precisely isomorphic with our actual linguistic behaviour, but rather of more generalized capacities, which are expressed in behaviour of that particular grammatical form only when set to work on the appropriate external tasks. In that case, we should no longer need to explain the universal forms of linguistic behaviour as reflecting corresponding innate forms in their underlying capacities and endowments. For those universal forms might simply reflect the universality of the 'objective functions' which language is everywhere used to perform. As with honey-bees, so with men: the universal forms in actual behaviour would be intelligible, only by considering how generalized 'innate propensities' find expression as applied to the specific problems of their actual lives. On this account, we could still retain a 'nativist' thesis about language, in the weakened form suggested by the honey-bee example; but the detailed grammatical regularities in human linguistic behaviour would be explained—at least in part—by appeal to the characteristic functions and tasks of language.

Supporters of the unitary view of language have paid little attention to this alternative, more generalized interpretation of 'deep grammar' and the 'native language capacity'. Chomsky himself has been inclined to dismiss 'functionalist' theories of grammatical form, as inconsistent with his view that the language capacity is 'innate', and as implying that language-learning consists entirely in 'conditioning' by environmental 'stimuli'.[1] But Chomsky's arguments have entirely disregarded the broader biological context of linguistics; and we can now see that—after all—there does exist room for reconciling a nativist account of our psycholinguistic capacities with a functionalist account of their behavioural expression. And it will be worthwhile spelling out a little further, at this point, the terms for achieving such a reconciliation between nativism and functionalism.

The nativist view leads to embarrassing biological consequences, only if we assume that the 'universal forms' of human grammar are represented, precisely and specifically, in our innate endowments or propensities. These embarrassments can be greatly reduced as a result of one small amendment, though at the price of abandoning the 'unitary' thesis. For let us merely suppose that the grammatical forms of language are represented, precisely and specifically, only in the actual behaviour through which the 'language capacity' is eventually given expression. The capacity to re-create language in every generation will remain, even then, something 'innate', and will depend upon the human beings concerned having suitable 'native propensities'. But all that we need now claim, in order to sustain the 'nativist' view against a 'behaviourist' account of language-learning, is that human beings possess at birth whatever native capacity, or capacities—singular or plural—are needed for generating behavioural end-products of the requisite form, in the presence of appropriate external tasks.

On this assumption, we can reopen two further questions. To begin with, we can enquire whether all of the various native capacities involved in language-use may not have their own

[1] Cf. Chomsky's review of Skinner's *Verbal Behaviour*; cf. above, p. 449, n. 1.

distinct physiological correlates; and, secondly, we can explore the possibility that the forms of the linguistic end-products reflect the nature of the external tasks on which those capacities are exercised, much more than they do the nature of the capacities themselves. The 'deep structures' of all language will then be the consequences, not of our 'native propensities' alone, but rather of the typical interplay between human beings—with all their inherited endowments and propensities—and the practical tasks on which they have occasion to exercise them.

The way in which this interplay could lead to an alternative theory of grammatical form can be indicated briefly by referring to the kind of approach exemplified in Karl Bühler's *Sprachtheorie*.[1] Trained in the Viennese neo-Kantian background of the early 1900s, Bühler was never tempted to support a naive empiricist theory of language-learning by straightforward 'conditioning'; on the contrary, language was for him the specialized possession of intelligent beings, having the innate capacity to recognize and act in accordance with 'rules'. On Bühler's account, the fundamental capacity underlying all higher cognitive behaviour was a generalized *Regelsbewusstsein*, or consciousness of rules; and this 'rule-awareness' was expressed in language, with its characteristic grammatical structure, as a result of being put to a particular kind of use.[2]

As for all the specific linguistic forms in which these 'native capacities' were expressed, those remained, for Bühler, something to be explained in functional terms, by reference to the different tasks in which language was developed and put to use. Do all languages embody, for instance, a differentiation between those expressions which are used as 'noun-phrases' and those which are used as 'verb-phrases'? If so, the objective external tasks of language give us evident reasons for expecting just such a feature. For, in its practical use, it is a matter of sheer efficiency to indicate which of our linguistic symbols are serving as 'indexical' terms, i.e., to draw attention to the things we are talking about, and which are serving as 'descriptive' terms, i.e. to say something about the subjects so indicated. A language which

[1] K. Bühler, *Sprachtheorie* (Jena, 1934).
[2] For the views of Bühler and the other Würzburg psychologists on 'rule-consciousness' and 'meaning-consciousness', see also G. Humphrey, *Thinking* (London, 1951), Chapters 2 and 3.

lacked any symbolic device to mark this functional distinction would end by being needlessly ambiguous and confusing; so we might expect it soon to develop some convention for differentiating those two classes of terms or symbols . . .

Having mistakenly associated functionalism in linguistics with empiricism in the theory of language-learning, the transformational grammarians have seen their position as essentially opposed to that of Bühler and other more traditional linguistic theorists. Yet, in certain respects, their own programme for linguistics can be carried through intelligibly only by implicitly adopting some elements from the functionalist position—especially in the theory of meaning. We need give only one illustration. Arguments for 'deep structure' frequently rely on demonstrating that superficially similar patterns of words in the same language have quite different *meanings*, and require to be construed as implying different 'deep' grammatical structures; while the same *meaning* may be expressed in different languages by the use of quite different superficial word-patterns. The sentence, 'I hope that so-and-so is the case', for example, looks very like the sentence, 'I assert that so-and-so is the case'; but it turns out that their meanings—and so their 'deep structures' —are very different. Before we go on to make any generalizations about universal grammar, however, there is one further question to ask: viz., 'On what conditions can such a distinction, as explained in terms of a single language, be applied first to other languages individually, and subsequently extended into a general truth about the grammatical structures of all languages whatever?'

Evidently, before we can make any such generalization, we must adopt some test for identifying words and sentences in other languages which correspond in meaning to (e.g.) our words 'hope' and 'assert', or our forms, 'I hope that . . .' and 'I assert that . . .' Failing such a test, we shall have no way of deciding just which sentences from different languages are to be taken as manifesting the 'same' grammatical forms on the 'deep' level. Yet, at this point, even supporters of the strong unitary view can scarcely identify equivalent items in different languages, except by using functional criteria. For how else are we to recognize 'equivalent' words and sentences in different languages, except by reference to similarities between the occasions on which, the tasks for which, and the manners in which, they are customarily employed? Deep structures can count as 'linguistic universals', that is, only when they are recognizable in sentences from different languages having 'equivalent' meanings; and, to the extent that the criteria for recognizing these 'equivalent' sentences themselves involve the same tests as are used to identify the structures of 'deep grammar', that claim will become empty. For it will then tell us only that equivalent sentences from different languages, when selected so as to have the same meaning and 'deep structure', universally turn out to have the same meaning and 'deep structure'!

The functional reanalysis of 'linguistic universals' hinted at here calls for a much longer argument than we have room to expound at this point. To sum up: if the differences in 'deep grammar' between statements of the forms, 'I hope that . . .' and 'I assert that . . .', are to be compared with any corresponding features of a language other than English, we must *already* have the means of identifying statements in that other language having corresponding meanings; and the discovery of 'universals' will then be a genuinely empirical discovery, only if the criteria of 'semantic equivalence' are independent of the deep-grammatical differences themselves. By contrast, if the rules for identifying 'deep structures' themselves turn out to rely on semantic criteria—as may well be the case—the existence of 'linguistic universals' will then be merely a formal consequence of the connections between the grammatical procedures for identifying 'deep structure', and our functional criteria for comparing 'meanings' in different languages.

This concession to functionalism should not seriously worry the advocates of transformational grammar; yet hitherto they do not seem to have reconciled themselves to it. Thus, Chomsky has brusquely dismissed Wittgenstein's account of meaning in terms of 'language games', on the ground that it implies an excessively pragmatic attitude towards language.[1] Such an account (he protests) treats the complex structures of propo-

---

[1] See Chomsky's dismissive remarks about Wittgenstein's conception of 'language-games' in *Cartesian Linguistics*, esp. pp. 21–2 and note 40. In fairness to Chomsky, it should be added that the Skinnerians, whose theories of language-learning he attacks, make the same mistake, in claiming Wittgenstein as implicitly supporting their 'behaviorism'.

sitional grammar as a mere extension of unstructured cries, commands, and exclamations. And, certainly, Chomsky's associates originally hoped to build an adequate 'semantics' (or theory of meaning) as an extension of 'syntactics' (or theory of grammatical form) rather than the other way around.[1] Yet this dismissal of the language-game notion was wholly premature. Wittgenstein came from the same general neo-Kantian background as Bühler, and was quite as opposed as Bühler to naive empiricism. Both men were preoccupied with the notion of 'rules of meaning'; and both of them ended by taking 'occasions' and 'manner of use' as a basis for recognizing expressions having the same meaning in different languages. Though first introduced for philosophical purposes, Wittgenstein's phrase 'language-games' could thus be used in theoretical linguistics for semantic purposes also.[2]

The issues raised in the contemporary linguistic debate—about the relationship between the theoretical 'deep structure' and the facts of actual linguistic behaviour—are reminiscent of those raised by Frege about the relationship between the theoretical 'core' of 'pure' number-concepts and the facts of actual arithmetical behaviour.[3] In either case, the central relationship at issue is one with which we are by now familiar, viz., that between the idealized articulation of a formal system and its empirical application in actual concept-use. This parallel has been made quite explicit by Eric Lenneberg, who has directly compared the human capacity for language with that for performing arithmetical calculations.[4] And Lenneberg is surely right to see significant analogies between the two capacities; many people, indeed, would regard the arithmetical capacity as a special form of the more general linguistic capacity. With Frege in mind, however, let us begin by distinguishing the specific human capacity for arithmetic from the more general

---

[1] The original position of Chomsky and his followers on semantics was set out by J. A. Fodor and J. J. Katz in a famous paper reprinted in *Readings in the Philosophy of Language* (Englewood Cliffs, 1963), Chapters 19 and 20: Fodor for one, it now appears, no longer considers himself bound by the position set out in that paper.

[2] On 'language-games', see L. Wittgenstein, *Philosophical Investigations* (Oxford, 1953), paras. 23ff., pp. 11ff.

[3] See the quotations from Frege, pp. 55–6 above.

[4] Lenneberg, 'Brain Correlates of Language', above.

capacity for language, and afterwards go on to explore the parallels between them.

Taking arithmetic first: if we consider what 'native capacities' are expressed in adult human arithmetical behaviour, the same pattern of argument recurs. There are two possible views: one stronger, the other weaker. According to the weaker view, our capacity to grasp arithmetic demands only the possession of some generalized capacity, or constellation of capacities, sufficient for us to get hold of whatever forms and structures are typical of the arithmetical operations current in our own milieu, without the need for precise and elaborate conditioning. This weaker view accepts historical and cultural differences between the arithmetical activities of different communities at their face-value; it treats the formal, abstract arithmetic of twentieth-century mathematical philosophers as a highly-specialized end-product of intellectual history; and it invokes, as the physiological basis of arithmetical behaviour, only such neurological processes as can find expression—equally—through any of the whole recorded historico-cultural spectrum of varied arithmetical activities. The stronger view, by contrast, interprets the human capacity for arithmetic as a single, specific, and unitary capacity, which is expressed, either explicitly or implicitly, in one and only one formal system of arithmetical procedures. The forms in question are those which have been analysed explicitly during the last eight years by such mathematical philosophers as Frege, Peano, and Russell, but those same forms—it is suggested—were implicitly operative in the arithmetical behaviour of all other cultures and periods. Man's progressive success in 'stripping away the accretions' that veil the 'pure form' of arithmetical concepts from the 'eye of the mind', and developing our modern axiomatic arithmetic, is then to be understood as the final expression of our 'innate grasp' of Peano's abstract forms.

The same problems arise, however, about this unitary view of arithmetic as about the corresponding view of language. Just how are the universal forms of Peano's abstract arithmetic to be related to the varied modes of numbering and ordering behaviour current in one milieu or another? Do the arithmetical practices of different milieus really share enough formal similarities to count as alternative expressions of a single native capacity as unitary and specific as the Peano axiom-system itself? Or might they not equally—as with the honey-bee —be alternative expressions of much more general capacities which have given rise to novel and different mathematical operations in each new epoch or culture in response to a succession of novel external tasks and problem-situations?

In the latter case, Peano's formal system would no longer represent the 'universal native capacity' of Man the arithmetizing animal. Rather, it would be a philosophical idealization, helpful in a retrospective analysis of men's calculative skills, but incapable of serving as a system of 'generative principles'—still less, as a system of causally efficacious 'generative influences'— in the actual behaviour of human populations at all historical times.[1] Instead of treating the real-life historical precursors of modern arithmetic as primitive gropings towards Peano's philosophical ideal, we could then accept them for what they were in actual practice: viz., alternative calculating-procedures which merely approximated to the modern forms, and which can be said to have Peano's abstract ideals 'implicit' in them, only in a strained and artificial sense. Throughout most of history, indeed, 'doing arithmetic' has embraced a whole constellation of skills and competences. It has included, for instance, counting objects out and taking them away; placing people or objects in a numerical sequence; identifying the numbers of objects by eye, by ear, or by touch; recognizing an equality between (say) a given number of heard sounds and the same number of light-flashes; giving names to numerals; reciting series of numbers, and so on—quite apart from more elaborate, scholarly activities, such as multiplying, dividing, and taking square roots. Computation, as it is known to modern mathematicians, is thus a formal elaboration, arrived at historically as an extension or extrapolation from a whole constellation of more elementary skills; and, far from all arithmetical behaviour having a common neurophysiological basis operating specifically to a 'Peaniform' pattern, each of these skills presumably has its own system of neurological prerequisites.

At this point, the evidence of aphasia and acalculia is directly relevant. If the human capacity for arithmetic and its physio-

[1] Lenneberg, 'Brain Correlates of Language'; see above, p. 453, n. 1.

logical basis were both of them unitary and specific, one must expect to find individuals developing and losing their arithmetical competence all of a piece; correspondingly, brain lesions should either disrupt arithmetical skills all at once, or else leave them completely untouched. Yet the neurological evidence strongly suggests the opposite. If we consider the whole constellation of performances normally regarded as elements in arithmetical behaviour, the clinical evidence strongly supports a *pattern* view of the 'arithmetical capacity'. Not only are different elements in this constellation developed in different contexts and at different stages in life: they are also destroyed separately and selectively, following different kinds of brain damage. So we may be left with a man who is quite unable to carry out certain elementary arithmetical tasks, even though his other arithmetical performances are entirely normal.[1] (Similarly, with the language capacity; brain damage can destroy some unexpectedly specific and restricted elements in the whole constellation of linguistic skills. A man may, for instance, be able as a result to write passages from dictation correctly, yet totally unable to read them back immediately afterwards.) In the face of such evidence as this, one can keep the stronger view on its feet only by supplementary distinctions and arguments.

(1) In the first place, we might seek to distinguish 'genuinely arithmetical' operations—involving formal transformations conforming to Peano's axiom-system—and 'merely pragmatic' activities, such as counting, ordering, comparing numbers, and reciting numerals; or, as Chomsky himself does, between 'genuinely linguistic' operations—involving formal operations within a strict grammatical scheme—and 'merely pragmatic' commands, exclamations, and the like. As to this distinction, it might *in fact* prove that the capacities to do formal computations, and to perform grammatical transformations, were represented psychologically by quite distinct and separate systems from the other, more practical capacities commonly associated with them; and it might *in fact* prove that these two formal sets of psychological capacities also had their own special physiological representations, inherited and transmitted by distinctive genetic pathways. This might in fact be so, but at

[1] Geschwind, 'Disconnection Syndromes in Animals and Man': cf. p. 449, n.1.

present there appears no sufficient evidence that it is. And while, in the course of intellectual history, the formal refinements of arithmetic have tended to converge on the abstract forms analysed by Frege and Peano, these common forms can—like the basic forms of grammar—be explained equally well by the succession of objective external tasks with which mathematicians have historically been dealing: i.e. explained in terms of their 'functional adaptation'.

(2) Alternatively, we might be tempted to argue that the neurophysiological correlates of arithmetic must at any rate be separate from those of the different sensory systems, on the grounds that a grasp (of expression) of the computational capacity is—manifestly—not dependent on the use of any particular sensory modality.[1] Arithmetic is learned and applied using alternative senses, and men with the most varied sensory disabilities have learned it. This second argument, however also risks a Platonic fallacy. Thus, in the *Meno*, Socrates appeals to the fact that an uneducated slave-boy follows a demonstration of Pythagoras' Theorem as proving the independent existence of the soul; for, unless the slave boy were born with the 'native capacity' to re-cognize the validity of intellectual steps which he had already pre-cognized, how could he possibly follow them? In a similar way, Lenneberg appeals to the evidence that common arithmetical skills are exercised using different sensory modalities, as proving the independent existence of a separate, non-sensory mental capacity (together with the necessary neurological mechanisms) by which human beings having any combination of the five senses are enabled to recognize the single universal formalism of Peano's arithmetic. And Chomsky himself repeats the same Socratic argument, citing the fact that children learn to use language correctly as proving that they 'unconsciously grasp beforehand' the same implicit system of rules that professional grammarians subsequently make explicit, when they analyse the linguistic utterances of others.[2]

Yet, given the undoubted fact that arithmetical computations, or linguistic transformations, can be learned without bringing some *single* sensory modality (e.g. sight) into play, how

[1] Lenneberg, 'Brain Correlates of Language', as before.
[2] See *Aspects of the Theory of Syntax*, Chapter 1, Section 8.

much really follows from this? On the face of things, it implies only that the same forms can be learned and exercised through *any* sensory modality. In a word, the forms of the relevant competences are 'neutral' as between different sensory modalities. We can learn through sight, or through hearing, or through touch, indifferently; there is no single sense, or combination of senses, through which we must learn either language or arithmetic. (Taste and smell would, all the same, be inefficient channels for developing these competences.) Still, none of this implies that we could develop, exercise, or even *possess* a capacity for language and computation, if we lacked all sensory modalities whatever, nor that the neurophysiology of language and arithmetic is totally separate from the neurophysiology of all the sensory-motor systems.

One final factor can help to make the *Meno* argument deceptively attractive, and reinforce a belief that the specific abstract forms of our eventual arithmetical and linguistic activities must somehow exist 'in the mind' before the development of any actual counting-behaviour or speech. This is the doctrine that linguistic and arithmetical activities are 'private' mental processes, distinct from all overt behaviour, and going on essentially 'in our heads'. Where men like Vygotsky, Luria, and Wittgenstein have interpreted the forms of thought as 'internalizing' the public forms of communal procedures,[1] Chomsky and Lenneberg reverse this relationship, and treat the forms of public speech and computation as 'externalizing' pre-existing forms of 'inner' mental activity.[2] We shall be taking up this topic again later, in Part II: here, there is room for only one brief remark. At this point, our earlier honeycomb argument can be deployed once again, and used to drive a wedge between two different theses. Leaving aside the notorious philosophical difficulties involved in treating language and thought as essentially 'private' or 'inner',[3] we may question

---

[1] L. S. Vygotsky, *Thought and Language* (Cambridge, Mass., 1964), Chapter 5; A. R. Luria, *Higher Cortical Functions in Man* (New York, 1968); L. Wittgenstein, *Philosophical Investigations* (Oxford, 1953).

[2] See *Cartesian Linguistics*, pp. 32–3, 40, and *Language and Mind*; also Lenneberg, 'Brain Correlates of Language'.

[3] One central topic of Wittgenstein's *Philosophical Investigations* is, of course, the impossibility of defining 'language' in such a way that the primary 'reference' of words, or sentences, is to 'essentially private' sensations.

whether our pre-existing mental capacities for arithmetical or grammatical behaviour are, in any case, as specific and detailed as the elaborately structured behaviour which is their eventual end-product. It is one thing to claim that, during the pre-speech or pre-arithmetic phase, the growing child must already be employing 'thought processes', of kinds that have not yet found formal linguistic or mathematical expression; but it is quite another thing to conclude that these 'pre-existent thought-processes' implicitly possess—in some real sense—the same detailed, precise forms that are characteristic of our adult linguistic or mathematical performances. Only if we are committed to the stronger and more problematic view of the 'native capacities' for language and arithmetic need we explain the forms of adult behaviour as reflecting the 'determining power' of those capacities to impose identical forms on the 'inner thoughts' of young children.[1]

Biologically speaking, then, there are grave objections to the strong nativist thesis—viz., that the human language capacity is specific and unitary—which do not affect the alternative weaker thesis—viz., that this language capacity involves a complex constellation of 'innate capabilities', capable of being expressed through a broad range of signalling behaviour and associated with an equally complex constellation of neurological correlates. Either form of nativism gives us an effective answer to behaviourist theories of language-learning, yet many advocates of 'transformational generative grammar' have taken the stronger form as the philosophical foundation for their linguistics, despite all biological difficulties it faces. If we take these biological considerations to heart, however, we shall find that the weaker form of nativism then has three advantages. It permits us to state clear empirical questions about the origins of language; it dispenses with the need for an exact isomorphism

[1] In the John Locke Lectures, for instance, Chomsky repeatedly used the term 'generative' in a way that left it unclear whether the formalism of 'generative grammar' was (i) *only* a set of formal expressions *by the application of which* grammatical expressions could be derived; or, alternatively, (ii) some kind of 'inner psychic agency' *guaranteeing* the grammatical character of our utterances. For a sensitive and shrewd criticism of Chomsky's position—especially its inherent ambiguities—see the essay by George Steiner, 'The Tongues of Men', in *The New Yorker* (15 November 1969), pp. 217–36.

between adult linguistic behaviour and its underlying native propensities and mechanisms; and it readily makes sense of the evidence from clinical neurology, which underlines the independence of different elements in human linguistic behaviour.

The contemporary doctrine that any coherent 'language' presupposes a single universal system of grammatical forms evidently faces, from the standpoint of history and biology, the same objections that we found confronting the Kantian and Piagetian views: viz. that all coherent 'thought' conforms—or approximates progressively—to a single universal system of intellectual forms. In each case, healthy scepticism immediately replies, 'Why must it?' In each case, the allegedly 'universal' and 'ahistorical' structures turn out on closer inspection to be products of theoretical idealization, rather than of empirical discovery. And, in each case, an empirical account of actual human language—or actual rational thought—proves to be at once more functional, and more generous in the diversity of forms that it acknowledges, than the fully-fledged Kantian or Chomskian theory. So we may set 'universal grammar' on one side, with Frege's 'pure number-concepts', and Kant's '*a priori* forms of rationality', and return to the actual diversity of empirical language, mathematics, and thoughts. Instead of the *a priori* forms of rational thought, and the universal 'deep structure' of linguistic behaviour, being exceptions to our evolutionary account of the development of collective concept-use, we can bring these too within our scope. The specific forms of argument in arithmetic and geometry, grammar and causal reasoning, current in any milieu, then represent—quite as much as those of (say) physics—the provisional end-products of cultural evolution.

CHAPTER 8

Conclusion: The
Cunning of Reason

THIS first group of enquiries has had two main aims: one
theoretical, the other philosophical. The theoretical aim has
been to develop a general analysis of collective concept-use and
its historical development, in terms applicable to as broad a
range of human activities as possible. Now, in this concluding
chapter—anticipating the arguments of part III—we must
briefly consider the philosophical implications of our account:
hoping to show how, despite the diversity of concepts and rational
standards in different human enterprises, situations and milieus,
we can nevertheless—in appropriate cases, at least—define for
ourselves an impartial standpoint of rationality, and so escape
from the threats or temptations of relativism.

In science as much as in ethics (we have argued) the historical
and cultural diversity of our concepts gives rise to intractable
problems, only so long as we continue to think of 'rationality'
as a character of particular systems of propositions or concepts,
rather than in terms of the procedures by which men change
from one set of concepts and beliefs to another. This leads us to
confuse the rationality of scientific theorizing itself with the
logicality of the inferences within scientific theories. Instead of
theories being recognized as formal artefacts, or abstractions,
taken from a historically developing enterprise whose rationality
lies primarily in its procedures of conceptual change, philosophers have treated the history of science as a chronicle of
successive propositional systems, whose comparative logicality
has provided the only measure of rational acceptability. Given
this initial confusion, the grounds for choosing between successive theories based on different sets of concepts has become a
matter of pragmatic preferences, rather than of rational
judgement. When invited to explain how the procedures for
improving the conceptual repertory of a science can have their

own inherent rationality, most twentieth-century philosophers of science have been content to dismiss this very question with the epigram, 'There can be no *logic* of discovery!', and leave it at that.

In the long run, this response was bound to appear totally unsatisfactory, except to those who saw philosophy of science as requiring merely a formal extension of the propositional calculus so as to embrace 'inductive' arguments. So, in recent years, several attempts have been made to extend the notion of rationality beyond the scope of formal logic, and to find some way of reapplying it in situations involving conceptual changes. (One might refer, for example, to the writings of Karl Popper,[1] Thomas Kuhn,[2] and Imre Lakatos.[3]) At first sight, our present analysis might even appear to be a step in the same general direction. But, in a more fundamental respect, our own argument has been moving in precisely the opposite direction. Just because the arguments of Popper and his associates have all taken the formal logician's approach to science as a starting-point, conceptual change has continued to be more or less of an anomaly for them, to be dealt with by extending an analysis of 'rationality' which is still primarily formal, so as to account for the rationality of scientific procedures also. Our own analysis, by contrast, has inverted the whole relationship between formal logic and rational enquiry.

What has to be demonstrated is not that the rational procedures of scientific enquiry have, after all, a kind of 'logic' of their own: rather it is, how the formal structures and relations of propositional logic are put to work in the service of rational enterprises at all.

The rationality of natural science and other collective disciplines has nothing intrinsically to do with formal entailments and contradictions, inductive logic, or the probability calculus.

[1] K. R. Popper, *Logic der Forschung* (Vienna, 1935), published in English as *The Logic of Scientific Discovery* (London, 1959); also *Conjectures and Refutations* (London, 1963).

[2] T. S. Kuhn, *The Structure of Scientific Revolutions*, etc.; see above, pp. 96–130.

[3] I. Lakatos, 'Proofs and Refutations' (see above, p. 252, n. 1); 'Criticism and the Methodology of Scientific Research Programmes', *Proceedings of the Aristotelian Society*, 69 (1968), 149–86; and 'Falsification and the Methodology of Scientific Research Programmes', in *Criticism and the Growth of Knowledge*, ed. Lakatos and Musgrave (Cambridge, England, 1970).

Propositional systems and formal inferences are legitimate instruments, among others, for the purposes of rational investigation and scientific explanation, but they are no more than this. Instead, the nature of that rationality must be analysed in quite independent terms, before the question can properly be raised, how 'logical systems' are given an application in science.

This is where our 'ecological' approach comes into the picture. For it allows us to attack the central problems about the rationality of collective concept-use in a way that involves no appeal (such as Popper makes) to an arbitrary, *a priori* demarcation-criterion, as the definition of 'science'. Again, instead of leaving us wandering (as Lakatos does) in an abstract world of 'methodological research programmes' whose very names are inherited from the arguments of the formal logicians, it requires us to focus directly and in detail on the historically-developing problems and strategies with which our rational enterprises are concerned. At the same time, it gives us the means of distinguishing between the actual conceptual choices in fact made by professional scientists, technologists, or lawyers, and those which the genuine needs of their specific problem-situations would— if accurately judged—have demanded of them; so that we can acknowledge the proper roles of professional élites or 'reference groups', without running the risk (as Kuhn does) of bowing absolutely to the judgements of the currently authoritative groups.

Thus: (1) Popper's analysis of the 'logic of scientific research' has been concerned throughout with problems of formal proof and/or refutation. In opposition to the 'verification' criteria of his positivist contemporaries in Vienna, he began by insisting that 'falsifiability' was the universal, timeless test of a genuinely scientific hypothesis, and that progressive 'falsification' (or the elimination of hypotheses) was the correct procedure for rational advance in science. Later, he modified his position, softening the negative aspects of his earlier account, and conceding that certain hypotheses might, in some more positive sense, be 'better corroborated' or more probable than others. Yet, at both stages, Popper's questions have been framed as questions about the *acceptability of propositions*, rather than about the *applicability of concepts*; while his approved procedures of rational investigation—i.e. scientific testing, refutation, and/or corroboration—have continued to be variants on those of earlier propositional logicians.[1] Despite his increasing personal interest in historical examples and episodes, his philosophy of science has, as a result, remained (in Lakatos's words) 'drily abstract and highly ahistorical'.[2] The final source of his rational standards for judging scientific arguments and procedures has, correspondingly, remained throughout a set of general *a priori* conditions, which have been imposed on all scientific reasoning from outside, by his own—ultimately arbitrary—definition of what is to count as a 'scientific' hypothesis, theory, or concept.

If, having taken the formal logician's approach to science as our starting-point, this is really the best that we can do to account for the 'rationality' of conceptual innovation in science, then the most understandable reaction is perhaps that of Paul Feyerabend. To base our fundamental criteria of rational judgement in science ultimately on our own *a priori* definition of 'science' is (Feyerabend retorts) an arbitrary—even an authoritarian—procedure; and it can easily lead us into closing off beforehand, without any substantive justification, legitimate lines of theoretical advance that do not happen to conform to our prior demands. If that is what considerations of 'rationality' demand of us in science, then it is better (he argues) to renounce all such intellectual Puritanism, and preserve our freedom of speculation and conceptual innovation, even at the price of being frankly anarchistic and 'irrational'—refusing to restrict, by any prior *fiat*, the possible directions of future scientific development.[3]

(2) Once we are committed to a 'propositional' starting-point, the only constructive alternative to Popper's apriorism

---

[1] In his Aristotelian Society paper, Lakatos distinguishes three phases in the development of Popper's philosophy, which he designates 'Popper₁,' 'Popper₂', and 'Popper₃': the third view is one in which 'Popper' cuts himself loose from the presuppositions of earlier propositional logic, and at last takes a historical view of the criteria of scientific merit. In retrospect, however, Lakatos now admits that 'Popper₃' is, at best, a far-reaching extrapolation of tendencies discoverable within Popper's development, not a position that Popper himself has ever actually adopted. In fact, the view designated as 'Popper₃' is better regarded as one aspect of *Lakatos's own position* in philosophy of science.

[2] I. Lakatos, 'History and its Rational Reconstructions', in *Boston Studies in Philosophy of Science*, Vol. 8, ed. R. S. Cohen and R. Buck (New York and Dordrecht, 1971).

[3] P. K. Feyerabend, 'Against Method', *Minnesota Studies in Philosophy in Science*, Vol. 4 (Minneapolis, 1970).

is—it appears—to take the actual historical experience of scientists more seriously than he does, and find some way of mobilizing it so as to throw light on the rationality of scientific procedures and strategies. This is what both Lakatos and Kuhn have tried in different ways to do. Unfortunately, Lakatos himself has simply taken over Popper's own argument at the point where it breaks off.[1] As a result, while he is clearly feeling his way towards an account in which the locus of 'rationality' lies in the strategies of intellectual change employed within scientific disciplines, he is in no position to tackle directly any specific, concrete questions, either about the strategies of particular sciences, or about the manner in which these strategies relate to the detailed problem-situations arising within the sciences. Instead, he discusses the intellectual procedures of science in a general, abstract terminology taken over directly from the propositional logicians: as 'inductivist', 'falsificationist', and/or 'conventionalist'. And, instead of showing how we can recognize the substantive virtues of particular scientific strategies as 'ecological' responses to specific intellectual problem-situations, Lakatos still hopes to achieve knock-down dialectical victories. By paying sufficiently careful attention to historical examples (he believes) falsification can be falsified, inductivism can be inductively refuted, conventionalism proved merely conventional, and so on . . .[2]

In rejecting apriorism, and emphasizing the philosophical relevance of the historical record, Lakatos's new account certainly promises to be an advance on Popper's position. Lakatos even agrees with Kuhn in allowing some special weight to the judgements of authoritative reference-groups in science. Thus, in direct opposition to Popper, he says, 'If a demarcation criterion between what is, and is not, scientifically "rational" is inconsistent with the basic appraisals of the scientific élite, it should be rejected.'[3] Yet he leaves it quite obscure just how

[1] See Lakatos's *Boston Studies* paper, part 2(a). In a subsequent discussion of this paper at Jerusalem (January, 1971), Lakatos confirmed his commitment to the 'logicist' position, by insisting—like Frege and the logical empiricists—that 'methodology' is concerned only with questions about the *ex post facto* appraisal of justificatory arguments, while all questions about discovery and conjecture are matters of 'mere psychology'. The proceedings of this meeting, organized by the Van Leer Foundation in honour of S. Sambursky, are due to be published during the year 1971/2.

[2] *Boston Studies* paper, part 2(a).   [3] Ibid.

his 'historiography of methodological research programmes' can be applied, in actual instances, to pass judgement on the rationality of particular modifications in scientific theory. He leaves it obscure, for instance, just how we can stand back from the actual facts of the historical record, and achieve the impartial standpoint we need in order to avoid relativism. Under what circumstances and on what conditions, for instance, should we be justified in concluding that even 'the basic appraisals of the scientific élite' had been misjudged? On occasion (Lakatos recognizes) we must be prepared to reach this conclusion; he rejects any implication that we can recognize progress in science 'solely by the march of actual history'. On Kuhn's account of the matter (he argues) a mass-conversion of astronomers back to astrology would carry its own unanswerable authority. Instead of representing a catastrophic lapse into superstition, this 'paradigm-switch' would catapult astrology into a rationally-unassailable position.[1] Yet how Lakatos himself can discredit that claim to authority, in turn, is the crucial unanswered question about his own position.

(3) Faced with this accusation of relativism, T. S. Kuhn has —as we saw in an earlier chapter—restated his position, most recently, in 'evolutionistic' terms, designed to eliminate the 'rational discontinuities' which plagued his previous revolutionary thesis.[2] But, quite apart from its objectionable Lamarckian features, this new posture does not give him any real way of escaping from the fundamental dilemma about the rationality of conceptual change. For his final statement about the judgement of new ideas in science is simply that 'One scientific theory is not as good as another for doing *what scientists normally do*'; and this statement is patently ambiguous. Does it mean, '. . . what scientists *habitually* do'; or does it mean, '. . . what they *properly* do'? If the first alternative is intended, his final position remains as relativist as his first, and still gives us no basis for appealing against the actual procedures of the most influential scientists. And, if his true position is the second, then he has simply switched camps, back to the abstract, absolutist position of his 'logical empiricist' predecessors. His 'universal criteria of scientific advance'—maximum accuracy of predic-

---

[1] Ibid., Section 2(b), last para. See above, p. 322–3.
[2] 'Reflections on my Critics', in *Criticism and the Growth of Knowledge*, p. 264.

tions, etc.—have in that case been imposed on the intellectual procedures of scientists from outside, rather than discovered in the course of their actual experience in developing novel scientific concepts, strategies, and disciplines.

In this final chapter, we shall take up these points in turn, and show in outline how our ecological approach to the problem of scientific rationality escapes the difficulties that beset Kuhn, Lakatos, and Popper alike. This means tackling head-on our initial questions about the impartial standpoint of rationality, and showing how a sufficiently full and detailed understanding of the features that make alternative strategies and procedural innovations 'adaptive' to particular problem-situations puts us in a position to avoid the outstanding dilemmas about rationaality, by resisting any last attempt to drive the standards of judgement in collective enterprises, either back onto the absolute authority of some external, *a priori* demarcation-criterion, or onto the relative and transitory authority of some local reference-group.

8.1 : *The Standpoint of Rationality*

How, then, are we to reconcile the need for an impartial standpoint of rational judgement with the actual facts about conceptual diversity, and the variety of rational standards accepted as authoritative in different milieus? What progress has our analysis of collective concept-use and conceptual change made towards defining this 'impartial standpoint', from which genuinely rational comparisons can be made between the concepts, judgements, and beliefs accepted in different cultures and epochs? Certainly, our argument has done nothing to make these problems appear any less serious. For it is not just the individual concepts and theoretical generalizations accepted in our intellectual disciplines that vary from one country and period to another. A similar historico-cultural diversity evidently affects—even if less drastically—the methodologies, the collective ambitions, and disciplinary goals of the activities concerned. So what scope now remains for discussing the comparative merits of corresponding concepts and beliefs from different milieus, without reviving the assumption of 'fixed and necessary principles of understanding', which we renounced at the outset?

On examination, the actual situation is not quite as desperate as it appears. At this point in our argument, we must demonstrate only that the problem of defining an 'impartial standpoint of rationality' is soluble *in principle*; and much of the material we need for this purpose has already been assembled in the course of the preceding chapters. (We shall be raising later on, in part III, some more detailed questions, both about the concept of 'rationality' in general, and about the 'rational appraisal' of our concepts.) For the moment, our key move will be to spell out a little further the significance of the phrase 'intellectual ecology', and to explain more precisely the claim that, at a fundamental level, questions of collective rational judgement need to be considered in ecological rather than in formal terms.

The exposition will involve three steps. First, we shall refer back to our arguments against the idea that 'systematicity' is the essential element in 'rationality', and recall how this distinction helped to free us from the threat of relativism. After that, we shall indicate how an ecological approach enables us —at any rate, in suitable cases—to make the rational cross-comparisons for which practical occasion arises. And, finally, we shall show how it enables us, also, to dispense with any external 'demarcation criterion', and so escape finally from reliance on *a priori* standards of judgement. The tasks of our rational enterprises—as we shall then see—are not to be defined as a matter of (e.g.) 'doing what scientists normally do'. They are a matter of *doing what there is there to be done*; and that is something we can only discover for ourselves gradually, in the light of experience, as we go along.

Let me begin by summarizing the negative part of this whole argument. In the first two chapters, we showed how a preoccupation with the systematicity of geometry and formal logic makes a universal, quasi-mathematical absolutism in our intellectual methods and standards appear the only alternative to outright historico-cultural relativism; how this opposition leads, in turn, to a seemingly-inescapable choice between uniformitarian and revolutionary theories of conceptual change; and how the arguments that impose these distasteful choices on us both rest on the assumption that 'rationality' is primarily a matter of 'logicality'—i.e., that the tests and/or criteria for judging the rationality of intellectual methods and positions primarily involve questions about coherence, entailment, consistency, and other such formal characteristics. Once philosophers have made that assumption, all that remains for them to do is to decide which (if any) of the alternative conceptual and propositional 'systems' available in any milieu is 'authoritative'—i.e. which system provides the 'logical framework' for defining standards of rationality having authority over intellectual choices in that milieu. And that decision immediately resolves itself, in practice, into a choice between (1) allowing men to treat the particular conceptual and propositional systems current in their own milieus as locally sovereign—the relativist approach—and (2) imposing from outside, on all milieus alike, an abstract and ideal set of formal criteria, defined in terms of a universal, quasi-mathematical 'logical system'—the absolutist approach.

The move from a systematic to a populational analysis of collective human activities and enterprises, by contrast, opens up the whole problem of 'rationality' afresh. The moment we stop assuming that the ideas of any milieu form static 'propositional systems', and recognize that they constitute historically developing 'conceptual populations', we are free to abandon also the philosophers' traditional assumption that rationality is a sub-species of logicality. There are, in any event, other, more general considerations favouring this change. In non-intellectual contexts, for instance, we judge the rationality of a man's conduct by considering, not how he habitually behaves, but rather how far he modifies his behaviour in new and unfamiliar situations,[1] and it is arguable that the rationality of intellectual performances should be judged, correspondingly, by considering, not the internal consistency of a man's habitual concepts and beliefs, but rather the manner in which he modifies this intellectual position in the face of new and unforeseen experiences.

The essential problem is, then, to show how 'rational' considerations are brought to bear—at whatever level—on those changes of mind which involve replacing one set of concepts by another improved set. If questions of rationality were indeed

[1] Cf. J. Bennett, *Rationality* (London, 1964).

concerned with formal relations, there would be no way of dealing with this problem, since such relations can hold, strictly, only between propositions framed in terms of one and the same set of concepts. Yet we must insist that conceptual changes at all levels lend themselves to judgement in rational terms, regardless of all questions about the relation of rationality to formal validity. The central point was expressed in a memorable simile some twenty years ago by Gilbert Ryle. He compared 'making a formal inference' with taking a journey along an existing road; whereas 'justifying that form of inference' was to be compared, rather, with laying out the road in the first place.[1] Once we have an established network of roads in any area, the question 'Which is the *right* way from A to B?' acquires a determinate sense. At the earlier stage of surveying for the road network, by contrast, no such single-valued questions arise, and all of the operative questions are comparative ones—e.g. 'Which of the alternative lines for a road would give us a cheaper, faster, more direct, and/or environmentally less damaging way of linking A to B?' The tasks of constructing novel sets of concepts in any field of enquiry and refashioning existing concepts so as to go beyond the scope of currently established procedures likewise raise comparative questions, about what changes would be 'better' or 'worse', rather than single-valued ones, about what step is 'correct' or 'incorrect'.

Formal inferences, framed in terms of statements using a given set of concepts, are thus the linguistic expression of *stereotyped* intellectual procedures, and so—in the nature of the case—the historical product of previously completed conceptual changes. By contrast, the establishment of new or modified concepts involves—in the nature of the case—*non-stereotyped* procedures, which can be expressed linguistically only in non-formal arguments, framed in terms of 'meta-statements' about those novel or modified concepts. In cross-cultural or cross-historical judgements, particularly, our basic concepts and intellectual structures are inevitably up for reappraisal, so no stereotyped arguments are possible. And, whenever this happens, we are—once again, in the nature of the case—

[1] Cf. G. Ryle, *The Concept of Mind* (London, 1949), Chapter 5, esp. pp. 120ff., and S. Toulmin, *The Uses of Argument* (Cambridge, England, 1958).

compelled to dig behind all purely formal, single-valued criteria of 'correctness' or 'validity', and bring to light the underlying comparisons from which those criteria derive their own current relevance and justification.

The more positive and constructive aim of this whole argument has been to indicate just what 'digging behind' existing intellectual procedures implies, and what is involved in 'bringing to light the underlying comparisons' on which rational procedures rely for justification. This positive part of our account is encapsulated in the phrase 'intellectual ecology'. Normally it is evident, from the context in which our concepts are employed, just what rational enterprise or activity we are implicitly concerned with, and so what 'point of view' the discussion in question is adopting. Once this point of view is clearly identified, we can then use previous experience with the concepts and procedures involved to define the relevant reservoir of unsolved problems; recognize the outstanding intellectual or practical 'demands' which accordingly face us in the rational enterprise concerned; and compare the 'rational merits' of proposed conceptual changes by seeing how far, and in what respects, they would give us the means of solving the outstanding conceptual problems and meeting the actual demands of the current problem-situation.

Stated in such entirely general terms, our positive conclusions are of course too vague for practical application. By focusing our attention on more particular, restricted classes of situations, however, we can do something to sharpen them up. For example: wherever a rational enterprise becomes the domain of a compact, well-structured 'discipline', its procedures and problems are determined with an eye to well-defined collective goals, which express the shared ideals of the discipline—whether explanatory, practical, or judicial. Within a compact discipline, conceptual development is guided by accepted 'strategies'; and these in turn are the outcome of men's previous experience in handling conceptual problems from the 'point of view' of that discipline. From time to time, those strategies may themselves be called in question; when this happens, they must be viewed against the whole historical background of the developing rational enterprise concerned, and the men concerned will have to judge—by a critical and comparative analysis of previous experience in this field—what fresh strategy, or direction of advance, is most 'promising' in this particular area of investigation.

On that level, the operative questions finally cease to be formal ones, to be settled by appeal to existing procedures and strategies; instead, they can be discussed only in discursive terms. Indeed, the issues involved in a choice between alternative new strategies demand, in certain respects, an element of prophecy. They call, that is, for an appraisal of past and present disciplinary experience broad and penetrating enough to yield a well-grounded forecast, as to which new direction of advance can—at this point in historical time and disciplinary development—fulfil most completely the long-term ambitions of the rational enterprise involved. If the discipline in question is a scientific one, the phrase 'fulfilling long-term ambitions' will mean providing richer and more fruitful modes of explanation and understanding; if it is a technological discipline, the goals will be practical and technical; if it is a legal or judicial discipline, they will be concerned with justice, equity, and/or 'public policy'.

In order to bring home the force of these positive arguments, we drew attention to certain parallels between intellectual strategies in the natural sciences and judicial strategies in the practice of the common law. As Mr. Justice Holmes demonstrated so clearly, questions of judicial strategy take us—at the limit—beyond the reach of all merely formal reasoning, in which accepted rules, principles, and patterns of inference are applied to novel situations or sets of facts.[1] Strategic reappraisals, in law as in science, have—necessarily—to be arrived at in the light of longer-term views about how, in a novel socio-historical situation, changes in the interpretation of the common law can fulfil most completely the basic responsibilities of any legal system. And similar non-formal considerations arise, also, in cases where law-courts are required to draw judicial cross-comparisons between the procedures and decisions of different legal milieus, or 'jurisdictions'.

At this point, we can begin to see how an ecological analysis helps us to make philosophical sense of 'rational' cross-

[1] Cf. G. Gottlieb, *The Logic of Choice* (London, 1968).

490 *Rationality and Human Adaptation*

comparisons, also. Lawyers and judges operating within the common-law tradition behave in neither a 'relativist' nor an 'absolutist' way. They neither treat judicial decisions and procedures as holding good merely for one single jurisdiction, nor impose some abstract and universal 'juridical logic' on all jurisdictions alike. Instead they find it reasonable to 'adopt' decisions from other jurisdictions as guiding precedents—subject only to certain qualifications of a *substantive* kind. Whether or no such decisions have any bearing on a current case is then determined, not by formal or *a priori* tests, but by considering each precedent in its own social, historical, and juridical context. They are adopted as precedents (that is to say) in those respects that illuminate the specific demands of the present case and historical situation, and help to show how those demands can be met in a manner concordant with the fundamental purposes of the law.

In other situations where 'cross-cultural' comparisons are drawn—e.g., in social and cultural anthropology—their validity again depends on certain *substantive* conditions and qualifications. Anthropologists, for instance, may have occasion to compare peoples with profoundly different levels of economic development and conditions of life, property relations, and kinship systems. So, when they come to discuss (e.g.) questions about comparative personal status in different cultures, they may be quite unable to establish exact equivalences between the nearest corresponding institutions, offices, and relations.[1] Yet this does not make it totally impossible to draw intelligible comparisons between different institutions of 'marriage'—as they exist in (say) the late-twentieth-century United States, and a matrilineal, polyandrous community, with a mixed fishing and farming economy, on an archipelago in the Pacific. Before any well-grounded comparisons can be established between the institutions and patterns of relationships in markedly different cultures, however, it will first be necessary to consider, in quite specific detail, just what the customary procedures achieve, within the broader life of each culture, towards satisfying the basic needs and solving the fundamental problems of our common humanity.

[1] Cf. E. Nida, *Towards a Science of Translating* (Leiden, 1964) also G. W. Stocking jr., *Race, Culture and Evolution* (New York, 1968).

To make an historical point, in passing: the central theme in Vico's account of the 'historical understanding' was that our shared humanity alone makes it possible to understand the lives and actions of our fellow-men, and twentieth-century idealist philosophers of history, such as Croce and Collingwood, have put forward similar arguments more recently.[1] Empiricist critics have sometimes caricatured this doctrine, and suggested that it treats historical understanding as some kind of trans-temporal telepathy, empathy, or clairvoyance—as though the historian had retrospectively to 're-experience' the sensations and feelings, imagery and agitations, of the men whose story he is reconstructing.[2] Such objections demonstrably misread Collingwood's actual arguments,[3] and the Viconian approach lends itself equally to a far more innocent interpretation. This is: that the possibility of understanding the actions, customs and beliefs of men in other milieus rests on our sharing, not common 'sensations' or 'mental images', but rather common human *needs and problems*.[4] The cultural patterns and 'forms of life' of other peoples are then open to our understanding, in just those respects and to just the extent that they represent alternative ways of attacking shared human problems and meeting shared human needs.

With the experience of jurisprudence and anthropology in mind, we can at last hint at a general answer to our own central questions about rationality: (1) 'From what impartial standpoint is it possible to compare the concepts and beliefs of men in different milieus?', and (2) 'In what respects and to what extent can rational considerations be brought to bear on these comparisons?' Once again, the difficulty that faces us in

---

[1] Recall that Collingwood's first book was on the philosophy of Vico, and he himself expressed a great debt to Croce.

[2] Cf., for instance, W. H. Walsh, *An Introduction to the Philosophy of History* (London, 1953).

[3] In *The Idea of History*, part V, Section 3, 'Historical Evidence', Collingwood makes it clear that the procedures by which a historian reaches his conclusions involve hypotheses and the appraisal of evidence, in just the same way as those used by the other rational investigator: in Section 4, 'History as Re-enactment of Past Experience', by contrast, he speaks explicitly about the *outcome* of the historian's work, not about his *methods*. One really has to be somewhat wilful, in order to overlook the clear distinction between these two separate arguments.

[4] Recall the essay by Isaiah Berlin (p. 23, n. 1, above).

practice—and should therefore concern us in theory—is a *substantive* one: viz., that of establishing in what respects concepts or beliefs taken from quite different populations of ideas can be relevant to one another.

Such questions of mutual relevance require us to consider afresh in every case just how far the concepts and beliefs in question represent 'corresponding' ideas; just how far they have the same significance, serve similar functions, and so on . . . These issues involve the same substantive problems as those which anthropologists face, when identifying 'corresponding' institutions in different cultures; and the fact that such issues cannot be settled in a purely formal or 'logical' manner in no way means, of course, that they cannot be handled in 'rational' terms. Far from it: to the extent that men living in different milieus have faced similar collective problems, and developed comparable collective activities—or 'rational enterprises'—for tackling them in an organized manner, we can recognize those parallel enterprises as defining corresponding *forums of judgement*. It is by reference to these forums that we may judge the 'significance' or 'function' of the concepts and judgements under consideration, and we can legitimately identify and compare 'corresponding' ideas just to the extent that sufficiently similar forums of judgement are available.

Needless to say, the resulting comparisons will carry weight, only on condition that we pay attention, both to the similarities of function between 'corresponding' ideas from different milieus, and also to their limits. In extreme cases, these limits will mean that rational comparisons between the ideas concerned are, after all, impracticable—'To what extent did Lincoln's emancipation of the slaves represent an advance on Manchester *laissez-faire* economics?' But that is as it should be; we can rationally compare only sufficiently-like with sufficiently-like. But the very same considerations that make rational comparisons between sufficiently dissimilar notions illegitimate can also, within appropriate limits, make comparisons between sufficiently similar ideas legitimate. To go to the other extreme: we can establish close similarities of intellectual function between the concept of 'impetus' in fourteenth-century scholastic discussions of change, the concept of 'quantity of motion' in Newton's dynamics, and the concept of 'momentum', as used in Einstein's twentieth-century physics. This done, it is quite clearly legitimate—having regard to those functions —to ask how far, and in what respects, the relativistic concept of 'momentum' represents an intellectual advance on Newton's *quantitas motus* and Buridan's 'impetus'.

Our positive goal here has been to show, at any rate in general terms, that rational comparisons can meaningfully be drawn between concepts and judgements operative within different milieus; and to do so in terms which at the same time show why such attempted comparisons sometimes cease to be meaningful. We began by analysing the historical development of those collective rational enterprises (or 'disciplines') within which the operative populations of concepts change from one time to another on an evolutionary pattern; and we went on to interpret the 'adaptive functions' of those concepts in ecological terms. This ecological analysis has now given us, also, the means of drawing a practical boundary between those rational comparisons which can hope to be legitimate and impartial, and those which cannot. Within the limits defined by the respective enterprises and milieus, we may in each case consider—retrospectively—what was in fact achieved, by accepting the concepts under consideration, towards meeting the relevant demands of (say) physics or history, law or technology; or towards solving the everyday problems of dealing effectively and harmoniously with our fellow-men, animals, or inanimate objects, as the case may be. Within the same limits, we may consider also—prospectively—what light such comparisons throw on possible ways in which the proper goals of (say) scientific understanding or historical analysis, judicial administration or inter-personal relations, could be better formulated for the future. And the basic tasks of rational appraisal have to do, in the last resort, with the combination of well-digested experience and well-grounded forecasts mobilized in the course of such comparisons.

One final qualification is needed at this point. I have referred to our concepts, and the standpoints from which they are judged, as reflecting our common 'humanity': i.e., as reflecting the problems which men in all milieus have to face, in the world as it is, simply in virtue of being men and living the lives of men. Vague and general though it is, even this statement is in one respect over-restrictive. While, to the best of our present knowledge, men are the only creatures to harness language and methodical thought to the solution of their problems in a 'conceptual' manner, it does not (of course) follow from this that those problems are themselves necessarily—or specifically—problems for humans *qua* humans. We shall be considering later (in part II) on what conditions the terminology of higher mental functions can be extended from normal adult humans and applied also to infants, mental defectives, animals, lower organisms, and/or inanimate things. Yet there are certain quite obvious resemblances between the mental activities of men and those of other animals: to the extent that these serve comparable functions in the lives of different species, indeed, we can even cross-compare the effectiveness of human and animal performances and the efficacy of the nervous systems they call into play.[1] Beyond a certain point, however, it would be misleading —even, incorrect—to see 'conceptual' procedures in the behaviour of other animals, or to treat inter-specific comparisons between the behaviour and physiology of men and other animals as raising issues of 'rationality'.

Whatever their other similarities, the higher mental capacities of men are clearly differentiated from those of other animals in two crucial respects. Men have developed and harnessed to the problems of human life a 'language' which, in its richness and subtlety, has no known animal counterpart.[2] And they are capable, not just of operating with stereotyped perceptual and cognitive routines, but also of criticizing and changing their procedures in the light of experience, so as to achieve a more powerful intellectual or practical grasp on their situation. These two distinctive features of the human species are what make it appropriate to discuss human perception and cognition —as we have done in the present group of enquiries—in terms of 'rational enterprises' and 'conceptual improvements'. Even though some of the problems arising for men out of their common humanity extend to other animals also, therefore, we must at the same time recognize that our solutions have a peculiarly 'human' character. For it is men, and men alone, that have devised the 'rational' enterprises which provide the

---

[1] Cf. E. H. Lenneberg, 'Brain Correlates of Language'; see above, p. 453, n.1.
[2] See, e.g., Geschwind; see above, p. 449, n.1.

forums for the judgement of conceptual changes, and so make possible the progressive refinement of the repertories of procedures which men—as men—employ in dealing with their common problems.

8.2: *The Escape from the* A Priori

One last radical objection has still to be met, in conclusion. This is directed against the very form of our answer: in particular, against any claim that the standpoint for rational judgement established here is an impartial one. Supposing that we have demonstrated some basis for cross-cultural and cross-historical comparisons, can this, even so, be regarded as truly 'impartial'? For surely, even on our account, anyone who compares the intellectual merits, judicial significance or practical efficacy of corresponding concepts and procedures from different milieus must do so—inevitably—from a standpoint *within his own culture and epoch*; and the manner in which he approaches these questions will—inevitably—be conditioned by his particular standpoint. So something more needs to be done, if we are to sustain the claim that 'rational' comparisons, as characterized here, are at the same time 'impartial' comparisons.

The argument underlying this objection must be dealt with explicitly here, because it brings to light an ambiguity in the terms that we have used at some of the most delicate points in our discussion. And it is potentially damaging because, in the course of resolving that ambiguity, it threatens to force us back once again into the very dilemma we first set out to escape: the invidious choice between the arbitrariness of the absolutist and the defeatism of the relativist. The crux of the matter is the following. At two sensitive points in our analysis—in discussing changes of strategy, and in defining 'ecological functions'—we appealed to the 'basic tasks', or 'fundamental goals', which determine the nature of the problem-situations in any discipline and so its 'ecological demands', by speaking (e.g.) of 'scientific understanding in general', or of 'the responsibilities of any system of law, properly so-called'. And those definitions can themselves be criticized, as begging questions which—if pressed —would undercut our whole philosophical position. For suppose that we are asked, *from what standpoint* we are deciding what exactly should count as the defining goals of scientific understanding 'in general', or alternatively as a legal system 'properly so-called'. What kind of an answer can we then give? Have we not simply put ourselves straight back into the same old trap: of being forced to choose, either, a Popperian reliance on arbitrary, external 'demarcation-criteria', or else a thorough-going empiricism which lands us inescapably in a renewed relativism?

To spell the objection out further:[1] we seem, in the long run, to be confronted by an unpalatable choice. We may leave all decisions about what exactly counts as 'scientific' or 'legal' (say) to be taken afresh in every milieu, or we may impose universal, *a priori* definitions of 'science' and 'law' from outside, on all milieus alike; there seems no third way. Depending on which alternative we select, we shall end up—even though at a remove—in one or other of the positions we originally rejected. By allowing each separate culture and epoch to decide, by its own standards, what properly counts as 'scientific understanding' (or 'technical efficiency', or 'justice') we plunge ourselves back into relativism; once that is done, the very question, whether some new set of concepts promotes the funda-mental goals of 'scientific understanding properly so-called', will be understood in quite different senses in different milieus, and answered in correspondingly independent ways. By impos-ing universal, abstract definitions of the 'scientific' and the 'legal', from outside, we land ourselves equally in an arbitrary absolutism; once that is done, we are laying down *a priori* standards of rationality for anything we shall acknowledge as (say) 'science' or 'law', in advance of any consideration of the actual diversity to be found in those enterprises. And such *a priori* standards will immediately run up against the same familiar difficulties. From what source do they derive their supposedly universal authority? And how can we demonstrate their relevance and application to the specific problems of men in the whole range of actual disciplines and milieus?

If we take the first course, the criteria appealed to in making 'ecological' comparisons between concepts will evidently hold good only within a single historico-cultural milieu; this being

[1] The following argument was specifically directed against the present position by Joseph Agassi, at a meeting of the American Association for the Advancement of Science (Section L) in December 1969.

so, trans-historical and cross-cultural comparisons will once again be ruled out. If we take the second course, our definitions of 'science', 'law' etc. will be entirely general and abstract, and we shall simply trade in the problem of historical relativity for that of historical relevance. The philosophical embarrassments of relativism and absolutism are accordingly no less serious, when they arise—implicitly and indirectly—over the 'demarcation criteria' for recognizing what counts as 'scientific' or 'legal' at all, than when they arise—directly and explicitly— over particular concepts, beliefs, or standards of judgement. In either case, we end by veering between a relativist account of rationality, which concedes too much to intellectual history and cultural anthropology, and an absolutist analysis, whose relevance to specific problem-situations is left entirely unclear.

Faced with this final challenge, we must demonstrate, in return, that it is still possible to steer a middle way between the absolutist and relativist extremes. The objection can be answered, in fact, once we recognize that it remains genuinely damaging, only so long as we permit abstract matters of formal dialectic to be substituted for the more fundamental substantive questions. Certainly, any 'ecological comparisons' that we make between scientific or legal concepts from two different milieus will depend on judgements of 'similarity' between the explanatory or judicial aims and procedures, demands and strategies, operative in the two contexts. And, certainly, it is we ourselves who will finally decide just what specific similarities justify us in regarding the problems of two milieus as 'common' problems, their rational enterprises as 'comparable' enterprises, and their resulting intellectual or judicial procedures as 'corresponding' procedures. It is we ourselves who will finally decide what is, or is not, properly to be called a 'scientific' or 'legal' issue at all; and that decision—we must grant—colours all our subsequent discussion of these 'rational', and supposedly 'impartial' comparisons. Yet, having granted this much, we must not let ourselves be stared down. For the decisions in question are not ones that we can take *on the basis of formal definitions alone*; nor are the consequential questions ones that we can settle by analytical dialectic alone. Even these decisions are made for *substantive* reasons, and are continually open to correction *in the light of experience*.

So we simply refuse to be pressured into accepting any formal or fixed demarcation-criterion for 'science' or 'law'; whether this is presented in abstract and absolute, or in concrete and relative terms. When we use such categorial terms as 'science' and 'law', we do so to refer neither to the timeless pursuit of abstract ideals, defined without reference to our changing grasp of men's actual needs and problems, nor to what the men of each separate milieu themselves happen to give the names of 'science' and 'law'. Rather, we work with certain broad, 'open-textured' and historically developing conceptions of what the scientific and judicial enterprises *are there to achieve*.[1] These substantive conceptions are arrived at in the light of the empirical record, both about the goals which the men of different milieus have set themselves, in their own cultivation of those rational enterprises, and about the kinds of success they in fact achieved in the pursuit of those goals. And we make our 'rational comparisons' by considering how far, and in what respects, the alternative strategies employed in each milieu have actually fulfilled the historically developing purposes of the relevant enterprises.

Being substantive and empirical, this reply of course has to take a certain amount for granted. It assumes, for instance, that men's lives do face them, in certain significant respects, with some very general but common problems, regardless of the milieu; and that these shared problems call for the development of corresponding sorts of techniques, concepts, and procedures. It assumes, furthermore, that men's collective rational enterprises can legitimately be regarded as so many attacks— whether in parallel cultures or successive epochs—on those common problems. In the long enough run, these presuppositions may be valid only within limits, and subject to qualifications. In the long enough run (that is) 'rational comparability' may strictly be possible, only in so far as, and to the extent that, the collective enterprises of different milieus are *in fact* directed at sufficiently similar problems. But this, too, is a substantive question, and cannot be decided *a priori*, in advance of considering the actual facts about men's activities and enterprises, procedures and achievements. In every case,

---

[1] For the notion of 'open texture', see particularly the essay by F. Waismann, in *Logic and Language*, ed. A. G. N. Flew (Oxford, 1951), pp. 120ff.

the first necessity is to understand how the current practices of our different rational enterprises—and our present 'everyday conceptual frameworks' also—are related to the genealogy of men's intellectual and practical activities; and how the strategies of conceptual change guiding the development of our enterprises today are designed to crystallize the human experience of other cultures and earlier ages, and apply it in ways that are relevant to our current problem-situations.

At this stage in our argument, indeed, we are at a level on which the very attempt to distinguish sharply between the *a priori* and the *a posteriori* is itself a source of mere confusion. If our definitions of 'science' and 'law' were purely *a priori*, we might perhaps be unable to avoid absolutism; if they were purely *a posteriori*, we might be unavoidably thrust back into relativism. But we need no longer accept either of these alternatives. By this stage, we are past being concerned, either with matters of formal consistency and contradiction alone, or with reporting new empirical observations and precedents alone. Instead, our task is now to investigate what there is there to be meant by 'rationality' and its associated concepts, on what conditions rational questions become operative, and how the practical force of those questions reflects the conditions of their applicability. Once we have reached this level, any proposal that we should proceed on a purely *a priori* basis is a non-starter; there is, in fact, no longer anything left for our conclusions to be 'prior' or 'posterior' to. One could suppose otherwise, only by assuming that some unchanging 'rational framework' of discussion could be identified, capable of serving as a permanent starting-point of criticism, prior to all empirical understanding. And such a totally 'pre-empirical' framework could rest only on the sort of 'fixed principles of human understanding' we renounced at the start of our whole argument.

For our purposes, any such anti-historical assumption is unnecessary and unhelpful. We arrive at our own 'impartial standpoint of rational judgement', not in advance of any empirical understanding, but rather in the light of all available insights into the empirical matrix within which understanding has to be sought. In the earlier phases of cultural and intellectual history, men were in little better position than a new-born infant to recognize the legitimate purposes and strategies for the development of rational enterprises. They had to find out what scope existed for such enterprises as they went along; gradually coming to see what scope existed for devising and improving acceptable systems of justice or equity, for systematically developing novel explanatory concepts, for demonstrating the merits of improved technical arts, and all the other rational activities of intellectual and practical life. In all these respects, the verdict of human experience is, of course, never final. We, too, are all the time having to find out, as we go along, what additional scope our own dealings with the world still provide for refining our intellectual, judicial, and technical strategies and procedures still further. That being so, there is one basis, and one alone, on which our judgements of 'rationality' and conceptual 'merit' can truly be impartial. This is one that takes into account the experience which men have accumulated when dealing with the relevant aspects of human life—explanatory or judicial, medical or technological—in *all* cultures and historical periods.

By requiring us to accept testimony about human experience in any epoch or culture whatever as relevant to all others, this standpoint of judgement avoids the fallacies that come from allowing special authority to the judgements of any one milieu. So, for better or for worse, our argument at any rate gives us a solution to our original problem having the required form: showing us how to escape from the dilemma over the historical and cultural diversity of our concepts and judgements, without being impaled on either the abstract, formalist 'absolutism' of Frege, or the concrete but parochial 'relativism' of Collingwood. Our impartial rational standpoint is thus an 'objective' one, in the sense of being neutral as between the local and temporary views of different historico-cultural milieus; but its conclusions are always subject to reconsideration, and it does not divorce itself from the actual testimony of history and anthropology. And it goes without saying that judgements arrived at from this standpoint are still, in one respect, at the mercy of history. As our experience accumulates still further, our ideas about rational strategies and procedures for dealing with the problems in any field are always open to reconsideration, revision and refinement.

Judgements of rationality arrived at on the basis of relevant experience from all periods and cultures are, however, 'objective' also in another, more fundamental sense. And, in recognizing the significance of this deeper objectivity, we shall find ourselves coming close to a view of 'historical destiny' shared with so unexpected a combination of figures as Vico and Kant, Epicurus and Hegel. All these men saw the confrontation between Man and Nature as a contest in which the stakes were real, substantial, and in certain respects unpredictable. To some—but only some—extent men could bring their rational grasp of the current situation to bear on their future expectations and patterns of life, in such a way that they anticipated, and so were 'rationally pre-adapted' to, the novel problem-situations that would face them in the future. Either way—whether their anticipations were successful or no—the verdict of experience rewarded even-handedly those men whose rational procedures and innovations proved, in the event, to meet the actual demands of history most adequately; while the resulting discoveries in due course imposed themselves on men's practice, with a severe price for nonconformity.

Each of the four philosophers gave an account of social and cultural evolution in which historical destiny was personified in a different metaphor, and given a different name. What Epicurus called Necessity,[1] Vico called Providence;[2] what Kant called the Plan of Nature,[3] Hegel called the Cunning of Reason.[4] Yet, whether one entitled the agent of destiny Reason or Nature, Providence or Necessity, the practical significance of this doctrine was the same. No judgement on Man's success in the rational organization of his experience is ever final, or immune to reconsideration. However much we may seem to have achieved—however much we may *actually* have achieved—future problem-situations may always impose unforeseeable demands on our procedures, techniques and methods of

---

[1] For the Epicurean notion of Necessity, see particularly Diodorus Siculus, *Library of History*, Book I, Secs. 8, 9; Loeb Classical Library (London and Cambridge, Mass., 1946), Vol. I, pp. 29–35.

[2] On Vico's notion of Providence, see *La Scienza Nuova*, para. 432.

[3] On Kant's notion of 'The Plan of Nature', see Collingwood, *The Idea of History* (Oxford, 1946), pp. 94–8.

[4] On Hegel's notion of 'The Cunning of Reason', see (e.g.) Collingwood, op. cit., pp. 116ff.

thought; and the outcome of the resulting 'conceptual competition' can never be predicted with absolute certainty.

Our own discussion carries a similar moral. The professional
judgements of a 'Supreme Court Justice' in any science are
never (as we saw) totally free, capricious or subjective: no
standard, well-established procedure can guide the authoritative scientist in devising a new intellectual strategy, yet he is
obliged to stake his own reputation on the outcome of his
intellectual reappraisal. He is obliged (that is) to commit himself now to a present estimate of the future consequences of
alternative rational policies, or strategies, and the decision he
arrives at in any particular case may subsequently turn out
to have been misjudged. In every case, that is to say, the historical outcome of the recommended strategy will in due course
give its own retroactive verdict on the scientist's insight. Though
men may forever be doing their 'rational best' to mobilize their
past experience, in tackling their present and future problems,
there is no guarantee that even their best will, on every single
occasion, meet the actual demands of all fresh problems-
situations. In the course of history, we have learned a good deal
about the occasions on which, the ways in which, and the considerations in the light of which, intellectual procedures and
practical techniques need to be 'rationally adaptable' to novel
'ecological demands'. But here, as elsewhere, any striving after
finality in our procedures carries with it the risk of idolizing
obsolescent concepts and methods.

In the intellectual decisions and conceptual changes that
continually face us—whether in the development of our rational
enterprises, or in the refinement of our everyday conceptual
framework—we are forever compelled to lay bets about our own
futures, and we can ground the strategic estimates on which
rational changes of policy are based, only on a well-digested
appreciation of earlier achievements in those same enterprises.
Yet future experience will always have occasional surprises in
store for us; the detailed outcome of our strategic changes can
be sensed only dimly, not seen clearly beforehand. If this is so,
it should not disturb us. In intellectual as much as in practical
and political affairs, the working-out of historical development
is liable to reward well-judged changes of policy in ways that
could not have been foreseen in precise detail.

At the final, strategic level of a rational enterprise, then, we must sometimes be prepared to play for subsequent achievements, of kinds that we could never aim at explicitly or with confidence. We are always liable to come up against 'rational frontiers' at which, as fallible individuals acting in the name of the human enterprises we represent, we must deal with novel unforeseen problem-situations, by opening up new possibilities and procedures. If our bets are not based on the fullest understanding of all that is at issue in such a strategic change—if we persist in the use of (say) explanatory procedures or judicial methods, beyond the point at which experience has already begun to show their limitations—the subsequent development of intellectual or judicial history will in due course find us out.

In all our 'rational interactions', as Men dealing with the World in which we live, the stakes for which we are playing are real and substantial. Having set our hands to the conceptual task, of improving our intellectual and practical understanding, we no longer have any choice whether or not to play for those stakes. The burden of 'rationality' then consists in the fundamental obligation to continue reappraising our strategies in the light of fresh experience. Having accepted this burden, we must admit that our strategic estimates will not, in the event, prove 100 per cent precise and well-founded in all cases. Yet if, in return, we seek some assurance that the burden is worth undertaking, we can already see in outline what form that assurance can take. For, about this, one thing at least can now be said. As those 'rational transactions' to which we have committed ourselves continue to work themselves out in the course of subsequent history, the same verdict of historical experience which earlier thinkers called the Cunning of Reason (or Nature, or Providence, or Necessity) will, in the long run, penalize all those who—whether knowingly, or through negligence—continue playing according to out-dated strategies.

APPENDIX   Research
             Programme

As far as it goes, our present analysis of the collective understanding, and of its historical evolution, has the merit of satisfying
our initial demands: for a theory of human understanding which
recognizes that neither the world that we have to deal with, nor
the set of concepts, methods, and beliefs that we develop in the course
of those dealings, is a historical invariant. But the more substantial
tests will come later, when there has been time to explore the consequential questions flowing from our analysis, and its implications
for the various fields of enquiry on which it bears. The true measure
of the insight which any serious theory provides lies, above all, in
the richness and variety of the novel questions it forces on our
attention, and in its power to reveal significant connections between
elements, or fields of enquiry, that had previously appeared entirely
independent. This means not just the power of the theory to generate
additional questions for investigation, but also its capacity to discredit questions left over from earlier accounts and to replace them
by other, more operative questions.

So here: our analysis has posed novel questions in fields as diverse
as history of science and sociology of art, theoretical linguistics and
moral philosophy. At the same time, it has pointed to unfounded
assumptions underlying current ideas in several of these fields, and
suggested alternative ways of approaching them. As a postscript to
this argument, accordingly, we may usefully assemble an outline
agenda for future research, summarizing the implications of our
account for a few of the fields of investigation touched on in these
enquiries.

(1) We may begin with most general implications. Abandoning
the theoretical pattern of formal 'systems' characterized by static
interactions, in favour of the alternative schema of 'populations'
subject to historical evolution, forces methodological changes on
many of the human sciences. In philosophy of science, sociology or
political theory, cultural anthropology or theoretical linguistics, it is
then necessary to find explanations for unities that were previously
accepted as 'self-explanatory'. It is a point of fact that different

natural sciences, different languages, different societies, and institutions all maintain a certain specificity and continuity, in the face of historical change; so we must look and see what kinds of dynamic equilibrium between innovative and selective processes are responsible, in each case, for preserving this specific character in the face of variability, and for governing the directions of conceptual, social, cultural, or linguistic evolution.

We may, for instance, ask such questions as the following. What conditions favour either the preservation of long-term intellectual stasis in a scientific discipline, or its fragmentation into two or more successor-disciplines, or the generation of a new hybrid science from two or more precursors, or the disappearance of the scientific enterprise concerned, whether by loss of definition, by extinction or otherwise? (Similar questions arise, equally, about societies, cultures and languages.) In what circumstances, again, might we expect to find the conceptual or social, cultural or linguistic counterparts of the apomictic, polytypic or ring species already familiar in the taxonomy of organic populations? And what scope is there, in the human sciences, for studying the 'adaptive responses' of different human activities and institutions to the practical and intellectual demands of different situations? Can the use of phrases like 'intellectual ecology', 'cultural environment', and 'social niche' be carried beyond a programmatic level and given some real substance, by a study of actual cases?

(2) More particularly, our account has implications for the historiography of science. We can now, for instance, demolish the last barriers between 'internalist' and 'externalist' histories of science: that is, between studies of the intellectual sequences by which the content of a science develops, and investigations into the influence of its socio-economic and cultural context on (e.g.) the direction and vigour of research. It is legitimate, within limits, to study these two aspects of science separately in certain cases, but the resulting self-limited pictures of scientific change are less thorough and comprehensive than we can now demand. Instead, we should set about relating the 'life of ideas' to the lives and institutions of the men who conceive and transmit them; and so reintegrate the 'internal' (or disciplinary) and 'external' (or professional) aspects of science.

Specifically, we have opened up four features of disciplinary evolution for study: (a) the changing 'explanatory goals' directing the various scientific disciplines at different times, and the implications of these changing ambitions for the problematics of the sciences, (b) the channels of intellectual innovation, by which 'conceptual variants' enter the current scientific debate, (c) the selection-criteria, procedures and/or prejudices to which those variants are exposed, in competing for an established place in a particular science, and (d) the ways in which broader features of the social or cultural context influence disciplinary development in a science.

Is a complete consensus about explanatory goals required, for instance, if a scientific discipline is to deal effectively with its outstanding problems? Or does a certain variety of intellectual ambitions, between scientists in different countries, schools or traditions, help to maintain the vigour of a science? In the latter case, how is a proper balance struck, so as to avoid both the diffuseness of a 'would-be discipline' and the scholastic uniformity of a dogmatic clique? Again: can we learn anything of a general kind from the actual historical record about the changing theoretical ideals of the various sciences—(e.g.) about the alternation between atomistic and continuum models in physics, genetics and elsewhere? Or about the possible schemata for explaining historical origins, whether in social theory, zoology, or physical cosmology? Or about the methodological relations between formal, mathematical theories (Duhem's physics *à la française*) and substantial models (Duhem's physics *à l'anglaise*)?

As to intellectual innovation in science: what kinds of argument are used to get a conceptual variant recognized as a genuine possibility, or ruled out as worthless? What kinds of factors, either internal (e.g. ripeness) or external (e.g. patronage), encourage a higher rate of innovation in one area of science than in another? And in what respects do broader social attitudes and incentives affect the types and rates of conceptual innovation in a science?

As to the procedures of intellectual selection: do the authoritative reference-groups serve similar judicial functions in all sciences, or do differences in selection-criteria entail different selection-procedures also? How has the resulting distribution of intellectual authority in science changed from one phase of history to another? How do such changes affect the eventual rate and direction of conceptual evolution? And in what respects is the character of scientific development in different milieus influenced by differences in the rigour with which the selection-procedures are enforced?

Finally: by what channels do disciplines of different kinds (e.g. sciences and technologies) interact? How far, for instance, does the effective development of explanatory concepts depend on, and in turn influence, the effective development of practical techniques? In what respects is the range of conceptual variants considered in the science of any milieu affected by larger presuppositions about (say) philosophy, ethics, or theology? And how can one distinguish legitimate and necessary interactions, on this level, from those which are merely conservative or metaphysical?

(3) Turning, next, to the sociology of scientific enterprises: we can ask some new sociological questions, both about the organization and operation of scientific institutions, and about their interaction with society at large. Sociologists as much as historians have tended to accept a sharp division between questions about the scientific professions, as elements within larger social 'systems', and questions about the content of the disciplines which those professions carry. This abstraction is, once again, legitimate only up to a point, and a more comprehensive analysis should indicate also how the professional institutions and practices of a science reflect the intellectual character of its problems. The institutions of science may share certain general features with all social institutions, but in some respects they are intelligible only with an eye to their special disciplinary functions.

We can, accordingly, pose problems of four further kinds: about (a) the manner in which the professional organizations of science act 'in the name of' their disciplines, (b) the channels by which authoritative reference-groups and individuals, forums, and journals, exercise their critical influence and direct the development of a discipline, (c) the processes by which authority is transferred from one reference-group to another within a profession, and (d) the steps by which disciplinary speciation is reflected, institutionally, in the emergence of correspondingly differentiated organizations and reference-groups.

Many of the resulting questions have already been touched on in these enquiries. We have discussed (e.g.) the effect of 'generational' differences on the pattern of professional authority, and the role of periodicals and standard textbooks in the formation and transmission of a conceptual tradition. But many others could be added: e.g., about differences between the institutional structures of science in different countries and their effect on the efficacy of disciplinary development, or about functional differences between the professional institutions of (say) science and technology, medicine and law.

(4) We may also frame some further questions for the psychology of scientific research and discovery. It is now necessary, for example, to pose our problems about the personal motives of individual scientists (and their 'causes') with an eye to the disciplinary problems (and 'reasons') of the collective science to which they are professionally committed. These relationships may be viewed from either of two directions: we can ask, either (a) how the collective goals of a

discipline impose professional demands on the problem-solving activities of the scientists involved, or, alternatively, (b) how the personal preoccupations that lead different individuals to enter a particular scientific field manifest themselves in their disciplinary work and professional lives.

(5) If pursued far enough, our analysis of collective rational enterprises has both theoretical implications, for the historical and philosophical understanding of the scientific enterprise, and also practical implications, for its political and administrative handling. The current arrangements for supporting the development of (say) scientific and technological activities always take for granted certain general beliefs, about (e.g.) the effective conditions for fruitful creative activity in such fields. A better theory of disciplinary change could thus give us a better practical understanding, about (a) the manner in which the professional institutions of any discipline can be adapted to the needs of the enterprise concerned, and about (b) the external conditions that encourage either the development of a particular discipline, or the cross-fertilization of different enterprises.

(6) Finally, our analysis allows us to compare the patterns of historical change in different kinds of collective enterprises. The central feature of our account was a model of historical development in 'compact disciplines'; but we have seen that an understanding of the conditions required for the applicability of this model can throw light also on other fields of human activity, which are not fully disciplinable. If we contrast the compact disciplines of (e.g.) the physical sciences, in turn, with the more diffuse disciplines of the behavioural sciences, the quasi-disciplines of the fine arts, and such non-disciplinary enterprises as ethics and philosophy, this can help us to understand better, not only those collective human enterprises which are in fact disciplinable, but also those which are not.

Index